T0293683

Recent Topics in
Differential Geometry
and its Related Fields

Recent Topics in Differential Geometry and its Related Fields

Proceedings of the 6th International Colloquium on Differential Geometry and its Related Fields

Veliko Tarnovo, Bulgaria   4 – 8 September 2018

editors

Toshiaki Adachi
Nagoya Institute of Technology, Japan

Hideya Hashimoto
Meijo University, Japan

 **World Scientific**

NEW JERSEY · LONDON · SINGAPORE · BEIJING · SHANGHAI · HONG KONG · TAIPEI · CHENNAI · TOKYO

Published by

World Scientific Publishing Co. Pte. Ltd.

5 Toh Tuck Link, Singapore 596224

USA office: 27 Warren Street, Suite 401-402, Hackensack, NJ 07601

UK office: 57 Shelton Street, Covent Garden, London WC2H 9HE

British Library Cataloguing-in-Publication Data
A catalogue record for this book is available from the British Library.

Cover image:
  (front) Historical remains of the main church, Pliska
  (back) The horseman, Madara
Photographed by Toshiaki ADACHI

RECENT TOPICS IN DIFFERENTIAL GEOMETRY AND ITS RELATED FIELDS
Proceedings of the 6th International Colloquium on Differential Geometry and
its Related Fields

ISBN 978-981-120-668-9

For any available supplementary material, please visit
https://www.worldscientific.com/worldscibooks/10.1142/11454#t=suppl

PREFACE

The *6th International Colloquium on Differential Geometry and its Related Fields* (ICDG2018) was held on the campus of St. Cyril and St. Methodius University of Veliko Tarnovo, Bulgaria, during the period of 4–8 September, 2018. The academic program ICDG was originally based on the agreement for academic exchange and cooperation between Nagoya Institute of Technology, Japan and University of Veliko Tarnovo, Bulgaria. This time, the Institute of Mathematics and Informatics, Bulgarian Academy of Sciences also co-organized ICDG, and we have some new young comers from Sofia. In addition, there was a satellite meeting at the Technical University of Iaşi, Romania during 23–24 August, 2018.

We here briefly recall the history of our program. In 1992, an academic program "International Workshops on Complex Structures and Vector Fields" had been started by Professors Stancho Dimiev and Kouei Sekigawa. Geometers and mathematical physists from Bulgaria, Japan and some other European countries have met together and discussed their problems at Sofia, Pravetz, St. Konstantin, Golden Sands and Plovdiv in every 2 years. After 16 years, in 2008, our program ICDG started as a continuation of this. The editors consider that these programs play an important role in exchanging ideas between scientists during this quarter century, and in creating friendships amongst each other throughout the meetings. Sadly, the editors had received news that Professor Dimiev, one of the founders of the first generation, had passed on. With his wife Anna, he played quite an important role in our academic program on geometry and analysis between Bulgaria and Japan. May his soul rest in peace.

This volume contains contributions by the main participants in the meeting. These cover recent topics on Differential Geometry, especially on Einstein metrics on Lie groups, some aspects on Hermitian symmetric spaces, geometric structures on cotangent bundles and semi-Riemannian manifolds, real hypersurfaces in complex hyperbolic spaces, and also on Discrete Mathematics such as coding theory and graph theory. We hope this volume also provides a broad overview of differential geometry and its relationship to other fields in mathematics and physics.

We appreciate all participants very much for their contributions in the Colloquium and to this proceedings. We acknowledge also the scientific reviewers who read the articles carefully and gave many important suggestions to authors. We are grateful to the members of the local committee, especially to Professor Piperkov, for their hospitality. With their help, all participants could spend several nice days in the old cultural town of Veliko Tarnovo and have a fantastic trip to Shipka and Kazanrak. We thank also Professor Milousheva for her huge effort in providing us the information on Professor Dimiev's CV. Lastly, we would like to thank the publishing editor Ms. Lai Fun Kwong and the designer for their help.

The Editors
10 May, 2019

The 6th International Colloquium
on Differential Geometry and its Related Fields

4–8 September, 2018 – Veliko Tarnovo, Bulgaria

ORGANIZING COMMITTEE

T. Adachi	– Nagoya Institute of Technology, Nagoya, Japan
S. H. Bouyuklieva	– St. Cyril and St. Methodius University of Veliko Tarnovo, Veliko Tarnovo, Bulgaria
H. Hashimoto	– Meijo University, Nagoya, Japan
M. J. Hristov	– St. Cyril and St. Methodius University of Veliko Tarnovo, Veliko Tarnovo, Bulgaria
V. Milousheva	– Institute of Mathematics and Informatics, Bulgarian Adademy of Science, Sofia, Bulgaria
M. Ohashi	– Nagoya Institute of Technology, Nagoya, Japan

LOCAL COMMITTEE

G. Nakova & V. Bencheva	– University of Veliko Tarnovo
P. Piperkov	– Bulgarian Academy of Science
Y. Aleksieva	– University of Sofia, Bulgaria

SCIENTIFIC ADVISORY COMMITTEE

G. Ganchev	– Bulgarian Academy of Sciences, Sofia, Bulgaria
K. Sekigawa	– Niigata University, Niigata, Japan
T. Sunada	– Meiji University, Tokyo, Japan

PRESENTATIONS

at Hall 400, Building 5,
St. Cyril and St. Methodius University

1. **Yusuke Sakane** (Osaka Univ.),
   Homogeneous Einstein metrics on Seifelt manifolds

2. **Paskal Piperkov** (IMI-BAS & Univ. Veliko Tarnovo),
   About some transforms in coding theory related to Vilenkin-Chrestenson transform

3. **Osamu Ikawa** (Kyoto Inst. Tech.),
   A generalization of the duality for Riemannian symmetric spaces and its applications

4. **Mancho Manev** & **Veselina Tavkova** (Univ. Plovdiv),
   On almost paracontact almost paracomplex Riemannian manifolds

5. **Misa Ohashi** (Nagoya Inst. Tech.),
   Examples of totally geodesic surfaces in symmetric spaces of type A

6. **Toshiaki Adachi** (Nagoya Inst. Tech.),
   Eigenvalues of finite vertex transitive Kaehler graphs and their zeta functions

7. **Galya Nakova** (Univ. Veliko Tarnovo),
   Some half lightlike submanifolds of almost contact B-metric manifolds

8. **Nobutaka Boumuki** (Oita Univ.),
   An indecomposable representation and the vector space of holomorphic vector fields on a pseudo-Hermitian symmetric space

9. **Yana Aleksieva** (Sofia Univ.) & **Velichika Milousheva** (IMI-BAS),
   Minimal Lorentz surfaces in pseudo-Euclidean 4-space with neutral metric

10. **Velichika Milousheva** (IMI-BAS),
    Meridian surfaces with parallel normalized mean curvature vector field in pseudo-Euclidean 4-space with neutral metric

11. **Takahiro Hashinaga** (Kitakyusyu Col.),
    *Homogeneous Lagrangian submanifolds in Hermitian symmetric
    spaces of noncompact type*

12. **Hideya Hashimoto** (Meijo Univ.),
    Geometrical structures and the Calabi-Bryant formula of G_2

13. **Cornelia-Livia Bejan** (Tech. Univ. Iasi),
    Generalized paracontact manifolds

14. **Alexander Petkov** (Sofia Univ.) & **Stefan Ivanov** (Sofia Univ.),
    *The QC Yamabe problem on non-spherical quaternionic
    contact manifolds*

15. **Naoya Ando** (Kumamoto Univ.),
    Holomorphic quartic differentials on surfaces

16. **Viktoria Bencheva** & **Milen Hristov** (Univ. Veliko Tarnovo),
    On the geometry of curves and surfaces in barycentrics

17. **Hiroshi Matsuzoe** (Nagoya Inst. Tech.),
    Sequential structures of statistical manifolds

18. **Georgi Ganchev** (IMI-BAS),
    *Surfaces of constant mean curvature in $\mathbb{R}_1^3$ isometric to a given one
    and minimal surfaces in $\mathbb{R}^4$ strongly isometric to a given one*

Building 5, St. Cyril and St. Methodius University, September 2018

CONTENTS

Preface   v

Organizing, Local and Scientific Advisory Committees   vii

Presentations   ix

In Memory of Professor Stancho Dimiev   1

Einstein metrics on special unitary groups $SU(2n)$ 5
   Andreas ARVANITOYEORGOS, Yusuke SAKANE
     and Marina STATHA

Construction of noncompact Lagrangian orbits in some
Hermitian symmetric spaces of noncompact type   29
   Takahiro HASHINAGA

Complex curves and isotropic minimal surfaces in
hyperKähler 4-manifolds   45
   Naoya ANDO

(α, ε)-structures of general natural lift type on cotangent bundles   63
   Simona-Luiza DRUŢĂ-ROMANIUC

Eigenvalues of regular Kähler graphs having commutative
adjacency operators   83
   Toshiaki ADACHI

Semi-Riemannian D-general warping manifolds   107
   Cornelia-Livia BEJAN and Sinem GÜLER

Symmetric triad with multiplicities and generalized duality
with applications to Leung's classification theorems   119
   Osamu IKAWA

An indecomposable representation and the complex vector
space of holomorphic vector fields on a pseudo-Hermitian
symmetric space   139
   Nobutaka BOUMUKI

Non-flat totally geodesic surfaces of $SU(4)/SO(4)$ and fibre
bundle structures related to $SU(4)$ 149
   Hideya HASHIMOTO, Misa OHASHI and Kazuhiro SUZUKI

An algorithm for computing the weight distribution of a linear
code over a composite finite field with characteristic 2   163
   *Paskal PIPERKOV, Iliya BOUYUKLIEV
   and Stefka BOUYUKLIEVA*

Extrinsic shapes of trajectories on real hypersurfaces of type
(B) in a complex hyperbolic space   183
   Tuya BAO and Toshiaki ADACHI

Periodic surfaces of revolution in $\mathbb{R}^3$ and closed plane curves   203
   Yasuhiro NAKAGAWA and Hidekazu SATO

In Memory of Professor Stancho Dimiev

Professor Stancho Dimiev at his office in IMI-BAS, September, 2006

Professor Stancho Dimiev was a member of the organizers of the conference "The International Workshop on Complex Structures and Vector Fields", which is the former academic program on Analysis and Geometry between Bulgaria and Japan. We received news that he had passed away on January 30, 2018. We here devote several pages in memoriam of him, who was one of the founders of our academic exchanging program.

Professor Stancho Dimiev was born on May 12, 1932 in Popovo, Bulgaria as a son of Gencho Stanchev and Sevdalina Dimiev. Popovo is a town in northeastern Bulgaria and is located between Shumen and Veliko Tarnovo. After finishing his studies in elementary school of 8 years in Popovo he entered a high school of 4 years in Sofia in 1950. He studied mathematics at the University of Sofia, Bulgaria during 1950–1955, obtained a master degree in 1955, and became a teaching assistant at this university in 1955–1960. From 1960 till 1963, he studied at the Moscow State University as a postgraduate student in Cybernetics with his scientific advisor Alexey Andreevich Lyapunov. Coming back to Bulgaria, he was appointed as a mathematician in the Mathematical Institute with Computing Center (which is now the Institute of Mathematics and Informatics), Bulgarian

Academy of Science in 1964, and was promoted to assistant professor in 1967, to an associate professor in 1981 and to a professor in 1991. In 2001, he left the institute and became a professor till 2005 at the Department of Mathematics, University of Plovdiv, where he had been an invited professor during 1975–1980. During the period of 1968 till 1969, he visited the Institute of Henri Poincaré, Paris. He got his Doctor of Philosophy from the Bulgarian Academy of Science in 1975. The title of his thesis was "Complex preordered vector spaces and stretched mappings". He served as a Ph.D. thesis advisor to 8 students: Kalin Petrov, Hussein Mehmedov, Fam Cheng, Lilia Apostolova, and others.

His major is complex analysis. According to MathSci Net, American Mathematical Society, he has 68 publications. His papers were published in many scientific journals such as Comptes Rendus Hebdomadaires des Séances de l'Acadámie des Sciences, The Bulletin of the London Mathematical Society, Banach Center Publications, Annales Polonici Mathematici, Comptes Rendus de l'Académie Bulgare des Sciences, and so on.

From 1992, he worked also as an organizer of the international conferences on geometry and analysis. The conference "The International Workshop on Complex Structure and Vector Fields" was started by Professors Dimiev and Kouei Sekigawa (Niigata Univ.) under the proposal of Prof. Dimiev on the occasion of his first visit to Niigata University in the autumn of 1991. The first workshop "International Workshop on Almost-Complex Structures" was held at Sofia in 1992, and it was successful with Professors Karlheinz Spallek (Ruhr-Universität Bochum), Julian Lawrynowicz (Polish Academy of Sciences) and Hideya Hashimoto. The workshops were held every two years, at Pravetz in 1994, at St. Konstantin near Varna in 1996, 1998, 2000, at Golden Sands in 2002, at Plovdiv in

At the 8th International Workshop on Complex Structures & Vector Fields, August 2006

2004 and again at Sofia in 2006, 2008 and 2010. From the second workshop till the eighth, the titles were "International Workshops on Complex Structures and Vector Fields" and the last two were titled as "International Workshops on Complex Structures, Integrability and Vector Fields". During these workshops, his wife Anna helped him as a manager of these workshops. She arranged accommodations of the participants and workshop excursions; she is a very nice lady. It is not too much to say that the workshops continued with her aid. In later life, she had knee problems, so Professor Dimiev usually helped her.

In 1997, from April till August, he stayed at Niigata University as an assistant professor, and gave a lecture on complex analysis for graduated students. Niigata is famous for Sake, Japanese spirits. He enjoyed his life in Niigata by discussing with students of the doctorate course at the department and also at university cafeteria, and by taking dinner at

Prof. Dimiev at his home
with Hashimoto, August 2006

Japanese classical pubs with Professors Sekigawa and Hashimoto. During this period he attended some mathematical meetings held in Japan and deepened friendships with some Japanese researchers.

Prof. Dimiev, his wife Anna and Adachi
at IMI-BAS, September 2004

Before 2000, in Japan, though Bulgarian yoghurt is usually taken in daily life and some Sumo wrestlers came from Bulgaria, in the field of mathematics academic exchanges between Bulgaria and Japan were not commonly held. It is quite remarkable that our academic program has continued for more than a quarter of a century. We would like to express our hearty thanks to the founders Professors Dimiev and Sekigawa and their wives, and also to all the participants.

It is regretful to have lost a key influential person between Bulgarian and Japanese geometers and analysts. We wish Professor Dimiev to rest peacefully in heaven.

The Editors

EINSTEIN METRICS
ON SPECIAL UNITARY GROUPS SU(2n)

Andreas ARVANITOYEORGOS

University of Patras, Department of Mathematics,
GR-26500 Rion, Greece
E-mail: arvanito@math.upatras.gr

Yusuke SAKANE

Department of Pure and Applied Mathematics,
Graduate School of Information Science and Technology,
Osaka University, Suita, Osaka 565-0871, Japan
E-mail: sakane@math.sci.osaka-u.ac.jp

Marina STATHA

University of Patras, Department of Mathematics,
GR-26500 Rion, Greece
E-mail: statha@master.math.upatras.gr

We obtain new invariant Einstein metrics on the compact Lie group $SU(2n)$ $(n \geq 3)$ which are not naturally reductive. We view the compact Lie group $SU(2n)$ as total space over the generalized Wallach space $SU(2n)/U(n)$ and consider $\mathrm{Ad}(U(n))$-invariant metrics on the group $SU(2n)$ associated to the generalized Wallach space to obtain these invariant Einstein metrics.

Keywords: Homogeneous space; Einstein metric; special unitary group; isotropy representation; naturally reductive metric; Gröbner basis.

1. Introduction

A Riemannian manifold (M, g) is called Einstein if it has constant Ricci curvature, i.e. $\mathrm{Ric}_g = \lambda \cdot g$ for some $\lambda \in \mathbb{R}$. Expositions for invariant Einstein metrics on homogeneous spaces can be found in [17] and [18]. General existence results for compact homogeneous spaces are difficult to obtain and some methods are described in [7, 8] and also existence and non existence problem is investigated in [19].

Left invariant Einstein metrics on compact semi-simple Lie groups have been studied in [13] by J.E. D'Atri and W. Ziller who proved existence of several naturally reductive Einstein metrics. First non naturally reductive Einstein metrics on compact semi-simple Lie groups were obtained by K. Mori in [14] for $SU(n)$ for $n \geq 6$, by using a method of Riemannian submersions and generalized flag manifolds. Then more non naturally reductive Einstein metrics on compact simple Lie groups were found in [2] by investigating left invariant metrics associated to the generalized flag manifolds. Chen and Liang [11] found a non naturally reductive Einstein metric on the compact simple Lie group F_4. More recently, the authors have obtained non naturally reductive Einstein metrics on the compact classical Lie groups SO(n), Sp(n) and SU(n) in [3–5]. I. Chrysikos and the second author obtained more non naturally reductive Einstein metrics on the compact exceptional Lie groups in [12]. Further, by studying left invariant metrics associated to the generalized Wallach space, H. Chen, Z. Chen and S. Deng [9] obtained more non naturally reductive Einstein metrics on the compact exceptional Lie groups E_6, E_7, E_8 and F_4, and Z. Chen, S. Deng and L. Zhang [10] obtained for E_7.

In the present article we investigate invariant Einstein metrics on compact unitary groups SU($2n$) which are not naturally reductive, by studying left invariant metrics associated to the generalized Wallach space SU($2n$)/U(n). Our main results are the following:

Theorem 1.1. *The compact Lie groups* SU($2n$) *admit, besides a bi-invariant metric, seven* Ad(U(n))*-invariant Einstein metrics for* $n \geq 60$. *Three of them are naturally reductive Einstein metrics and the other four metrics are non naturally reductive Einstein metrics. For* $3 \leq n \leq 59$, *there are five* Ad(U(n))*-invariant Einstein metrics. Three of them are naturally reductive Einstein metrics and the other two metrics are non naturally reductive Einstein metrics.*

For $n = 2$ we see that all Ad(U(2))-invariant Einstein metrics on SU(4) are naturally reductive (see Remark 6.1).

In section 2 we refer a formula for Ricci tensor for compact Lie group in case when the metrics have more symmetry, that is, the metrics are associated with a homogeneous space. In section 3 we consider the generalized Wallach space SU($2n$)/U(n) due to Nikonorov [15] and a decomposition of $\mathfrak{su}(2n)$ associated. We compute Ricci tensor for the associated left invariant metrics in section 4. In section 5 we give conditions for the metric being naturally reductive. In section 6 we give new Einstein metrics on SU($2n$) by solving algebraic equations. We compute Gröbner basis with the aid of a computer. But the obtained Gröbner base are huge, so we carry the files in the Supplementary Material(s) webpage https://www.worldscientific.com/worldscibooks/10.1142/11454#t=suppl. By analyzing these polynomials, we prove existence of non naturally reductive Einstein metrics.

2. The Ricci tensor for compact Lie groups

We recall an expression for the Ricci tensor for a left invariant metric g on a compact Lie group G. The Ricci tensor r of the metric g is given as follows ([6] p. 381):

$$r(X,Y) = -\frac{1}{2}\sum_i g([X,X_i],[Y,X_i]) - \frac{1}{2}B(X,Y)$$
$$+\frac{1}{4}\sum_{i,j} g([X_i,X_j],X)g([X_i,X_j],Y), \qquad (1)$$

where $\{X_j\}$ is an orthonormal basis of $\mathfrak{g}$ with respect to the metric g and B is the Killing form of the Lie algebra $\mathfrak{g}$. In general, from the equation (1), it is not easy to obtain the condition for a left invariant metric being Einstein. We consider a subset of all left invariant metrics on G. Now we assume that G is a compact semi-simple Lie group and consider a connected closed subgroup H of G. Let $\mathfrak{g}$ and $\mathfrak{h}$ be the corresponding Lie algebras of G and H respectively. The Killing form B of $\mathfrak{g}$ is negative definite, so we can define an $\mathrm{Ad}(G)$-invariant inner product $-B$ on $\mathfrak{g}$. Let $\mathfrak{g} = \mathfrak{h} \oplus \mathfrak{p}$ be a reductive decomposition of $\mathfrak{g}$ with respect to $-B$, that is, let $\mathfrak{p}$ be the orthogonal complement of $\mathfrak{h}$, then we have $\mathrm{Ad}(H)\mathfrak{p} \subset \mathfrak{p}$.

We decompose the space $\mathfrak{p}$ into irreducible $\mathrm{Ad}(H)$-modules as follows:

$$\mathfrak{p} = \mathfrak{p}_1 \oplus \cdots \oplus \mathfrak{p}_q.$$

Let $\mathfrak{h} = \mathfrak{h}_0 \oplus \mathfrak{h}_1 \oplus \cdots \oplus \mathfrak{h}_p$ be the decomposition into ideals of $\mathfrak{h}$, where $\mathfrak{h}_0$ is the center of $\mathfrak{h}$ and $\mathfrak{h}_i$ $(i = 1,\ldots,p)$ are simple ideals of $\mathfrak{h}$. We assume that the Lie algebra $\mathfrak{g}$ can be decomposed into mutually non equivalent irreducible $\mathrm{Ad}(H)$-modules:

$$\mathfrak{g} = \mathfrak{h}_0 \oplus \mathfrak{h}_1 \oplus \cdots \oplus \mathfrak{h}_p \oplus \mathfrak{p}_1 \oplus \cdots \oplus \mathfrak{p}_q. \qquad (2)$$

For simplicity, we write the decomposition (2) as

$$\mathfrak{g} = \mathfrak{m}_0 \oplus \mathfrak{m}_1 \oplus \cdots \oplus \mathfrak{m}_p \oplus \mathfrak{m}_{p+1} \oplus \cdots \oplus \mathfrak{m}_{p+q}. \qquad (3)$$

Then an $\mathrm{Ad}(H)$-invariant inner product $\langle\ ,\ \rangle$ on $\mathfrak{g}$ which give rise to a left invariant metric on G can be expressed as

$$\langle\ ,\ \rangle = x_0(-B)|_{\mathfrak{m}_0} + x_1(-B)|_{\mathfrak{m}_1} + \cdots + x_{p+q}(-B)|_{\mathfrak{m}_{p+q}}, \qquad (4)$$

for positive real parameters $(x_0,x_1,\ldots,x_{p+q}) \in \mathbb{R}_+^{p+q+1}$. Moreover, the Ricci tensor r for the metric (4) is of the same form

$$r = z_0(-B)|_{\mathfrak{m}_0} + z_1(-B)|_{\mathfrak{m}_1} + \cdots + z_{p+q}(-B)|_{\mathfrak{m}_{p+q}}, \qquad (5)$$

for $(z_0,z_1,\ldots,z_{p+q}) \in \mathbb{R}^{p+q+1}$, since Ricci tensor r is also $\mathrm{Ad}(H)$-invariant.

Let $\{e_\alpha\}$ be a $(-B)$-orthonormal basis adapted to the decomposition of $\mathfrak{m}$, i.e. $e_\alpha \in \mathfrak{m}_i$ for some i, and $\alpha < \beta$ if $i < j$. We put $A_{\alpha\beta}^\gamma = -B([e_\alpha, e_\beta], e_\gamma)$ so that $[e_\alpha, e_\beta] = \sum_\gamma A_{\alpha\beta}^\gamma e_\gamma$ and set $\begin{bmatrix} k \\ ij \end{bmatrix} = \sum (A_{\alpha\beta}^\gamma)^2$, where the sum is taken over all indices α, β, γ with $e_\alpha \in \mathfrak{m}_i, e_\beta \in \mathfrak{m}_j, e_\gamma \in \mathfrak{m}_k$ (cf. [19]). Then the positive numbers $\begin{bmatrix} k \\ ij \end{bmatrix}$ are independent of the $(-B)$-orthonormal bases chosen for $\mathfrak{m}_i, \mathfrak{m}_j, \mathfrak{m}_k$, and $\begin{bmatrix} k \\ ij \end{bmatrix} = \begin{bmatrix} k \\ ji \end{bmatrix} = \begin{bmatrix} j \\ ki \end{bmatrix}$.

Let $d_k = \dim \mathfrak{m}_k$. Then, by the same argument as in [16], we have the following:

Lemma 2.1. *The components $r_0, r_1, \ldots, r_{p+q}$ of the Ricci tensor r of the metric $\langle\ ,\ \rangle$ of the form (4) on G are given by*

$$r_k = \frac{1}{2x_k} + \frac{1}{4d_k} \sum_{j,i} \frac{x_k}{x_j x_i} \begin{bmatrix} k \\ ji \end{bmatrix} - \frac{1}{2d_k} \sum_{j,i} \frac{x_j}{x_k x_i} \begin{bmatrix} j \\ ki \end{bmatrix} \tag{6}$$

$(k = 0, 1, \ldots, p+q)$, *where the sum is taken over $i, j = 0, 1, \ldots, p+q$. Moreover, for each k we have* $\sum_{i,j} \begin{bmatrix} j \\ ki \end{bmatrix} = d_k$.

By Lemma 2.1, it follows that left invariant Einstein metrics on G corresponding to $\mathrm{Ad}(H)$-invariant inner products on $\mathfrak{g}$ are exactly the positive solutions $(x_0, x_1, \ldots, x_{p+q}) \in \mathbb{R}_+^{p+q+1}$ of the system of equations $\{r_0 = \lambda, r_1 = \lambda, \ldots, r_{p+q} = \lambda\}$, where $\lambda \in \mathbb{R}_+$ is the Einstein constant.

3. A decomposition of $\mathfrak{su}(2n)$

We recall the notion of generalized Wallach spaces due to Nikonorov [15]. Let G/H be an almost effective compact homogeneous space G/H with connected compact semisimple Lie group G and its compact subgroup H. Denote by $\mathfrak{g}$ and $\mathfrak{h}$ Lie algebras of G and H respectively. Let $\mathfrak{p}$ be the $(-B)$-orthogonal complement to $\mathfrak{h}$ in $\mathfrak{g}$. A compact homogeneous space G/H is called a *generalized Wallach space* if the module $\mathfrak{p}$ is decomposed as a direct sum of three $\mathrm{Ad}(H)$-invariant irreducible modules pairwise orthogonal with respect to $-B$, that is,

$$\mathfrak{p} = \mathfrak{p}_1 \oplus \mathfrak{p}_2 \oplus \mathfrak{p}_3 \tag{7}$$

such that

$$[\mathfrak{p}_i, \mathfrak{p}_i] \subset \mathfrak{h} \quad \text{for}\ \ i \in \{1, 2, 3\}. \tag{8}$$

The inclusion (8) implies that for all $i \in \{1, 2, 3\}$

$$\mathfrak{k}_i = \mathfrak{h} \oplus \mathfrak{p}_i \qquad (9)$$

is a subalgebra of $\mathfrak{g}$, and the pair $(\mathfrak{k}_i, \mathfrak{h})$ is irreducible symmetric (it could be non-effective). From (7) and (8) we easily get that $[\mathfrak{p}_j, \mathfrak{p}_k] \subset \mathfrak{p}_i$ for pairwise distinct i, j, k. Therefore,

$$[\mathfrak{p}_j \oplus \mathfrak{p}_k, \mathfrak{p}_j \oplus \mathfrak{p}_k] \subset \mathfrak{h} \oplus \mathfrak{p}_i = \mathfrak{k}_i, \qquad \{i, j, k\} = \{1, 2, 3\}$$

and all the pairs $(\mathfrak{g}, \mathfrak{k}_i)$ are also irreducible symmetric.

We consider the generalized Wallach space $G/H = \mathrm{SU}(2n)/\mathrm{U}(n)$. Then the tangent space $\mathfrak{p}$ of G/H at eH decomposes into three non equivalent irreducible $\mathrm{Ad}(H)$-submodules

$$\mathfrak{p} = \mathfrak{p}_1 \oplus \mathfrak{p}_2 \oplus \mathfrak{p}_3,$$

where $\mathfrak{k}_1 = \mathfrak{u}(n) \oplus \mathfrak{p}_1 = \mathfrak{so}(2n)$, $\mathfrak{k}_2 = \mathfrak{u}(n) \oplus \mathfrak{p}_2 = \mathfrak{sp}(n)$ and $\mathfrak{k}_3 = \mathfrak{u}(n) \oplus \mathfrak{p}_3 = \mathfrak{s}(\mathfrak{u}(n) \oplus \mathfrak{u}(n))$.

The Lie algebra $\mathfrak{su}(2n)$ splits into $\mathrm{Ad}(\mathrm{U}(n))$-irreducible modules which are non equivalent each other as follows:

$$\mathfrak{su}(2n) = \mathfrak{h}_0 \oplus \mathfrak{su}(n) \oplus \mathfrak{p}_1 \oplus \mathfrak{p}_2 \oplus \mathfrak{p}_3 \qquad (10)$$

where $\mathfrak{h}_0$ is the center of $\mathfrak{u}(n)$. We rewrite the decompositions (10) as follows:

$$\mathfrak{su}(2n) = \mathfrak{m}_0 \oplus \mathfrak{m}_1 \oplus \mathfrak{m}_2 \oplus \mathfrak{m}_3 \oplus \mathfrak{m}_4. \qquad (11)$$

This is an orthogonal decomposition with respect to $-B$. Note that $d_0 = \dim \mathfrak{m}_0 = 1$, $d_1 = \dim \mathfrak{m}_1 = n^2 - 1$, $d_2 = \dim \mathfrak{m}_2 = n^2 - n$, $d_3 = \dim \mathfrak{m}_3 = n^2 + n$ and $d_4 = \dim \mathfrak{m}_4 = n^2 - 1$.

Lemma 3.1. *The submodules in the decomposition (11) satisfy the following bracket relations:*

$$[\mathfrak{m}_0, \mathfrak{m}_1] = (0), \qquad [\mathfrak{m}_0, \mathfrak{m}_j] \subset \mathfrak{m}_j, (j = 2, 3, 4), \qquad [\mathfrak{m}_1, \mathfrak{m}_1] = \mathfrak{m}_1,$$

$$[\mathfrak{m}_1, \mathfrak{m}_i] \subset \mathfrak{m}_i, (i = 2, 3, 4), \qquad [\mathfrak{m}_2, \mathfrak{m}_3] \subset \mathfrak{m}_4.$$

Therefore, we see that the possibility of non zero numbers (up to permutation of indices) are

$$\begin{bmatrix} 2 \\ 02 \end{bmatrix}, \begin{bmatrix} 3 \\ 03 \end{bmatrix}, \begin{bmatrix} 4 \\ 04 \end{bmatrix}, \begin{bmatrix} 1 \\ 11 \end{bmatrix}, \begin{bmatrix} 2 \\ 12 \end{bmatrix}, \begin{bmatrix} 3 \\ 13 \end{bmatrix}, \begin{bmatrix} 4 \\ 14 \end{bmatrix}, \begin{bmatrix} 2 \\ 34 \end{bmatrix}. \qquad (12)$$

In order to compute the triplets $\begin{bmatrix} i \\ jk \end{bmatrix}$ we need the following lemma (see Lemma 4.1 in [1], Lemma 5.2 in [4]).

Lemma 3.2. *Let* $\mathfrak{q}$ *be a simple subalgebra of* $\mathfrak{g} = \mathfrak{su}(2n)$. *Consider an orthonormal basis* $\{f_j\}$ *of* $\mathfrak{q}$ *with respect to* $-B$ *(negative of the Killing form of* $\mathfrak{su}(2n)$), *and denote by* $B_{\mathfrak{q}}$ *the Killing form of* $\mathfrak{q}$. *Then, for* $i = 1, \ldots, \dim \mathfrak{q}$, *we have*

$$\sum_{j,k=1}^{\dim \mathfrak{q}} \left(- B([f_i, f_j], f_k) \right)^2 = \alpha^{\mathfrak{q}}_{\mathfrak{su}(2n)},$$

where $\alpha^{\mathfrak{q}}_{\mathfrak{su}(2n)}$ *is the constant determined by* $B_{\mathfrak{q}} = \alpha^{\mathfrak{q}}_{\mathfrak{su}(2n)} \cdot B|_{\mathfrak{q}}$.

Lemma 3.3. *The triplets* $\begin{bmatrix} i \\ jk \end{bmatrix}$ *are given by:*

$$\begin{bmatrix} 2 \\ 02 \end{bmatrix} = \frac{n-1}{2n}, \quad \begin{bmatrix} 3 \\ 03 \end{bmatrix} = \frac{n+1}{2n}, \quad \begin{bmatrix} 4 \\ 04 \end{bmatrix} = 0,$$

$$\begin{bmatrix} 1 \\ 11 \end{bmatrix} = \frac{n^2-1}{4}, \quad \begin{bmatrix} 2 \\ 12 \end{bmatrix} = \frac{(n^2-1)(n-2)}{4n}, \quad \begin{bmatrix} 3 \\ 13 \end{bmatrix} = \frac{(n^2-1)(n+2)}{4n},$$

$$\begin{bmatrix} 4 \\ 14 \end{bmatrix} = \frac{n^2-1}{4}, \quad \begin{bmatrix} 4 \\ 23 \end{bmatrix} = \frac{n^2-1}{4}.$$

Proof. First note that Killing forms of $\mathfrak{so}(2n)$, $\mathfrak{sp}(n)$, $\mathfrak{su}(m)$ are given as follows:

$$B_{\mathfrak{so}(2n)}(X,Y) = -(2n-2)\operatorname{tr}(XY), \quad B_{\mathfrak{sp}(n)}(X,Y) = -2(n+1)\operatorname{tr}(XY),$$
$$B_{\mathfrak{su}(m)}(X,Y) = -2m\operatorname{tr}(XY).$$

Note also that $\mathfrak{u}(n)$ is imbedded in $\mathfrak{so}(2n) \subset \mathfrak{su}(2n)$ by

$$A + iB \to \begin{pmatrix} A & B \\ -B & A \end{pmatrix}$$

where A is a real skew symmetric matrix and B is a real symmetric matrix. Thus, for each $f_i \in \mathfrak{su}(n) \subset \mathfrak{su}(2n)$, we have that

$$\sum_{j,k=1}^{\dim \mathfrak{su}(n)} \left(- B([f_i, f_j], f_k) \right)^2 = \frac{1}{2}\alpha^{\mathfrak{su}(n)}_{\mathfrak{su}(2n)} = \frac{1}{2} 2n/(4n) = 1/4$$

and that $\begin{bmatrix} 1 \\ 11 \end{bmatrix} = (n^2-1)/4$.

Now, for each $f_i \in \mathfrak{so}(2n)$, we have

$$\sum_{j,k=1}^{\dim \mathfrak{so}(2n)} \left(- B([f_i, f_j], f_k) \right)^2 = (2n-2)/(4n) = (n-1)/(2n).$$

In particular, we see that $\begin{bmatrix} 2 \\ 02 \end{bmatrix} = (n-1)/(2\,n)$, since $\mathfrak{m}_0 = \mathrm{span}\{f_1\}$.

Now we have that

$$\begin{bmatrix} 1 \\ 11 \end{bmatrix} + \begin{bmatrix} 2 \\ 12 \end{bmatrix} = \sum_{i=1}^{\dim\,\mathfrak{su}(n)} \sum_{j,k=1}^{\dim\,\mathfrak{so}(2n)} \left(-B([f_i, f_j], f_k) \right)^2 = (n^2-1)(n-1)/(2\,n).$$

Hence, we obtain

$$\begin{bmatrix} 2 \\ 12 \end{bmatrix} = (n^2-1)(n-1)/(2\,n) - (n^2-1)/4 = (n^2-1)(n-2)/(4n).$$

For each $f_i \in \mathfrak{sp}(n)$, we have

$$\sum_{j,k=1}^{\dim\,\mathfrak{sp}(n)} \left(-B([f_i, f_j], f_k) \right)^2 = 2(n+1)/(4n) = (n+1)/(2\,n).$$

and

$$\sum_{i=1}^{\dim\,\mathfrak{su}(n)} \sum_{j,k=1}^{\dim\,\mathfrak{sp}(n)} \left(-B([f_i, f_j], f_k) \right)^2 = (n^2-1)(n+1)/(2\,n).$$

We also see that

$$\sum_{i=1}^{\dim\,\mathfrak{su}(n)} \sum_{j,k=1}^{\dim\,\mathfrak{sp}(n)} \left(-B([f_i, f_j], f_k) \right)^2 = \begin{bmatrix} 1 \\ 11 \end{bmatrix} + \begin{bmatrix} 3 \\ 13 \end{bmatrix}.$$

Hence, we obtain

$$\begin{bmatrix} 3 \\ 13 \end{bmatrix} = (n^2-1)(n+1)/(2\,n) - (n^2-1)/4 = (n^2-1)(n+2)/(4n).$$

Note that

$$\begin{bmatrix} 1 \\ 11 \end{bmatrix} + \begin{bmatrix} 2 \\ 12 \end{bmatrix} + \begin{bmatrix} 3 \\ 13 \end{bmatrix} + \begin{bmatrix} 4 \\ 14 \end{bmatrix} = n^2 - 1.$$

Thus we see that $\begin{bmatrix} 4 \\ 14 \end{bmatrix} = (n^2-1)/4$. Since

$$\begin{bmatrix} 2 \\ 20 \end{bmatrix} + \begin{bmatrix} 0 \\ 22 \end{bmatrix} + \begin{bmatrix} 1 \\ 22 \end{bmatrix} + \begin{bmatrix} 2 \\ 21 \end{bmatrix} + \begin{bmatrix} 4 \\ 23 \end{bmatrix} + \begin{bmatrix} 3 \\ 24 \end{bmatrix} = n^2 - n,$$

we see that $\begin{bmatrix} 4 \\ 23 \end{bmatrix} = (n^2-1)/4$. $\qquad\square$

4. Ricci tensor of the compact Lie group SU(2n)

We now consider the compact Lie group $G = \mathrm{SU}(2n)$, the decomposition (10) of its Lie algebra and a subset of all left invariant metrics on G, determined by the $\mathrm{Ad}(H) = \mathrm{Ad}(\mathrm{U}(n)))$-invariant scalar products $\langle\ ,\ \rangle$ on $\mathfrak{g}$ given by

$$x_0(-B)|_{\mathfrak{h}_0} + x_1(-B)|_{\mathfrak{su}(n)} + x_2(-B)|_{\mathfrak{p}_1} + x_3(-B)|_{\mathfrak{p}_2} + x_4(-B)|_{\mathfrak{p}_3} \quad (13)$$

where $x_i > 0$ $(i = 0, 1, 2, 3, 4)$.

We use the formula for the Ricci tensor in Lemma 2.1 to obtain the following:

Lemma 4.1. *The components of the Ricci tensor r for the $\mathrm{Ad}(H)$-invariant scalar product $\langle , \rangle$ on $G = \mathrm{SU}(2n)$ defined by (13) are given as follows:*

$$r_0 = \frac{1}{4}\left(\begin{bmatrix}0\\22\end{bmatrix}\frac{x_0}{x_2{}^2} + \begin{bmatrix}0\\33\end{bmatrix}\frac{x_0}{x_3{}^2} + \begin{bmatrix}0\\44\end{bmatrix}\frac{x_0}{x_4{}^2}\right),$$

$$r_1 = \frac{1}{4d_1 x_1}\begin{bmatrix}1\\11\end{bmatrix} + \frac{1}{4d_1}\left(\begin{bmatrix}1\\22\end{bmatrix}\frac{x_1}{x_2{}^2} + \begin{bmatrix}1\\33\end{bmatrix}\frac{x_1}{x_3{}^2} + \begin{bmatrix}1\\44\end{bmatrix}\frac{x_1}{x_4{}^2}\right),$$

$$r_2 = \frac{1}{2x_2} + \frac{1}{2d_2}\begin{bmatrix}2\\34\end{bmatrix}\left(\frac{x_2}{x_3 x_4} - \frac{x_3}{x_2 x_4} - \frac{x_4}{x_2 x_3}\right) - \frac{1}{2d_2}\left(\begin{bmatrix}0\\22\end{bmatrix}\frac{x_0}{x_2{}^2} + \begin{bmatrix}1\\22\end{bmatrix}\frac{x_1}{x_2{}^2}\right),$$

$$r_3 = \frac{1}{2x_3} + \frac{1}{2d_3}\begin{bmatrix}2\\34\end{bmatrix}\left(\frac{x_3}{x_2 x_4} - \frac{x_2}{x_3 x_4} - \frac{x_4}{x_2 x_3}\right) - \frac{1}{2d_3}\left(\begin{bmatrix}0\\33\end{bmatrix}\frac{x_0}{x_3{}^2} + \begin{bmatrix}1\\33\end{bmatrix}\frac{x_1}{x_3{}^2}\right),$$

$$r_4 = \frac{1}{2x_4} + \frac{1}{2d_4}\begin{bmatrix}2\\34\end{bmatrix}\left(\frac{x_4}{x_2 x_3} - \frac{x_3}{x_2 x_4} - \frac{x_2}{x_4 x_3}\right) - \frac{1}{2d_4}\left(\begin{bmatrix}0\\44\end{bmatrix}\frac{x_0}{x_4{}^2} + \begin{bmatrix}1\\44\end{bmatrix}\frac{x_1}{x_4{}^2}\right).$$

Then by taking into account Lemma 3.3 we obtain the following:

Proposition 4.1. *The components of the Ricci tensor r for the $\mathrm{Ad}(H)$-invariant scalar product $\langle\ ,\ \rangle$ on $G = \mathrm{SU}(2n)$ defined by (13) are given as follows:*

$$r_0 = \frac{1}{4}\left(\frac{n-1}{2n}\frac{x_0}{x_2{}^2} + \frac{n+1}{2n}\frac{x_0}{x_3{}^2}\right),$$

$$r_1 = \frac{1}{16x_1} + \frac{1}{16}\left(\frac{(n-2)}{n}\frac{x_1}{x_2{}^2} + \frac{(n+2)}{n}\frac{x_1}{x_3{}^2} + \frac{x_1}{x_4{}^2}\right),$$

$$r_2 = \frac{1}{2x_2} + \frac{n+1}{8n}\left(\frac{x_2}{x_3 x_4} - \frac{x_3}{x_2 x_4} - \frac{x_4}{x_2 x_3}\right) - \frac{1}{4n^2}\left(\frac{x_0}{x_2{}^2} + \frac{(n-2)(n+1)}{2}\frac{x_1}{x_2{}^2}\right),$$

$$r_3 = \frac{1}{2x_3} + \frac{n-1}{8n}\left(\frac{x_3}{x_2 x_4} - \frac{x_2}{x_3 x_4} - \frac{x_4}{x_2 x_3}\right) - \frac{1}{4n^2}\left(\frac{x_0}{x_3{}^2} + \frac{(n+2)(n-1)}{2}\frac{x_1}{x_3{}^2}\right),$$

$$r_4 = \frac{1}{2x_4} + \frac{1}{8}\left(\frac{x_4}{x_2 x_3} - \frac{x_2}{x_3 x_4} - \frac{x_3}{x_2 x_4}\right) - \frac{1}{8}\frac{x_1}{x_4{}^2}.$$

We consider the system of equations

$$r_0 = r_1, \quad r_1 = r_2, \quad r_2 = r_3, \quad r_3 = r_4. \tag{14}$$

Then finding Einstein metrics of the form (13) reduces to finding positive solutions of the system (14).

5. Naturally reductive metrics on the compact Lie group SU(2n)

We recall the basic results of D'Atri and Ziller in [13], where they have studied naturally reductive metrics on compact Lie groups and gave a complete classification in the case of simple Lie groups. Let G be a compact, connected semisimple Lie group, L a closed subgroup of G and let $\mathfrak{g}$ be the Lie algebra of G and $\mathfrak{l}$ the subalgebra corresponding to L. Let $\mathfrak{g} = \mathfrak{l} \oplus \mathfrak{m}$ be a reductive decomposition of $\mathfrak{g}$ with respect to the negative of the Killing form $-B$ so that $[\mathfrak{l}, \mathfrak{m}] \subset \mathfrak{m}$ and $\mathfrak{m} \cong T_o(G/L)$. Let $\mathfrak{l} = \mathfrak{l}_0 \oplus \mathfrak{l}_1 \oplus \cdots \oplus \mathfrak{l}_p$ be a decomposition of $\mathfrak{l}$ into ideals, where $\mathfrak{l}_0$ is the center of $\mathfrak{l}$ and $\mathfrak{l}_i$ $(i = 1, \ldots, p)$ are simple ideals of $\mathfrak{l}$. Let $A_0|_{\mathfrak{l}_0}$ be an arbitrary metric on $\mathfrak{l}_0$.

Theorem 5.1 ([13], Theorem 1, p. 9 and Theorem 3, p. 14). *Under the notations above, a left invariant metric on G of the form*

$$\langle\,,\,\rangle = x \cdot (-B)|_{\mathfrak{m}} + A_0|_{\mathfrak{l}_0} + u_1 \cdot (-B)|_{\mathfrak{l}_1} + \cdots + u_p \cdot (-B)|_{\mathfrak{l}_p}, \tag{15}$$

with positive $x, u_1, \cdots, u_p$ is naturally reductive with respect to $G \times L$, where $G \times L$ acts on G by $(g, l)y = gyl^{-1}$.

Moreover, if a left invariant metric $\langle\,,\,\rangle$ on a compact simple Lie group G is naturally reductive, then there is a closed subgroup L of G and the metric $\langle\,,\,\rangle$ is given by the form (15).

Proposition 5.1. *If a left invariant metric $\langle\,,\,\rangle$ of the form (13) on $\mathrm{SU}(2n)$ $(n \geq 2)$ is naturally reductive with respect to $\mathrm{SU}(2n) \times L$ for some closed subgroup L of $\mathrm{SU}(2n)$, then one of the following holds:*

1) *for $n \geq 3$, $x_0 = x_1 = x_2$, $x_3 = x_4$ and for $n = 2$, $x_0 = x_2$, $x_3 = x_4$,*
2) *$x_0 = x_1 = x_3$, $x_2 = x_4$,*
3) *$x_1 = x_4$, $x_2 = x_3$,*
4) *$x_2 = x_3 = x_4$.*

Conversely, if one of the conditions 1), 2), 3), 4) *is satisfied, then the metric* $\langle \, , \, \rangle$ *of the form* (13) *is naturally reductive with respect to* $\mathrm{SU}(2n) \times L$ *for some closed subgroup* L *of* $\mathrm{SU}(2n)$.

Proof. Let $\mathfrak{l}$ be the Lie algebra of L. Then we have that $\mathfrak{l} \subset \mathfrak{h}$ or $\mathfrak{l} \not\subset \mathfrak{h}$. First we consider the case of $\mathfrak{l} \not\subset \mathfrak{h}$. Let $\mathfrak{k}$ be the subalgebra of $\mathfrak{su}(2n)$ generated by $\mathfrak{l}$ and $\mathfrak{h}$. Then the metric $\langle \, , \, \rangle$ of the form (13) is ad($\mathfrak{k}$)-invariant. Since $\mathfrak{su}(2n) = \mathfrak{h}_0 \oplus \mathfrak{su}(n) \oplus \mathfrak{p}_1 \oplus \mathfrak{p}_2 \oplus \mathfrak{p}_3$ is an irreducible decomposition as Ad(H)-modules, we see that the Lie algebra $\mathfrak{k}$ contains at least one of $\mathfrak{p}_1$, $\mathfrak{p}_2$, $\mathfrak{p}_3$. We first consider the case that $\mathfrak{k}$ contains $\mathfrak{p}_1$. Since $[\mathfrak{p}_1, \mathfrak{p}_1] \subset \mathfrak{h}_0 \oplus \mathfrak{su}(n) = \mathfrak{h}$ and $\mathfrak{h}_0 \oplus \mathfrak{su}(n) \oplus \mathfrak{p}_1 = \mathfrak{so}(2n)$, we see that $\mathfrak{k}$ contains $\mathfrak{so}(2n)$. Note that, for $n \geq 3$, $\mathfrak{so}(2n)$ is a simple Lie algebra and $\mathfrak{so}(4) = \mathfrak{su}(2) \oplus \mathfrak{su}(2)$. Thus the metric $\langle \, , \, \rangle$ satisfies the property of 1). Similarly, if $\mathfrak{k}$ contains $\mathfrak{p}_2$, then $\mathfrak{k}$ contains $\mathfrak{sp}(n)$ and the metric $\langle \, , \, \rangle$ satisfies the property of 2). If $\mathfrak{k}$ contains $\mathfrak{p}_3$, then $\mathfrak{k}$ contains $\mathfrak{h}_0 \oplus \mathfrak{su}(n) \oplus \mathfrak{p}_3$. Since the pair $(\mathfrak{su}(n) \oplus \mathfrak{p}_3, \mathfrak{su}(n))$ is irreducible symmetric and the metric $\langle \, , \, \rangle$ is ad($\mathfrak{su}(n) \oplus \mathfrak{p}_3$)-invariant, we obtain the property of 3).

In case of $\mathfrak{l} \subset \mathfrak{h}$, noting that the orthogonal complement $\mathfrak{l}^\perp$ of $\mathfrak{l}$ with respect to $-B$ contains the orthogonal complement $\mathfrak{h}^\perp$ of $\mathfrak{h}$, we see that $x_2 = x_3 = x_4$ by Theorem 5.1.

The converse is a direct consequence of Theorem 5.1.   □

6. New invariant Einstein metrics on $\mathrm{SU}(2n)$

We normalise our equations by setting $x_2 = 1$ and obtain the following system of equations from (14) for the variables x_0, x_1, x_3 and x_4:

$$
\left\{
\begin{aligned}
g_1 &= 2nx_0x_1x_3^2x_4^2 + 2nx_0x_1x_4^2 - nx_1^2x_3^2x_4^2 - nx_1^2x_3^2 \\
&\quad -nx_1^2x_4^2 - nx_3^2x_4^2 - 2x_0x_1x_3^2x_4^2 + 2x_0x_1x_4^2 \\
&\quad +2x_1^2x_3^2x_4^2 - 2x_1^2x_4^2 = 0, \\
g_2 &= 3n^2x_1^2x_3^2x_4^2 + n^2x_1^2x_3^2 + n^2x_1^2x_4^2 + 2n^2x_1x_3^3x_4 \\
&\quad -8n^2x_1x_3^2x_4^2 + 2n^2x_1x_3x_4^3 - 2n^2x_1x_3x_4 + n^2x_3^2x_4^2 \\
&\quad -4nx_1^2x_3^2x_4^2 + 2nx_1^2x_4^2 + 2nx_1x_3^3x_4 + 2nx_1x_3x_4^3 \\
&\quad -2nx_1x_3x_4 + 4x_0x_1x_3^2x_4^2 - 4x_1^2x_3^2x_4^2 = 0, \\
g_3 &= 2 - n^2x_1x_3^2x_4 + n^2x_1x_4 - 2n^2x_3^3 + 4n^2x_3^2x_4 - 4n^2x_3x_4 \\
&\quad +2n^2x_3 + nx_1x_3^2x_4 + nx_1x_4 - 2nx_3x_4^2 - 2x_0x_3^2x_4 \\
&\quad +2x_0x_4 + 2x_1x_3^2x_4 - 2x_1x_4 = 0, \\
g_4 &= n^2x_1x_3^2 - n^2x_1x_4^2 + 2n^2x_3^3x_4 - 4n^2x_3^2x_4 - 2n^2x_3x_4^3 \\
&\quad +4n^2x_3x_4^2 - nx_1x_4^2 - nx_3^3x_4 + nx_3x_4^3 + nx_3x_4 \\
&\quad -2x_0x_4^2 + 2x_1x_4^2 = 0.
\end{aligned}
\right.
\tag{16}
$$

By solving equation $g_4 = 0$ in the system (16), we obtain that

$$x_0 = \frac{-1}{2x_4{}^2}\left(-n^2x_1x_3{}^2 + n^2x_1x_4{}^2 - 2n^2x_3{}^3x_4 + 4n^2x_3{}^2x_4 + 2n^2x_3x_4{}^3\right.$$
$$\left. - 4n^2x_3x_4{}^2 + nx_1x_4{}^2 + nx_3{}^3x_4 - nx_3x_4{}^3 - nx_3x_4 - 2x_1x_4{}^2\right). \tag{17}$$

By substituting (17) into the system (16), we obtain the system of equations:

$$\begin{cases}
f_1 = n^2x_1{}^2x_3{}^4 - n^2x_1{}^2x_3{}^2x_4{}^2 + n^2x_1{}^2x_3{}^2 - n^2x_1{}^2x_4{}^2 + 2n^2x_1x_3{}^5x_4 \\
\quad -4n^2x_1x_3{}^4x_4 - 2n^2x_1x_3{}^3x_4{}^3 + 4n^2x_1x_3{}^3x_4{}^2 + 2n^2x_1x_3{}^3x_4 \\
\quad -4n^2x_1x_3{}^2x_4 - 2n^2x_1x_3x_4{}^3 + 4n^2x_1x_3x_4{}^2 - nx_1{}^2x_3{}^4 \\
\quad +nx_1{}^2x_3{}^2 - 2nx_1{}^2x_4{}^2 - 3nx_1x_3{}^5x_4 + 4nx_1x_3{}^4x_4 \\
\quad +3nx_1x_3{}^3x_4{}^3 - 4nx_1x_3{}^3x_4{}^2 + 2nx_1x_3{}^3x_4 - 4nx_1x_3{}^2x_4 \\
\quad -nx_1x_3x_4{}^3 + 4nx_1x_3x_4{}^2 + nx_1x_3x_4 + 2x_1{}^2x_3{}^2x_4{}^2 - x_1{}^2x_3{}^2 \\
\quad +x_1x_3{}^5x_4 - x_1x_3{}^3x_4{}^3 - 2x_1x_3{}^3x_4 + x_1x_3x_4{}^3 + x_1x_3x_4 \\
\quad -x_3{}^2x_4{}^2 = 0, \\
f_2 = 2nx_1{}^2x_3{}^4 + nx_1{}^2x_3{}^2x_4{}^2 + nx_1{}^2x_3{}^2 + nx_1{}^2x_4{}^2 + 4nx_1x_3{}^5x_4 \\
\quad -8nx_1x_3{}^4x_4 - 4nx_1x_3{}^3x_4{}^3 + 8nx_1x_3{}^3x_4{}^2 + 2nx_1x_3{}^3x_4 \\
\quad -8nx_1x_3{}^2x_4{}^2 + 2nx_1x_3x_4{}^3 - 2nx_1x_3x_4 + nx_3{}^2x_4{}^2 \\
\quad -6x_1{}^2x_3{}^2x_4{}^2 + 2x_1{}^2x_4{}^2 - 2x_1x_3{}^5x_4 + 2x_1x_3{}^3x_4{}^3 + 4x_1x_3{}^3x_4 \\
\quad +2x_1x_3x_4{}^3 - 2x_1x_3x_4 = 0, \\
f_3 = nx_1x_3{}^3 - nx_1x_3 + 2nx_3{}^4x_4 - 4nx_3{}^3x_4 - 2nx_3{}^2x_4{}^3 \\
\quad +4nx_3{}^2x_4{}^2 - 4nx_3x_4{}^2 + 4nx_3x_4 + 2nx_4{}^3 - 2nx_4 - 2x_1x_3x_4{}^2 \\
\quad -x_3{}^4x_4 + x_3{}^2x_4{}^3 + 2x_3{}^2x_4 + x_4{}^3 - x_4 = 0.
\end{cases} \tag{18}$$

We consider a polynomial ring $R = \mathbb{Q}[n][z, x_1, x_3, x_4]$ and an ideal J generated by $\{f_1, f_2, f_3, z\,x_1\,x_3\,x_4 - 1\}$ to find non zero solutions of equation (18). We take the lexicographic order $>$ with $z > x_1 > x_3 > x_4$ for a monomial ordering on R. Then, by an aid of computer, we see that a Gröbner basis for the ideal J contains the polynomial

$$(x_4 - 1)\left((n-1)x_4 - 3n - 1\right)\left((3n^2 - 1)\,x_4 - n^2 - 1\right)h_4(x_4),$$

where the polynomial $h_4(x_4)$ of x_4 with degree 28 is given by

$$h_4(x_4) = 64(n-2)^4(n-1)^3n^6(2n-1)^4\left(3n^2-1\right)^3\left(5n^3-8n^2+n-2\right)\times$$
$$\left(37n^3+24n^2+n-2\right)x_4{}^{28} - 128(n-2)^4(n-1)^2n^5(2n-1)^3\left(3n^2-1\right)^2\times$$
$$\left(5402n^{11}-8575n^{10}-1806n^9+6748n^8+196n^7-236n^6+206n^5-332n^4\right.$$
$$\left.-202n^3+63n^2+12n-4\right)x_4{}^{27} - 64(n-2)^3(n-1)n^3(2n-1)^2\left(3n^2-1\right)\times$$
$$\left(8880n^{19}-372166n^{18}+1406947n^{17}-1404788n^{16}-954207n^{15}\right.$$
$$+2493752n^{14}-918039n^{13}-803592n^{12}+647455n^{11}-28276n^{10}$$
$$-161047n^9+70196n^8+26327n^7-23512n^6+1167n^5+2384n^4-711n^3$$

$$-38n^2+60n-8)x_4{}^{26} + 64(n-2)^3n^2(2n-1)\big(562992n^{24}-10179984n^{23}$$
$$+37870292n^{22}-41813566n^{21}-34568911n^{20}+116422630n^{19}-74282195n^{18}$$
$$-40586431n^{17}+75112281n^{16}-20174005n^{15}-22624145n^{14}+16116529n^{13}$$
$$+1909107n^{12}-5118393n^{11}+823905n^{10}+908023n^9-384989n^8-25975n^7$$
$$+55897n^6-14607n^5-1880n^4+1935n^3-298n^2-28n+8\big)x_4{}^{25}$$
$$+16(n-2)^2\big(142080n^{29}-25591680n^{28}+338032396n^{27}-1433782024n^{26}$$
$$+2332931107n^{25}+186738481n^{24}-5888175167n^{23}+7658364979n^{22}$$
$$-1670881346n^{21}-4985782370n^{20}+4803263959n^{19}-358165949n^{18}$$
$$-1866287065n^{17}+909062087n^{16}+252424389n^{15}-313833243n^{14}$$
$$+20927521n^{13}+63636213n^{12}-22320015n^{11}-4691423n^{10}+4991058n^9$$
$$-656476n^8-479194n^7+171508n^6-2283n^5-5055n^4+408n^2-144n$$
$$+16)x_4{}^{24} - 16(n-2)^2n\big(1619712n^{28}-104206400n^{27}+986995624n^{26}$$
$$-3203504120n^{25}+3134184574n^{24}+4915900156n^{23}-14938710031n^{22}$$
$$+11528212742n^{21}+4416730912n^{20}-13360845878n^{19}+6786275180n^{18}$$
$$+2756271154n^{17}-4122583224n^{16}+724373410n^{15}+887429806n^{14}$$
$$-393021858n^{13}-98038716n^{12}+95350174n^{11}+1040292n^{10}-18350098n^9$$
$$+2823914n^8+2539082n^7-775207n^6-66316n^5+20780n^4+11248n^3$$
$$-5936n^2+1408n-64)x_4{}^{23} + 8(n-2)n\big(19007488n^{29}-703916896n^{28}$$
$$+6078800176n^{27}-20966478212n^{26}+25964166104n^{25}+26008617679n^{24}$$
$$-117919125843n^{23}+125733694429n^{22}+2521124625n^{21}-124560503688n^{20}$$
$$+104352864628n^{19}-2958514968n^{18}-46491008740n^{17}+25088828534n^{16}$$
$$+2904825486n^{15}-6935748174n^{14}+1421062086n^{13}+714246264n^{12}$$
$$-160000884n^{11}-159312908n^{10}+14181428n^9+45033347n^8-10269459n^7$$
$$-1984215n^6-232079n^5+515624n^4-164184n^3+31216n^2-1392n$$
$$+384)x_4{}^{22} - 8(n-2)n^2\big(76139776n^{28}-1889261168n^{27}+12978605344n^{26}$$
$$-34908386576n^{25}+20305347196n^{24}+91164179983n^{23}-207659196319n^{22}$$
$$+131259413696n^{21}+104529220382n^{20}-224622063278n^{19}+120341519778n^{18}$$
$$+32511504994n^{17}-73090955406n^{16}+29984642344n^{15}+5083504164n^{14}$$
$$-8330701526n^{13}+2450916174n^{12}-378417626n^{11}+233828974n^{10}$$
$$+95800022n^9-174430806n^8+28692169n^7+10898315n^6+3287662n^5$$
$$-3141252n^4+676088n^3-116544n^2+1920n-1152)x_4{}^{21}+n^2\big(1856516864n^{29}$$
$$-37749347840n^{28}+255528647360n^{27}-746933346816n^{26}+641382587100n^{25}$$
$$+1755993735488n^{24}-5353628548937n^{23}+4994978457987n^{22}$$
$$+1052948331766n^{21}-6503829409034n^{20}+5710479496942n^{19}$$
$$-1099603073026n^{18}-1894684780114n^{17}+1762642528742n^{16}$$
$$-527995735636n^{15}-113976683820n^{14}+165383672298n^{13}-100217401638n^{12}$$
$$+45119135602n^{11}+768247010n^{10}-11434127106n^9+2414447106n^8$$
$$+608228301n^7+279214521n^6-234977000n^5+32236040n^4-2163536n^3$$
$$-388336n^2-246912n+51456)x_4{}^{20} - 4n^3\big(1140065856n^{28}-18591182912n^{27}$$

$+104213236184n^{26}-234768521320n^{25}+18241401874n^{24}+945809905156n^{23}$
$-1805947061742n^{22}+970070124129n^{21}+1152399036963n^{20}$
$-2253473512839n^{19}+1455335394115n^{18}-55497515191n^{17}$
$-637797379025n^{16}+520475656321n^{15}-184512803145n^{14}+5578915195n^{13}$
$+45727773057n^{12}-47056771693n^{11}+19548688185n^{10}+1357191859n^{9}$
$-2301471093n^{8}-40586145n^{7}-141335693n^{6}+148174224n^{5}-10137392n^{4}$
$-232880n^{3}+162416n^{2}+151552n-33920\big)x_4{}^{19}+2n^2\big(4683840960n^{29}$
$-63112846320n^{28}+291183605024n^{27}-460181493020n^{26}-529046635232n^{25}$
$+2957838161646n^{24}-3825310252429n^{23}+463033658295n^{22}$
$+4035328061696n^{21}-4972842516820n^{20}+2491270394000n^{19}$
$+338053149940n^{18}-1540831405616n^{17}+1217051832888n^{16}$
$-539198397978n^{15}+110946104346n^{14}+110221460120n^{13}$
$-132818172260n^{12}+46320263464n^{11}-4577417968n^{10}+3131558016n^{9}$
$-27080918n^{8}-926425857n^{7}-31953641n^{6}+53582776n^{5}+572504n^{4}$
$+1356752n^{3}+105008n^{2}+6016n+7680\big)x_4{}^{18}-4n^3\big(4125803840n^{28}$
$-46625945528n^{27}+173768896384n^{26}-148312205598n^{25}-629943045046n^{24}$
$+1815721286288n^{23}-1525437775604n^{22}-688751891053n^{21}$
$+2481657327201n^{20}-2307861961849n^{19}+939225618905n^{18}$
$+391434743123n^{17}-907271789563n^{16}+698988012631n^{15}$
$-363325497855n^{14}+101241249493n^{13}+62001906787n^{12}-68808726427n^{11}$
$+25425724147n^{10}-12335474457n^{9}+4903931509n^{8}+322550293n^{7}$
$-22446153n^{6}-139456836n^{5}+8714040n^{4}-6112560n^{3}-437520n^{2}-409664n$
$-21504\big)x_4{}^{17}+n^3\big(25369911296n^{28}-241857552448n^{27}+703327077064n^{26}$
$+1421408832n^{25}-3798867522230n^{24}+7169222676054n^{23}$
$-3238141071909n^{22}-5375999746973n^{21}+9690104135726n^{20}$
$-7851134014402n^{19}+2669877312270n^{18}+2450325349798n^{17}$
$-4063141358710n^{16}+3075801315594n^{15}-1819901374572n^{14}$
$+538440535372n^{13}+198143288330n^{12}-186410664974n^{11}$
$+126602399178n^{10}-92850345926n^{9}+19132759460n^{8}-1686993576n^{7}$
$+2646746113n^{6}-187639775n^{5}+83770192n^{4}-21608616n^{3}+3354352n^{2}$
$+219664n+148992\big)x_4{}^{16}-8n^4\big(4311454756n^{27}-34656183284n^{26}$
$+74255564393n^{25}+83150043370n^{24}-546453975358n^{23}+715192498187n^{22}$
$-68192931055n^{21}-735374364043n^{20}+1026255323921n^{19}-833584949643n^{18}$
$+209227893551n^{17}+378936215857n^{16}-485137563379n^{15}+395716183163n^{14}$
$-258810788897n^{13}+71999497015n^{12}-903746169n^{11}-6868075325n^{10}$
$+20451910948n^{9}-10512608327n^{8}+1947558693n^{7}-1394412202n^{6}$
$+266746132n^{5}-10043136n^{4}+14402768n^{3}-57248n^{2}+1041600n+20992\big)x_4{}^{15}$
$+4n^4\big(10512645416n^{27}-71096950440n^{26}+102554231558n^{25}$
$+285091226776n^{24}-1013758586128n^{23}+932051858275n^{22}$

$$+202070248099n^{21}-1199699711980n^{20}+1688483967184n^{19}$$
$$-1381830149504n^{18}+130182363564n^{17}+681457171424n^{16}$$
$$-767399789592n^{15}+781977876302n^{14}-470094490362n^{13}$$
$$+145492500992n^{12}-79459420408n^{11}-6216558648n^{10}+33533433078n^{9}$$
$$-7419359048n^{8}+6766278104n^{7}-3111674489n^{6}+82241279n^{5}$$
$$-151084644n^{4}+14322000n^{3}-4899032n^{2}-715600n-104416\big)x_4^{14}$$
$$-8n^{5}\big(5849742192n^{26}-33403427048n^{25}+28882977878n^{24}$$
$$+157902297360n^{23}-390072904610n^{22}+266156850682n^{21}$$
$$+98818585287n^{20}-437913195047n^{19}+700975144797n^{18}-474014683367n^{17}$$
$$-19752293155n^{16}+183873705881n^{15}-305264374175n^{14}$$
$$+331844601029n^{13}-152719099961n^{12}+94946457555n^{11}-51123311841n^{10}$$
$$-5928257253n^{9}-167489423n^{8}-1836595781n^{7}+3884298229n^{6}$$
$$-385113595n^{5}+243295982n^{4}-53525648n^{3}+2113504n^{2}-1971088n$$
$$-191392\big)x_4^{13}+n^{4}\big(48651259052n^{27}-238925831680n^{26}+122523871383n^{25}$$
$$+1064994157195n^{24}-2084392005828n^{23}+1315499472780n^{22}$$
$$+320254219290n^{21}-2716514460102n^{20}+4064241578578n^{19}$$
$$-1898683606166n^{18}+219690238560n^{17}+582629183656n^{16}$$
$$-2017194780694n^{15}+1419891516170n^{14}-804690414642n^{13}$$
$$+792045244142n^{12}-175496837318n^{11}+71074521278n^{10}-71874870871n^{9}$$
$$-29649451891n^{8}+5399119442n^{7}-1375760382n^{6}+2387248056n^{5}$$
$$-124895976n^{4}+40437840n^{3}+4762432n^{2}+1919936n+219648\big)x_4^{12}$$
$$-4n^{5}\big(12104155728n^{26}-53087875612n^{25}+23763944950n^{24}$$
$$+184188234181n^{23}-332997545373n^{22}+248365103509n^{21}+40414700575n^{20}$$
$$-514527437127n^{19}+549589646287n^{18}-244342483407n^{17}$$
$$+152021635003n^{16}+166407704131n^{15}-267647070379n^{14}$$
$$+112061709211n^{13}-178143775643n^{12}+84918714555n^{11}-16615917523n^{10}$$
$$+40213049391n^{9}-2904611205n^{8}-493836924n^{7}-2380539428n^{6}$$
$$-1018583860n^{5}+84225488n^{4}-30588928n^{3}+13872416n^{2}-142144n$$
$$-65920\big)x_4^{11}+2n^{5}\big(23381008024n^{26}-95099964890n^{25}+52146038553n^{24}$$
$$+232558270041n^{23}-445141358362n^{22}+361926989794n^{21}$$
$$+108550635144n^{20}-596976332228n^{19}+503174689800n^{18}$$
$$-414490567200n^{17}+108280508586n^{16}+298303224358n^{15}$$
$$-107684805348n^{14}+174437128704n^{13}-162990916720n^{12}-10168005468n^{11}$$
$$-64197462016n^{10}+24466467930n^{9}+4244961029n^{8}+12961343065n^{7}$$
$$-39109954n^{6}-349192514n^{5}-92830944n^{4}-47064504n^{3}+2645152n^{2}$$
$$-2448640n+58368\big)x_4^{10}-4n^{6}\big(10851364160n^{25}-41133090836n^{24}$$
$$+24533650540n^{23}+75806407369n^{22}-145449292843n^{21}+103244940115n^{20}$$
$$+32566507397n^{19}-123305614489n^{18}+162597878013n^{17}-130557731209n^{16}$$
$$-32985835579n^{15}+34751037737n^{14}-38776872473n^{13}+72619664893n^{12}$$

$$+10492158431n^{11}+2962608245n^{10}-9724126265n^9-9075025927n^8$$
$$-4838126909n^7+1730681362n^6+375479840n^5+521762324n^4-29048920n^3$$
$$-24662496n^2-1242592n+39488)x_4^9 + n^6(37547824328n^{25}$$
$$-128336123028n^{24}+62918987891n^{23}+201471972131n^{22}-313784049782n^{21}$$
$$+207370351874n^{20}-37398839542n^{19}-247511986334n^{18}+435898216442n^{17}$$
$$-131060722806n^{16} + 12901626344n^{15} + 7828925616n^{14} - 234847592954n^{13}$$
$$+54248360766n^{12}+2494255630n^{11}+37366533318n^{10}+54427066998n^9$$
$$-12368761054n^8-3724147771n^7-7075162323n^6-2400109896n^5$$
$$+775737416n^4-16127800n^3+50810072n^2+5234816n+164192)x_4^8$$
$$-16n^7(1823987834n^{24}-5310281985n^{23}+1352896898n^{22}+7072303732n^{21}$$
$$-7921891744n^{20}+8906538106n^{19}-4241962924n^{18}-11868328702n^{17}$$
$$+8925683772n^{16}-3523534964n^{15}+6898559848n^{14}+5672790666n^{13}$$
$$-4170407840n^{12}+225399506n^{11}-4290193276n^{10}-697783546n^9$$
$$+1111905610n^8-74129195n^7+813706214n^6-92209142n^5-25345888n^4$$
$$-20932268n^3-2210216n^2+3850992n-181600)x_4^7 + 4n^6(n^2+1)\times$$
$$(4933450284n^{23}-11310248899n^{22}-5360404660n^{21}+21712808970n^{20}$$
$$-6737837040n^{19}+5151656898n^{18}+6100731680n^{17}-25943190458n^{16}$$
$$-2800369716n^{15}+2584622244n^{14}+6031414400n^{13}+8993028762n^{12}$$
$$+72336972n^{11}+384904862n^{10}-3919399424n^9-1444180814n^8+897579752n^7$$
$$-140036961n^6+266513284n^5-33731196n^4-24235836n^3-3407840n^2$$
$$-877680n+162528)x_4^6 - 16n^7(n^2+1)^2(705296664n^{20}-1132454665n^{19}$$
$$-1740662210n^{18}+2576283269n^{17}+898325446n^{16}-780522851n^{15}$$
$$+1257751202n^{14}-1145958947n^{13}-1827418074n^{12}+102754827n^{11}$$
$$+772372046n^{10}+315507375n^9+26230610n^8+156756247n^7-151136762n^6$$
$$-86700137n^5+33702746n^4+2102858n^3+1394124n^2-75688n-14000)x_4^5$$
$$+4n^7(n^2+1)^3(1324763915n^{18}-1197266391n^{17}-4323463484n^{16}$$
$$+2931908280n^{15}+4116945405n^{14}-2381180009n^{13}+357520258n^{12}$$
$$+1351269450n^{11}-2472882679n^{10}-1081714701n^9+1337859200n^8$$
$$+311756356n^7-300514101n^6+123078449n^5+1922282n^4-27508806n^3$$
$$+7272228n^2+885276n-8880)x_4^4 - 32n^8(n+1)(n^2+1)^4(61565680n^{14}$$
$$-75687447n^{13}-140049561n^{12}+176611806n^{11}+75048672n^{10}-145185458n^9$$
$$+45224444n^8+48501614n^7-65534498n^6+3009583n^5+23416645n^4$$
$$-6521964n^3-1425070n^2+367290n+4440)x_4^3 + 16n^8(n+1)(n^2+1)^5\times$$
$$(34369898n^{12}-20811891n^{11}-92685753n^{10}+60668294n^9+81076152n^8$$
$$-69130940n^7-14882080n^6+33786470n^5-11135956n^4-5216145n^3$$
$$+3749737n^2-73260n-132830)x_4^2 - 64n^9(n+1)^2(2n-1)(n^2+1)^6\times$$
$$(815184n^7-459199n^6-1570232n^5+997103n^4+805068n^3-615325n^2$$
$$-56980n+89725)x_4 + 2960n^8(n+1)^2(2n-1)^2(3n+1)(n^2+1)^7(289n^4$$
$$-458n^2+185).$$

At first we consider the solutions of $h_4(x_4) = 0$. Note that the coefficient of highest degree of the polynomial $h_4(x_4)$ is given by

$$(n-2)^4(n-1)^3 n^6 (2n-1)^4 (3n^2-1)^3 (5n^3-8n^2+n-2)(37n^3+24n^2+n-2).$$

Thus, for large value of $x_4 = a$, we see $h_4(a) > 0$. In fact, we see $h_4(5) > 0$ for $n \geq 3$ (see the file "SU(2n)_solution_of_eq_h4(x4)=0.pdf" page 7).

For $x_4 = 0$ we see that

$$h_4(0) = 2960 n^8 (n+1)^2 (2n-1)^2 (3n+1)(n^2+1)^7 (289 n^4 - 458 n^2 + 185) > 0$$

and for $x_4 = 1$ we see that

$$h_4(1) = -16(n-1)^3(n+1)^3(n+2)^4 (5n^3+8n^2+n+2)^2$$
$$\times (5n^3+11n^2-9n+1)^2 (65n^5+5n^4+8n^3+6n^2+3n+1) < 0.$$

Now we have

$$h_4\left(1 - \frac{2}{n} + \frac{70}{n^2}\right)$$

$$= \frac{16}{n^{50}} \big(1936896000 n^{75} - 262442957360 n^{74} + 14017427210100 n^{73}$$
$$- 448833495256664 n^{72} + 10737091614743005 n^{71}$$
$$- 200261879904600229 n^{70} + 3085132101074824126 n^{69}$$
$$- 46258411978676442296 n^{68} + 704268578219772454337 n^{67}$$
$$- 12271292624170498870243 n^{66} + 216825391035342259273920 n^{65}$$
$$- 3765386297599693450710168 n^{64} + 60038773891613692764264298 n^{63}$$
$$- 889859006568473741799472914 n^{62}$$
$$+ 12346554217449262027149633860 n^{61}$$
$$- 165175176275059371260494292576 n^{60}$$
$$+ 2180671369605096673118810821658 n^{59}$$
$$- 28930147543136316209266973664550 n^{58}$$
$$+ 385523054948967269901696622580380 n^{57}$$
$$- 5104286302980660050380934455740760 n^{56}$$
$$+ 66003906322069462558493450191268185 n^{55}$$
$$- 822892201649562887221898667142855657 n^{54}$$
$$+ 9802042963893413069266516508300576206 n^{53}$$
$$- 111104205937972421787983975868265765832 n^{52}$$
$$+ 1196487361570043078098335531606705226213 n^{51}$$
$$- 12245538496345620938254097994121101170839 n^{50}$$
$$+ 119238409875487788489013524245814475984160 n^{49}$$
$$- 1106201804945086654294694469739608899867576 n^{48}$$
$$+ 9792742257633450073961393877654691020515616 n^{47}$$
$$- 82837343661383903513259845863448375487886160 n^{46}$$

$+670447332106186483063530598820604311905483584n^{45}$

$-519724502642631490273635533698551659320208976 0n^{44}$

$+386220093339387707421227137591689062253230526 72n^{43}$

$-275312817515933260456449816311176565154981250112n^{42}$

$+188338283587575592227304699938225548101833787 2896n^{41}$

$-123669335328813514478881049734739469541034740 73088n^{40}$

$+779452138509833772622835432713435096748043452 83584n^{39}$

$-471419370407997867406577902605957209081193353 293056n^{38}$

$+273457714221032199945969267743429116757178972 9856000n^{37}$

$-152014597246091909109010019861669259673607986 23022592n^{36}$

$+808922588249879081848460766996947246375611053 67851520n^{35}$

$-411442621430985011339381454086662439855282727 995748864n^{34}$

$+199655132445727559406931726328366241007010648 2845937664n^{33}$

$-922148059379456871670579132691811751170662192 8053932032n^{32}$

$+404231674206168696965664327730115705088543438 3222616064n^{31}$

$-167590302385224407056842869753003222287163209 387196940288n^{30}$

$+654365376487757620789352216608257224374428775 577557327872n^{29}$

$-239387812976011385843180716123523762158910852 6642870329344n^{28}$

$+815372723481184782802302542475029636935723009 6570832388096n^{27}$

$-256568733488485574806964825204166000914434570 72487780663296n^{26}$

$+738628310960615082928214321620831634054373697 56173806141440n^{25}$

$-192172049779352843781723873456754981500907163 942635174297600n^{24}$

$+444719333807516929259266464182101422906555382 603851669504000n^{23}$

$-896140792688418896430867785613497018835272917 493014364160000n^{22}$

$+152581104539338889087588722737558677245196588 7881692774400000n^{21}$

$-209435773784356588752423067063277896473645979 1060248576000000n^{20}$

$+212227532662219866815228338686231672227504643 0265943040000000n^{19}$

$-123152760828113092031607923331598531921951257 5163161600000000n^{18}$

$-268482955923874757056073191297750156250083175 745536000000000n^{17}$

$+134986796230011226934043576920097966083147719 8883840000000000n^{16}$

$-125483538421279009531215314813524635259052575 2473600000000000n^{15}$

$+376660704352429162876367335259546399035361714 944000000000000n^{14}$

$+294545993818482702979085252339357451036841418 240000000000000n^{13}$

$-352520080270027081719547900028079148031975449 60000000000000000n^{12}$

$+124786573683778414962770521362997366068601344 00000000000000000n^{11}$

$+254656510848158506998876315500014736824422400 00000000000000000n^{10}$

$-451482022146773741883329364607577044000896000 00000000000000000n^{9}$

$+193574187235096196474503295843461904051840000 00000000000000000n^{8}$

$-518205458155456490709890411817109135360000000 00000000000000000n^{7}$

$$-3987753740960889920409717766122066233600000000000000000000000n^6$$
$$+2034555191740759358806792634612489344000000000000000000000000n^5$$
$$-1871083591831209149108071193463737600000000000000000000000000n^4$$
$$-2136417282776382763823945227850496000000000000000000000000000n^3$$
$$+9780276719807503921871508976902400000000000000000000000000000n^2$$
$$-17461854945633880975430706918400000000000000000000000000000n$$
$$+1177565533554534300100610560000000000000000000000000000000\Big).$$

We can see that

$$\begin{cases} h_4\Big(1 - \dfrac{2}{n} + \dfrac{70}{n^2}\Big) > 0 & \text{for } n \geq 60, \\[2mm] h_4\Big(1 - \dfrac{2}{n} + \dfrac{70}{n^2}\Big) < 0 & \text{for } 8 \leq n \leq 59 \end{cases}$$

(see the file "SU(2n)_solution_of_eq_h4(x4)=0.pdf" from page 9 to page 22).
    We also see that

$$
\begin{aligned}
h_4&\Big(1 - \frac{2}{n}\Big) \\
&= -\frac{16(n-1)^2}{n^{24}}\big(88270000n^{48} - 808606075n^{47} - 2504544915n^{46} \\
&\quad + 33761710845n^{45} + 58424075719n^{44} - 1473324513352n^{43} \\
&\quad + 4714334856996n^{42} + 2104706447808n^{41} + 3168758230476n^{40} \\
&\quad - 661160551049798n^{39} + 6027133183599726n^{38} - 32900745038233470n^{37} \\
&\quad + 135818296545860322n^{36} - 461992592727670064n^{35} \\
&\quad + 1345722292227854524n^{34} - 3420937721728495528n^{33} \\
&\quad + 7669950613538195196n^{32} - 15255090821100344263n^{31} \\
&\quad + 26949927770356830413n^{30} - 42122000755190819751n^{29} \\
&\quad + 57666675586248284655n^{28} - 67930205958690371584n^{27} \\
&\quad + 66753404123670639224n^{26} - 51388493313864215024n^{25} \\
&\quad + 25689211749651961840n^{24} + 735282215618519296n^{23} \\
&\quad - 17849678074631147264n^{22} + 21232232457576938496n^{21} \\
&\quad - 14071471102642907648n^{20} + 3887277637566840832n^{19} \\
&\quad + 2916173828952522752n^{18} - 4413362747029266432n^{17} \\
&\quad + 2596461049957064704n^{16} - 422421517856407552n^{15} \\
&\quad - 571689357837008896n^{14} + 529872799268143104n^{13} \\
&\quad - 189074347228725248n^{12} - 13399724475285504n^{11} \\
&\quad + 48340358307250176n^{10} - 23911759231320064n^9 \\
&\quad + 4639981602603008n^8 + 953293324943360n^7 \\
&\quad - 963743668240384n^6 + 338962140889088n^5 - 72620649218048n^4 \\
&\quad + 10175582830592n^3 - 911338373120n^2 + 47244640256n - 1073741824\big).
\end{aligned}
$$

We can see that $h_4(1-2/n) < 0$ for $n \geq 8$ and that $h_4(3/4) < 0$ for $3 \leq n \leq 7$ (see the file "SU(2n)_solution_of_eq_h4(x4)=0.pdf" page 22). Thus we see that the equation $h_4(x_4) = 0$ has four positive solutions $x_4 = \alpha, \beta, \gamma, \delta$ for $n \geq 60$ with

$$0 < \alpha < 1 - 2/n < \beta < 1 - 2/n + 70/n^2 < \gamma < 1 < \delta < 5$$

and two positive solutions $x_4 = \alpha, \delta$ for $3 \leq n \leq 59$ with

$$0 < \alpha < 1 < \delta < 5.$$

Moreover, if $(x_4 - 1)\left((n-1)x_4 - 3n - 1\right)\left((3n^2 - 1)x_4 - n^2 - 1\right) \neq 0$, the Gröbner basis for the ideal J (with above lexicographic order $z > x_1 > x_3 > x_4$ as a monomial ordering on $R = \mathbb{Q}[n][z, x_1, x_3, x_4]$) contains polynomials of the form $b(n)x_1 - X(x_4)$ and $c(n)x_3 - Y(x_4)$, where $b(n)$ is a polynomial of n of the form $b(n) = 8n^7(n+1)(n^2+1)^5 b_1(n)$ where $b_1(n)$ is a polynomial of n with integer coefficients of the degree 507, $c(n) = 32(n-1)^3 n^7 (n+1)(n+2)^3(2n+1)(3n-1)(n^2+1)^6(5n^2-1)c_1(n)$ where $c_1(n)$ is a polynomial of n with integer coefficients of the degree 490, and $X(x_4)$ and $Y(x_4)$ are polynomials of x_4 of degree 27 with (huge) coefficients in $\mathbb{Z}[n]$ (cf. the files "x1=g(x4)forSU(2n).pdf" and "x3=f(x4)forSU(2n).pdf"). We also can see that $b(n)$ and $c(n)$ are not zero for integers $n \geq 3$ (for $b(n)$, see page 882 in the file "x1=g(x4)forSU(2n).pdf" and, for $c(n)$, see page 877 in the file "x3=f(x4)forSU(2n).pdf"). Thus, if x_4 are reals, so are the solutions x_1 and x_3.

Now we take the lexicographic order $>$ with $z > x_1 > x_4 > x_3$ for a monomial ordering on R. Then, by an aid of computer, we see that a Gröbner basis for the ideal J contains the polynomial

$$(x_3 - 1)\left((n-1)x_3 - 3n - 1\right)\left((3n - 1)x_3 - n - 1\right)h_3(x_3),$$

where $h_3(x_3)$ is a polynomial of x_3 with degree 28 with coefficients in $\mathbb{Z}[n]$ (cf. the files "SU(2n)_eq_h3(x3)=0.pdf"). We can see that, for $n \geq 3$, the coefficients of even degree of the polynomial of $h_3(x_3)$ of x_3 are positive and odd degree are negative (cf. the files "SU(2n) equation h3(x3)=0" page 8). Thus we obtain that real solutions of $h_3(x_3) = 0$ are always positive.

We also take another lexicographic order $>$ with $z > x_3 > x_4 > x_1$ for a monomial ordering on R. Then, by an aid of computer, we see that a Gröbner basis for the ideal J contains the polynomial

$$(x_1 - 1)((3n - 1)x_1 - n - 1)\left((3n^2 - 1)x_1 - n^2 - 1\right)h_1(x_1),$$

where $h_1(x_1)$ is a polynomial of x_1 with degree 28 with coefficients in $\mathbb{Z}[n]$ (see the file "SU(2n)_eq_h1(x1)=0.pdf"). We can see that, for $n \geq 17$, the coefficients of even degree of the polynomial of $h_1(x_1)$ of x_1 are positive and odd degree are negative for the degree d between $0 \leq d \leq 24$ and the coefficients of degrees are positive for the degree d between $25 \leq d \leq 28$ (see the file "SU(2n)_eq_h1(x1)=0.pdf" from page 12 to page 102). But we can show that real solutions of $h_1(x_1) = 0$ are always positive for $n \geq 19$ by considering the polynomial $H[x_1, n]$ of degree 4 defined from the part of degrees between $25 \leq d \leq 28$ in $h_1(x_1)$ (see the file "SU(2n)_eq_h1(x1)=0.pdf" from page 102 to page 110). We also can see that, for $3 \leq n \leq 18$, the real solutions of equation $h_1(x_1) = 0$ are positive (see the file "SU(2n)_eq_h1(x1)=0.pdf" page 111).

Note that, from $g_1 = 0$ in the system (16), we see that

$$x_0 = \frac{(n-2)x_1{}^2 x_3{}^2 x_4{}^2 + n x_1{}^2 x_3{}^2 + (n+2)x_1{}^2 x_4{}^2 + n x_3{}^2 x_4{}^2}{2 x_1 x_4{}^2 \left((n-1)x_3{}^2 + n + 1\right)}.$$

In particular, for positive values x_1, x_3, x_4, we see the value x_0 is also positive. Thus we obtain four Einstein metrics for $n \geq 60$ and two Einstein metrics for $3 \leq n \leq 59$.

We consider the case $(x_4 - 1)\left((n-1)x_4 - 3n - 1\right)\left((3n^2 - 1)x_4 - n^2 - 1\right) = 0$ for $n \geq 2$. Then, with an aid of computer, by computing the Gröbner basis for the ideal J, we see that the solutions for (16) with $x_0 x_1 x_3 x_4 \neq 0$ are given as following (cf. the file "Einstein_metrics_on_SU(2n)_for_the_cases_x4=fraction.pdf"):

(1) If $x_4 = 1$, then $x_0 = x_1 = x_2 = x_3 = 1$,
   or $x_0 = x_1 = x_3 = \dfrac{n+1}{3n-1}$ and $x_2 = x_4 = 1$;

(2) If $x_4 = \dfrac{3n+1}{n-1}$, then $x_0 = x_1 = x_2 = 1$ and $x_3 = x_4$;

(3) If $x_4 = \dfrac{n^2+1}{3n^2-1}$, then $x_0 = \dfrac{5n^4 - 2n^2 + 1}{(3n^2-1)(n^2+1)}$, $x_1 = x_4$ and $x_2 = x_3 = 1$.

Note that these Einstein metrics are naturally reductive by Proposition 5.1.

Further, with an aid of a computer, by computing the Gröbner basis for the ideal J, we obtain the following proposition.

Proposition 6.1. *For the system of equations* (14), *if two of variables* $\{x_0, x_1, x_2, x_3, x_4\}$ *are equal, then, by normalizing* $x_2 = 1$, *the solutions for* (16) *with* $x_0 x_1 x_3 x_4 \neq 0$ *are one of the followings* (cf. the file "SU(2n)_xi=xj.pdf"):

(1) $x_0 = x_1 = x_3 = x_4 = 1$,

(2) $x_0 = x_1 = x_3 = \dfrac{n+1}{3n-1}$, $x_4 = 1$,

(3) $x_0 = x_1 = 1$, $x_3 = x_4 = \dfrac{3n+1}{n-1}$,

(4) $x_0 = \dfrac{5n^4 - 2n^2 + 1}{(3n^2-1)(n^2+1)}$, $x_1 = x_4 = \dfrac{n^2+1}{3n^2-1}$, $x_3 = 1$.

Thus we see that the real solutions $\{x_0, x_1, x_3, x_4\}$ for the system of equations (18) obtained from the solutions of $h_4(x_4) = 0$ satisfy the property that $x_i \neq x_j$ whenever $i \neq j$. Thus we see that these solutions are non naturally reductive Einstein metrics on $SU(2n)$ by Proposition 5.1. Thus we have proved Theorem 1.1.

For $n = 2$ we have the follwing:

Remark 6.1. All $\mathrm{Ad}(U(2))$-invariant Einstein metrics on $SU(4)$ are naturally reductive. Besides a bi-invariant metric, there are five naturally reductive metrics on $SU(4)$.

In fact, for $n = 2$, we see that $h_4(x_4) = 20480 \left(5x_4^2 - 24x_4 + 15\right) k_4(x_4)$, where

$$
\begin{aligned}
k_4(x_4) = {}& 4253356800x_4{}^{18} - 32501387520x_4{}^{17} + 112370936992x_4{}^{16} \\
& - 206248875056x_4{}^{15} + 263687860667x_4{}^{14} - 313019641568x_4{}^{13} \\
& + 212196662751x_4{}^{12} + 473282418384x_4{}^{11} - 887913037555x_4{}^{10} \\
& + 614757956896x_4{}^{9} - 685517477633x_4{}^{8} - 117254939648x_4{}^{7} \\
& + 2725578596856x_4{}^{6} - 3864264762784x_4{}^{5} + 3581601367650x_4{}^{4} \\
& - 3919355953200x_4{}^{3} + 3279264058875x_4{}^{2} - 1487268180000x_4 \\
& + 325283765625.
\end{aligned}
$$

We see that $k_4(x_4) = 0$ has no real solutions. Thus $h_4(x_4) = 0$ has only two real solutions $x_4 = (12 - \sqrt{69})/5$ and $x_4 = (12 + \sqrt{69})/5$. We also see that, for $5x_4^2 - 24x_4 + 15 = 0$, $x_3 = x_4$, $5 + 5x_1 - 8x_4 = 0$, $x_0 = 1$ by computing the Gröbner basis. Thus we see that these Einstein metrics are naturally reductive by Proposition 5.1.

Acknowledgments

The second author was supported by JSPS KAKENHI (No. 16K05130).

References

[1] A. Arvanitoyeorgos, V. V. Dzhepko and Yu. G. Nikonorov, Invariant Einstein metrics on some homogeneous spaces of classical Lie groups, *Canad. J. Math.* **61**(6) (2009), 1201–1213.

[2] A. Arvanitoyeorgos, K. Mori and Y. Sakane, Einstein metrics on compact Lie groups which are not naturally reductive, *Geom. Dedicata*
160(1) (2012), 261–285.

[3] A. Arvanitoyeorgos, Y. Sakane and M. Statha, New Einstein metrics
on the Lie group SO(n) which are not naturally reductive, *Geom.
Imaging Comput.* **2**(2) (2015), 77–108.

[4] A. Arvanitoyeorgos, Y. Sakane and M. Statha, Einstein metrics on the
symplectic group which are not naturally reductive, *Current developments in differential geometry and its related fields*, 1–22, World Sci.
Publ., Hackensack, NJ, 2016.

[5] A. Arvanitoyeorgos, Y. Sakane and M. Statha, Homogeneous Einstein
metrics on complex Stiefel manifolds and special unitary groups, *Contemporary perspectives in differential geometry and its related fields*,
1–20, World Sci. Publ., Hackensack, NJ, 2018.

[6] A. L. Besse, *Einstein Manifolds*, Springer-Verlag, Berlin, 1986.

[7] C. Böhm, Homogeneous Einstein metrics and simplicial complexes, *J.
Diff. Geom.* **67**(1) (2004), 79–165.

[8] C. Böhm, M. Wang and W. Ziller, A variational approach for compact
homogeneous Einstein manifolds, *Geom. Func. Anal.* **14** (4) (2004),
681–733.

[9] H. Chen, Z. Chen and S. Deng, New non-naturally reductive Einstein
metrics on exceptional simple Lie groups, *J. Geom. Phys.* **124** (2018),
268–285.

[10] Z. Chen, S. Deng and L. Zhang, New Einstein metrics on E_7, *Diff.
Geom. Appl.* **51** (2017), 189–202.

[11] Z. Chen and K. Liang, Non-naturally reductive Einstein metrics on
the compact simple Lie group F_4, *Ann. Glob. Anal. Geom.* **46** (2014),
103–115.

[12] I. Chrysikos and Y. Sakane, Non-naturally reductive Einstein metrics
on exceptional Lie group, *J. Geom. Phys.* **116** (2017), 152–186.

[13] J. E. D'Atri and W. Ziller, Naturally reductive metrics and Einstein
metrics on compact Lie groups, *Memoirs A.M.S.* **19** (215) (1979).

[14] K. Mori, Left Invariant Einstein Metrics on $SU(N)$ that are not Naturally Reductive, Master Thesis (in Japanese) Osaka University 1994,
English Translation: *Osaka University RPM* 96010 (preprint series)
1996.

[15] Yu. G. Nikonorov, Classification of generalized Wallach spaces, *Geom.
Dedicata* **181**(1) (2016), 193–212.

[16] J-S. Park and Y. Sakane, Invariant Einstein metrics on certain homogeneous spaces, *Tokyo J. Math.* **20**(1) (1997), 51–61.

[17] M. Wang, Einstein metrics from symmetry and bundle constructions, in *Surveys in Differential Geometry: Essays on Einstein Manifolds, Surv. Diff. Geom.* VI, Int. Press, Boston, MA 1999.

[18] M. Wang, Einstein metrics from symmetry and bundle constructions: A sequel, in *Differential Geometry : Under the Influence of S.-S. Chern, Advanced Lectures in Mathematics*, vol. 22, 253–309, Higher Education Press/International Press, 2012.

[19] M. Wang and W. Ziller, Existence and non-existence of homogeneous Einstein metrics, *Invent. Math.* **84** (1986), 177–194.

Received February 15, 2019
Revised February 20, 2019

Recent Topics in Differential Geometry
and its Related Fields   29 – 44

CONSTRUCTION OF NONCOMPACT LAGRANGIAN ORBITS IN SOME HERMITIAN SYMMETRIC SPACE OF NONCOMPACT TYPE

Takahiro HASHINAGA

National Institute of Technology, Kitakyushu College,
Kitakyushu, Fukuoka, 802-0985, Japan
E-mail: hashinaga@kct.ac.jp

In this paper, we construct noncompact homogeneous Lagrangian submanifolds in some Hermitian symmetric spaces of noncompact type. Our noncompact homogeneous Lagrangian submanifolds are obtained as connected closed subgroups of the solvable part of the Iwasawa decomposition.

Keywords: Lagrangian submanifolds; Hermitian symmetric spaces; Iwasawa decomposition; solvable models.

1. Introduction

Let (M, ω) be a $2n$-dimensional symplectic manifold with symplectic form ω. A submanifold L of (M, ω) is called *Lagrangian* if it has dimension n and the restriction of the symplectic form ω to L vanishes. When $M = (M, g, J, \omega)$ is a Kähler manifold, we call a Lagrangian submanifold L of M *homogeneous* if L is obtained by an orbit of a connected closed subgroup H of the automorphism group of the Kähler manifold M. Furthermore, if we take H to be a compact subgroup, we say L is *compact* homogeneous Lagrangian submanifold. Homogeneous Lagrangian submanifolds in a Kähler manifold are investigated in several contexts (cf. [7, 8, 10, 12]).

Homogeneous Lagrangian submanifolds in a specific Kähler manifold have been constructed and classified by some authors. For instance, in [1, 13], they constructed compact homogeneous Lagrangian submanifolds in $\mathbb{C}^n$ or $\mathbb{C}P^n$. Beduli and Gori [2] classified compact homogeneous Lagrangian submanifolds in the complex projective space $\mathbb{C}P^n$ when H is a compact simple Lie group by using the theory of prehomogeneous vector spaces. Ma and Ohnita [9] gave the complete classification of compact homogeneous Lagrangian submanifolds in the complex hyperquadric $\mathcal{Q}_n(\mathbb{C})$ by applying a geometry of compact homogeneous hypersurfaces in the sphere. In [5], Kajigaya and the author classified noncompact homogeneous Lagrangian submanifolds in the complex hyperbolic space $\mathbb{C}H^n$ which can be obtained by the actions of connected closed subgroups of the solvable part of the Iwasawa decomposition.

Due to McDuff's result, any Hermitian symmetric space of noncompact type G/K is diffeomorphic to the standard symplectic vector space $(\mathbb{C}^n, \omega_0)$ as a symplectic manifold, namely there exists a symplectic diffeomorphism $\Phi : G/K \to \mathbb{C}^n$ (see [11]). Since the automorphism groups of Hermitian symmetric spaces of noncompact type are also noncompact, it would be expected to provide many examples of homogeneous Lagrangian submanifolds not only compact ones but also noncompact ones in $\mathbb{C}^n$ through the study of them in Hermitian symmetric spaces of noncompact type. Note that McDuff's result was generalized by Deltour in [3].

In the present paper, we study noncompact homogeneous Lagrangian submanifolds in Hermitian symmetric spaces of noncompact type $M = G/K$. As well known, any Hermitian symmetric space of noncompact type is identified with certain solvable Lie group equipped with left-invariant structures, called a solvable model. In this paper, we construct closed connected subgroups of the solvable model admitting a Lagrangian orbit. More precisely, we construct Lagrangian subalgebras $\mathfrak{s}'$ with $\dim_{\mathbb{R}} \mathfrak{s}' = \frac{1}{2} \dim_{\mathbb{R}} M$, which are Lie subalgebras of the Lie algebra of the solvable model satisfying the condition $\langle J(\mathfrak{s}'), \mathfrak{s}' \rangle = 0$ (see Definition 3.1).

In § 2, we briefly review the solvable model of an irreducible Hermitian symmetric space of noncompact type M, and describe a Lagrangian subalgebra in the solvable model. In § 3 and § 4, we study Lagrangian subalgebras in the cases of $M = \mathrm{SO}_0(2, n)/\mathrm{SO}(2) \times \mathrm{SO}(n)$ and $M = \mathrm{Sp}(n, \mathbb{R})/\mathrm{U}(n)$ respectively.

2. The solvable model of a Hermitian symmetric space of noncompact type

In this subsection, we recall fundamental facts on a solvable model of an irreducible Hermitian symmetric space of noncompact type. We refer to [6].

Let $M = G/K$ be an irreducible Hermitian symmetric space of noncompact type, where G is the identity component of isometry group, and K is the isotropy subgroup of G at some point o, called the *origin*. Denote by $(\mathfrak{g}, \mathfrak{k}, \theta)$ the corresponding symmetric pair, where $\mathfrak{g}$ and $\mathfrak{k}$ the Lie algebras of G and K, respectively, and θ is the Cartan involution. Then the eigenspace decomposition $\mathfrak{g} = \mathfrak{k} \oplus \mathfrak{p}$ with respect to the Cartan involution θ is called the *Cartan decomposition*. Note that $\mathfrak{p}$ is the (-1)-eigenspace, and it is well known that $\mathfrak{p} \cong T_o(G/K)$.

First of all, we recall the definition of the Iwasawa decomposition. Let us fix $\mathfrak{a}$ as a maximal abelian subspace of $\mathfrak{p}$, and denote by $\mathfrak{a}^*$ the dual space of $\mathfrak{a}$. Then we define

$$\mathfrak{g}_\lambda := \{X \in \mathfrak{g} \mid \text{ad}(H)X = \lambda(H)X \ (\forall H \in \mathfrak{a})\}$$

for each $\lambda \in \mathfrak{a}^*$. We call $\lambda \in \mathfrak{a}^* \setminus \{0\}$ a *restricted root* with respect to $\mathfrak{a}$ if $\mathfrak{g}_\lambda \neq 0$. Denote by Σ the set of restricted roots, which is called *the restricted root system*. Let Λ be a set of simple roots of Σ, and denote by Σ^+ the set of positive roots associated with Λ. Recall that $[\mathfrak{g}_\lambda, \mathfrak{g}_\mu] \subset \mathfrak{g}_{\lambda+\mu}$ holds for each $\lambda, \mu \in \Sigma \cup \{0\}$. Let us define

$$\mathfrak{n} := \bigoplus_{\lambda \in \Sigma^+} \mathfrak{g}_\lambda.$$

Note that $\mathfrak{n}$ is a nilpotent Lie algebra, and $\mathfrak{a} \oplus \mathfrak{n}$ is a solvable Lie algebra.

Definition 2.1. The decomposition $\mathfrak{g} = \mathfrak{k} \oplus \mathfrak{a} \oplus \mathfrak{n}$ is called the *Iwasawa decomposition* of $\mathfrak{g}$, and $\mathfrak{a} \oplus \mathfrak{n}$ is called the *solvable part* of the Iwasawa decomposition.

Let S be the connected Lie subgroup of G whose Lie algebra is $\mathfrak{a} \oplus \mathfrak{n}$. Then S is simply-connected and acts simply transitively on G/K. In particular, the following map is a diffeomorphism:

$$\Phi : S \to G/K = M : g \mapsto [g] = g.o.$$

A solvable model is defined by the solvable part of the Iwasawa decomposition $\mathfrak{a} \oplus \mathfrak{n}$ with certain geometric structures. We here recall the geometric structures on $M = G/K$. Note that $\mathfrak{p} \cong T_o(G/K)$, and every K-invariant geometric structures on $\mathfrak{p}$ can be extend to G-invariant geometric structures on M. Let B be the Killing form of $\mathfrak{g}$, $k > 0$, and define a positive definite inner product on $\mathfrak{g}$ by

$$\langle X, Y \rangle_\mathfrak{g} := -kB(\theta X, Y).$$

Since $\langle X, Y \rangle_\mathfrak{g}|_{\mathfrak{p} \times \mathfrak{p}}$ is K-invariant, this gives a G-invariant Riemannian metric on M. Since M is irreducible, by the Schur lemma, we see that every G-invariant Riemannian metric on M can be obtained in this way. Let $C(\mathfrak{k})$ be the center of $\mathfrak{k}$. Then there exists $Z \in C(\mathfrak{k})$ uniquely (up to sign), such that $(\text{ad}(Z)|_\mathfrak{p})^2 = -\text{id}_\mathfrak{p}$. Let us denote $J_\mathfrak{p} := \text{ad}(Z)|_\mathfrak{p}$. This $J_\mathfrak{p}$ defines a complex structure on M. For the uniqueness of the complex structure, see

[6, Theorem 4.5, Ch. VIII]). As well known, $\omega_{\mathfrak{p}} := \langle J_{\mathfrak{p}} \cdot, \cdot \rangle_{\mathfrak{g}}|_{\mathfrak{p} \times \mathfrak{p}}$ is a compatible symplectic form on $\mathfrak{p}$, which also defines a G-invariant Kähler form on M.

Geometric structures on $\mathfrak{a} \oplus \mathfrak{n}$ are induced by the geometric structures on $\mathfrak{p}$ via the differential of the diffeomorphism $\Phi : S \to G/K$ at the identity e defined by

$$(d\Phi)_e : \mathfrak{a} \oplus \mathfrak{n} \to \mathfrak{p} : X \mapsto X_{\mathfrak{p}} = (1/2)(X - \theta X).$$

Here $\mathfrak{p}$-subscript means the orthogonal projection.

Definition 2.2. Let $\mathfrak{a} \oplus \mathfrak{n}$ be the solvable part of the Iwasawa decomposition. Then the triplet $(\mathfrak{a} \oplus \mathfrak{n}, \langle\,,\,\rangle, J)$ defined by the following is called the *solvable model* of $M = G/K$:

$$\begin{aligned}
\langle X, Y \rangle &:= \langle (d\Phi)_e X, (d\Phi)_e Y \rangle_{\mathfrak{g}} \\
&= k \langle X_{\mathfrak{a}}, Y_{\mathfrak{a}} \rangle_{\mathfrak{g}} + (k/2) \langle X_{\mathfrak{n}}, Y_{\mathfrak{n}} \rangle_{\mathfrak{g}}, \quad (k > 0, \ X, Y \in \mathfrak{a} \oplus \mathfrak{n}),
\end{aligned}$$

where $\mathfrak{a}$- and $\mathfrak{n}$-subscripts mean the orthogonal projections respectively, and

$$J := (d\Phi)_e^{-1} \circ \mathrm{ad}(Z) \circ (d\Phi)_e.$$

The solvable model $(\mathfrak{a} \oplus \mathfrak{n}, \langle\,,\,\rangle, J)$ induces the left-invariant Riemannian metric and the complex structure on S. Denote by $(S, \langle\,,\,\rangle, J)$ the corresponding triple of $(\mathfrak{a} \oplus \mathfrak{n}, \langle\,,\,\rangle, J)$. We also call $(S, \langle\,,\,\rangle, J)$ the *solvable model* of M. It is known that the solvable model $(S, \langle\,,\,\rangle, J)$ is isomorphic to $M = G/K$ as a Kähler manifold.

3. Lagrangian subalgebras in the solvable model

Let $M = G/K$ be an irreducible Hermitian symmetric space of noncompact type, $(\mathfrak{a} \oplus \mathfrak{n}, \langle\,,\,\rangle, J)$ be the solvable model of M, and $\mathfrak{s}'$ be a subspace of $\mathfrak{a} \oplus \mathfrak{n}$.

Definition 3.1. A subspace $\mathfrak{s}'$ of $\mathfrak{a} \oplus \mathfrak{n}$ is called a *Lagrangian subalgebra* if the following conditions hold:
(1) $\mathfrak{s}'$ is a Lie subalgebra of $\mathfrak{a} \oplus \mathfrak{n}$;
(2) $\dim_{\mathbb{R}} \mathfrak{s}' = \frac{1}{2} \dim_{\mathbb{R}}(\mathfrak{a} \oplus \mathfrak{n})$;
(3) $\langle J(\mathfrak{s}'), \mathfrak{s}' \rangle = 0$.

Let $(S, \langle\,,\,\rangle, J)$ be the corresponding triplet of the solvable model $(\mathfrak{a} \oplus \mathfrak{n}, \langle\,,\,\rangle, J)$, and S'' be a connected closed subgroup of S. Suppose that the action of S'' on M admits a Lagrangian orbit $S'' \cdot p$ for some point $p \in M$.

If necessary, we replace S'' by a conjugate subgroup S' of S'' in S so that the orbit through the origin $S' \cdot o$ is Lagrangian. Then the Lie algebra $\mathfrak{s}'$ of S' is a Lagrangian subalgebra. Conversely, if $\mathfrak{s}'$ is a Lagrangian subalgebra, then the action of $S' = \exp_{\mathfrak{s}} \mathfrak{s}'$ admits a Lagrangian orbit $S' \cdot o$ in S.

Let $2n$ be the dimension of M. Since $(M, g, J) \cong (S, \langle\,,\,\rangle, J)$ is a complex manifold, we can take an orthonormal basis $\{X_1, \ldots, X_n, Y_1, \ldots, Y_n\}$ of the solvable model $(\mathfrak{a} \oplus \mathfrak{n}, \langle\,,\,\rangle, J)$ so that $J(X_i) = Y_i$ for each $i \in \{1, \ldots, n\}$. We choose $V_i := \cos \varphi_i X_i + \sin \varphi_i Y_i$ for each $i \in \{1, \ldots, n\}$, and put

$$\mathfrak{s}' := \mathrm{span}\{V_1, \ldots, V_n\}.$$

Lemma 3.1. *The subspace $\mathfrak{s}'$ of $\mathfrak{a} \oplus \mathfrak{n}$ satisfies that the conditions (2) and (3) in Definition 3.1.*

In the following sections, we investigate Lie subalgebra conditions of $\mathfrak{s}'$ for some irreducible Hermitian symmetric spaces of noncompact type. We also investigate orbit equivalent classes of S' with Lagrangian subalgebra $\mathfrak{s}'$. We here recall the definition of the orbit equivalence.

Definition 3.2. *Two isometric actions $S_1 \curvearrowright M$ and $S_2 \curvearrowright M$ are said to be orbit equivalent if there exists an isometry Ψ on M such that $\Psi(S_1 \cdot p) = S_2 \cdot \Psi(p)$ for any point $p \in M$.*

It is easily seen that two isometric actions S_1 and S_2 in S are orbit equivalent if the corresponding Lie algebras $\mathfrak{s}_1$ and $\mathfrak{s}_2$ are conjugate in $\mathfrak{a} \oplus \mathfrak{n}$, namely there exists $g \in \mathfrak{a} \oplus \mathfrak{n}$ such that $\mathrm{Ad}(\exp(g))\mathfrak{s}_1 = \mathfrak{s}_2$.

4. The case of $M = \mathrm{SO}_0(2, n)/(\mathrm{SO}(2) \times \mathrm{SO}(n))$

In this section, we construct Lagrangian subalgebras in the solvable model of $M = \mathrm{SO}_0(2, n)/(\mathrm{SO}(2) \times \mathrm{SO}(n))$.

First of all, we introduce the solvable model of $\mathrm{SO}_0(2, n)/(\mathrm{SO}(2) \times \mathrm{SO}(n))$. For details, we refer to [4].

Definition 4.1. *Let $c > 0$ and $n \geq 3$. We call $(\mathfrak{s}(c), \langle\,,\,\rangle, J)$ the solvable model of $M = \mathrm{SO}_0(2, n)/(\mathrm{SO}(2) \times \mathrm{SO}(n))$ if this triplet satisfies the following conditions:*
(1) $\mathfrak{s}(c) := \mathrm{span}_{\mathbb{R}}\{A_1, A_2, X_0, Y_1, \ldots, Y_n, Z_1, \ldots, Z_n, W_0\}$ is a $(2n + 4)$-dimensional Lie algebra whose bracket relations are defined by

- $[A_1, X_0] = cX_0$, $[A_1, Y_i] = -(c/2)Y_i$, $[A_1, Z_i] = (c/2)Z_i$,
- $[A_2, W_0] = cW_0$, $[A_2, Y_i] = (c/2)Y_i$, $[A_2, Z_i] = (c/2)Z_i$,

- $[X_0, Y_i] = cZ_i$, $[Y_i, Z_i] = cW_0$,
- and the other relations vanish;

(2) $\langle \, , \, \rangle$ is an inner product on $\mathfrak{s}(c)$ so that the above basis is orthonormal;
(3) J is a complex structure on $\mathfrak{s}(c)$ given by

$$J(A_1) = -X_0, \quad J(A_2) = W_0, \quad J(Y_i) = Z_i.$$

Let $S(c)$ be the connected and simply-connected Lie group with Lie algebra $\mathfrak{s}(c)$. The solvable model $(S(c), \langle \, , \, \rangle, J)$ is isomorphic to $\mathrm{SO}_0(2, n)/(\mathrm{SO}(2) \times \mathrm{SO}(n))$ with minimal sectional curvature $-c^2$.

Let us denote that

$$T_1 := \cos\varphi_1 A_1 + \sin\varphi_1 X_0 \quad (\varphi_1 \in [0, \pi/2]),$$
$$T_2 := \cos\varphi_2 A_2 + \sin\varphi_2 W_0 \quad (\varphi_2 \in [0, \pi/2]),$$

and

$$V_i := \cos\psi_i Y_i - \sin\psi_i Z_i \quad (\psi_i \in [0, \pi/2]),$$

for each $i = 1, \ldots, n$. We also define

$$\mathfrak{s}'(\varphi_1, \varphi_2, \psi_1, \ldots, \psi_n) := \mathrm{span}_{\mathbb{R}}\{T_1, T_2, V_1, \ldots, V_n\}.$$

Lemma 4.1. *The subspace* $\mathfrak{s}'(\varphi_1, \varphi_2, \psi_1, \ldots, \psi_n)$ *is a Lagrangian subalgebra if and only if one of the following conditions holds:*

(1) $\varphi_1 = \psi_1 = \cdots \psi_n = \frac{\pi}{2}$,
(2) $\varphi_1 = 0$ *and* $\psi_i \in \{0, \frac{\pi}{2}\}$ *for each* $i \in \{1, \ldots, n\}$,
(3) $\varphi_1 \in (0, \frac{\pi}{2})$ *and* $\psi_i \in \{\varphi_1, \frac{\pi}{2}\}$ *for each* $i \in \{1, \ldots, n\}$.

Proof. For simplicity, we put $\mathfrak{s}' := \mathfrak{s}'(\varphi_1, \varphi_2, \psi_1, \ldots, \psi_n)$. Lemma 3.1 states that $\mathfrak{s}'$ satisfies that the conditions (2) and (3) in Definition 3.1. Hence we only need to check the Lie subalgebra conditions in $\mathfrak{s}(c)$. For each $i, j \in \{1, \ldots, n\}$, direct calculations give that

$$[T_1, T_2] = 0 \in \mathfrak{s}',$$
$$[T_1, V_i] = -\frac{c}{2}\cos\varphi_1\cos\psi_i Y_i + \left(-\frac{c}{2}\cos\varphi_1\sin\psi_i + c\sin\varphi_1\cos\psi_i\right)Z_i,$$
$$[T_2, V_i] = \frac{c}{2}\cos\varphi_2\cos\psi_i Y_i + \frac{c}{2}\cos\varphi_2\sin\psi_i Z_i = \frac{c}{2}\cos\varphi_2 V_i \in \mathfrak{s}',$$
$$[V_i, V_j] = 0 \in \mathfrak{s}',$$

which yield that $\mathfrak{s}'$ is a Lagrangian subalgebra if and only if $[T_1, V_i] \in \mathfrak{s}'$ for each $i = 1, \ldots, n$. We first consider the cases of $\varphi_1 \in (0, \frac{\pi}{2})$. We take any $\varphi_1 \in (0, \frac{\pi}{2})$ and fix it. If $[T_1, V_i] \in \mathfrak{s}'$, then there exists $k \in \mathbb{R}$ such that $[T_1, V_i] = kV_i$. By comparing the coefficients of the both sides of $[T_1, V_i] = kV_i$, we have

$$\begin{cases} -\frac{c}{2}\cos\varphi_1\cos\psi_i = k\cos\psi_i, \\ -\frac{c}{2}\cos\varphi_1\sin\psi_i + c\sin\varphi_1\cos\psi_i = -k\sin\psi_i. \end{cases}$$

When $\psi_i = \frac{\pi}{2}$, $[T_1, V_i] = -\frac{c}{2}\cos\varphi_1 V_i \in \mathfrak{s}'$. When $\psi_i \neq \frac{\pi}{2}$, $[T_1, V_i] \in \mathfrak{s}'$ if and only if $k = -\frac{c}{2}\cos\varphi_1$ and $0 = c\sin(\psi_i - \varphi_1)$. Since $\varphi_1 \in (0, \frac{\pi}{2})$ and $\psi_i \in [0, \frac{\pi}{2}]$, we obtain $\psi_i = \varphi_1$.

By similar arguments, for each $i = 1, \ldots, n$, it is easy to see that

$$[T_1, V_i] = c\cos\psi_i Z_i \in \mathfrak{s}' \iff \psi_i = \frac{\pi}{2}$$

when $\varphi_1 = \frac{\pi}{2}$, and

$$[T_1, V_i] = -\frac{c}{2}\cos\psi_i Y_i - \frac{c}{2}\sin\psi_i Z_i \in \mathfrak{s}' \iff \psi_i \in \left\{0, \frac{\pi}{2}\right\}$$

when $\varphi_1 = 0$. We complete the proof. $\square$

Next we investigate the orbit equivalent classes of the connected closed subgroups S' of $S(c)$ with Lagrangian subalgebras $\mathfrak{s}'$. We first prove the following lemma.

Lemma 4.2. *Put* $U = a_1 A_1 + a_2 A_2 + x_0 X_0 + w_0 W_0 + \displaystyle\sum_{i=1}^{n}(y_i Y_i + z_i Z_i)$. *Then we have*

$$\mathrm{Ad}(\exp(tX_0))(U) = U - ta_1 cX_0 + t\sum_{i=1}^{n} y_i cZ_i,$$

$$\mathrm{Ad}(\exp(tW_0))(U) = U - ta_2 cW_0.$$

Proof. We show them directly using Baker-Campbell-Hausdorff formula

$$\mathrm{Ad}(\exp(A))B = B + [A, B] + \frac{1}{2!}[A, [A, B]] + \frac{1}{3!}[A, [A, [A, B]]] \cdots.$$

By Lie structures of $\mathfrak{s}(c)$ (see (1) in Definition 4.1), one can see that $[X_0, [X_0, U]] = 0$, and $[W_0, [W_0, U]] = 0$ for any $U \in \mathfrak{s}(c)$. Easy computations give that

$$\mathrm{Ad}(\exp(tX_0))U = U + [tX_0, U]$$

$$= U - ta_1 cX_0 + t\sum_{i=1}^{n} y_i cZ_i,$$

$$\mathrm{Ad}(\exp(tW_0))U = U + [tW_0, U]$$

$$= U - ta_2 cW_0,$$

which complete the proof. $\square$

We define

$$\mathfrak{s}_0'(\varphi_1, \varphi_2) := \mathrm{span}\{T_1, T_2, Z_1, \ldots, Z_n\},$$
$$\mathfrak{s}_b'(\varphi_1, \varphi_2) := \mathrm{span}\{T_1, T_2, Y_1, \ldots Y_b, Z_{b+1}, \ldots, Z_n\} \quad (b \in \{1, \ldots, n-1\}),$$
$$\mathfrak{s}_n'(\varphi_1, \varphi_2) := \mathrm{span}\{T_1, T_2, Y_1, \ldots, Y_n\},$$

where $\varphi_1, \varphi_2 \in [0, \frac{\pi}{2}]$. Let us denote by $S'(\varphi_1, \varphi_2, \psi_1, \ldots, \psi_n)$ and $S_b'(\varphi_1, \varphi_2)$ the connected closed subgroup of $S(c)$ with Lie algebra $\mathfrak{s}'(\varphi_1, \varphi_2, \psi_1, \ldots, \psi_n)$ and $\mathfrak{s}_b'(\varphi_1, \varphi_2)$ respectively.

Theorem 4.1. *Let* $S' := S'(\varphi_1, \varphi_2, \psi_1, \ldots, \psi_n)$ *be a connected closed subgroup of* $S(c)$ *with Lie subalgebra* $\mathfrak{s}' := \mathfrak{s}'(\varphi_1, \varphi_2, \psi_1, \ldots, \psi_n)$. *Suppose that* $\mathfrak{s}'$ *is a Lagrangian subalgebra. Then the action of* S' *is orbit equivalent to the action of one of the following:*

- $S_0'(\frac{\pi}{2}, 0)$,
- $S_0'(\frac{\pi}{2}, \frac{\pi}{2})$,
- $S_b'(0, 0)$, *where* $b \in \{0, 1, \ldots, n\}$,
- $S_b'(0, \frac{\pi}{2})$, *where* $b \in \{0, 1, \ldots, n\}$.

Proof. It follows from Lemma 4.1 that the Lie algebra $\mathfrak{s}'$ coincides with one of the following:
(1) $\mathfrak{s}'(\frac{\pi}{2}, \varphi_2, \frac{\pi}{2}, \ldots, \frac{\pi}{2})$, where $\varphi_2 \in [0, \frac{\pi}{2}]$,
(2) $\mathfrak{s}'(0, \varphi_2, \psi_1, \ldots, \psi_n)$, where $\varphi_2 \in [0, \frac{\pi}{2}]$, and $\psi_i \in \{0, \frac{\pi}{2}\}$,
(3) $\mathfrak{s}'(\varphi_1, \varphi_2, \psi_1, \ldots, \psi_n)$, where $\varphi_1 \in (0, \frac{\pi}{2})$, $\varphi_2 \in [0, \frac{\pi}{2}]$, and $\psi_i \in \{\varphi_1, \frac{\pi}{2}\}$.
As we mentioned at the end of Section 2, isometric actions S_1 and S_2 are orbit equivalent if the Lie algebras $\mathfrak{s}_1$ and $\mathfrak{s}_2$ are conjugate. We thus investigate conjugate classes of the above Lagrangian subalgebras.
(1) The cases of $\mathfrak{s}' = \mathfrak{s}'(\frac{\pi}{2}, \varphi_2, \frac{\pi}{2}, \ldots, \frac{\pi}{2})$, where $\varphi_2 \in [0, \frac{\pi}{2}]$. When $\varphi_2 = \frac{\pi}{2}$, it is obvious that $\mathfrak{s}' = \mathfrak{s}_0'(\frac{\pi}{2}, \frac{\pi}{2})$. Therefore S' is orbit equivalent to $S_0'(\frac{\pi}{2}, \frac{\pi}{2})$. When $\varphi_2 \neq \frac{\pi}{2}$, Lemma 4.1 and Lemma 4.2 yield that

$$\mathrm{Ad}\left(\exp\left(\frac{\tan\varphi_2}{c} W_0\right)\right)\mathfrak{s}' = \mathfrak{s}'(\pi/2, 0, \pi/2, \ldots, \pi/2) = \mathfrak{s}_0'(\pi/2, 0),$$

which means that S' is orbit equivalent to $S_0'(\frac{\pi}{2}, 0)$.
(2) The cases of $\mathfrak{s}' = \mathfrak{s}'(0, \varphi_2, \psi_1, \ldots, \psi_n)$ where $\varphi_2 \in [0, \frac{\pi}{2})$, $\psi_1, \ldots, \psi_n \in \{0, \frac{\pi}{2}\}$. By Lemma 4.1 and Lemma 4.2, we have that

$$\mathrm{Ad}\left(\exp\left(\frac{\tan\varphi_2}{c} W_0\right)\right)\mathfrak{s}' = \mathfrak{s}'(0, 0, \psi_1, \ldots, \psi_n).$$

Let b be a number of i such that $\psi_i = 0$. Then one can easily see that there exists isometric automorphism between $\mathfrak{s}'(0, 0, \psi_1, \ldots, \psi_n)$ and $\mathfrak{s}'_b(0, 0)$ since $\psi_i \in \{0, \frac{\pi}{2}\}$ for each $i \in \{1, \ldots, n\}$. Moreover it makes an isometric automorphism on S, and which shows that $S'(0, 0, \psi_1, \ldots, \psi_n)$ is orbit equivalent to $S'_b(0, 0)$. Hence S' is orbit equivalent to $S'_b(0, 0)$.

(3) The cases of $\mathfrak{s}' = \mathfrak{s}'(\varphi_1, \varphi_2, \psi_1, \ldots, \psi_n)$ where $\varphi_1 \in (0, \frac{\pi}{2}), \varphi_2 \in [0, \frac{\pi}{2}], \psi_1, \ldots \psi_n \in \{\varphi_1, \frac{\pi}{2}\}$. By Lemma 4.1 and Lemma 4.2, we have that

$$\mathrm{Ad}\left(\exp\left(\frac{\tan\varphi_2}{c} W_0\right)\right)\left(\mathrm{Ad}\left(\exp\left(\frac{\tan\varphi_1}{c} X_0\right)\right)\mathfrak{s}'\right) = \mathfrak{s}'(0, 0, \psi_1, \ldots, \psi_n)$$

when $\varphi_2 \in [0, \frac{\pi}{2})$, and

$$\mathrm{Ad}\left(\exp\left(\frac{\tan\varphi_1}{c} X_0\right)\right)\mathfrak{s}' = \mathfrak{s}'(0, \pi/2, \psi_1, \ldots, \psi_n)$$

when $\varphi_2 = \frac{\pi}{2}$. By the same argument in the proof of (2), one can prove that S' is orbit equivalent to $S'_b(0, 0)$ when $\varphi_2 \in [0, \frac{\pi}{2})$ and $S'_b(0, \frac{\pi}{2})$ when $\varphi_2 = \frac{\pi}{2}$. We complete the proof. $\qquad\square$

5. The case of $M = \mathbf{Sp}(n, \mathbb{R})/\mathbf{U}(n)$

In this section, we consider Lagrangian subalgebras in the solvable model of $M = \mathrm{Sp}(n, \mathbb{R})/\mathrm{U}(n)$.

We first introduce the solvable model of $M = \mathrm{Sp}(n, \mathbb{R})/\mathrm{U}(n) =: G/K$. Recall that matrix expressions of $\mathfrak{g}$ and $\mathfrak{k}$ are given by

$$\mathfrak{g} = \mathfrak{sp}(n, \mathbb{R}) = \left\{\left(\begin{array}{c|c} X_1 & X_2 \\ \hline X_3 & -{}^t X_1 \end{array}\right) \in M_{2n}(\mathbb{R}) \;\middle|\; X_1 \in M_n(\mathbb{R}), X_2, X_3 \in \mathrm{Sym}_n(\mathbb{R})\right\},$$

$$\mathfrak{k} = \mathfrak{u}(n) \cong \left\{\left(\begin{array}{c|c} A & -B \\ \hline B & A \end{array}\right) \in M_{2n}(\mathbb{R}) \;\middle|\; A \in \mathrm{Alt}_n(\mathbb{R}), B \in \mathrm{Sym}_n(\mathbb{R})\right\},$$

and the Cartan involution θ is given by

$$\theta : \mathfrak{g} \to \mathfrak{g} : X \mapsto -{}^t X.$$

Let us put

$$\mathfrak{p} := \left\{\left(\begin{array}{c|c} X & Y \\ \hline Y & -X \end{array}\right) \in M_{2n}(\mathbb{R}) \;\middle|\; X, Y \in \mathrm{Sym}_n(\mathbb{R})\right\}.$$

Then, $\mathfrak{g} = \mathfrak{k} \oplus \mathfrak{p}$ is the Cartan decomposition with respect to θ. We denote by $E_{i,j}$ the usual matrix unit, and choose a maximal abelian subspace $\mathfrak{a}$ in $\mathfrak{p}$ as follows:

$$\mathfrak{a} := \mathrm{span}_{\mathbb{R}}\left\{H_i := (E_{i,i} - E_{i+n,i+n}) \;\middle|\; i = 1, \ldots, n\right\} \subset \mathfrak{g}.$$

Define $\varepsilon_i \in \mathfrak{a}^*$ $(i = 1, \ldots, n-1)$ by

$$\varepsilon_i : \mathfrak{a} \to \mathbb{R} : \sum_{j=1}^{n} a_j H_j \mapsto a_i,$$

and $\Sigma := \{\pm(\varepsilon_i \pm \varepsilon_j) \mid 1 \le i < j \le n\} \cup \{\pm 2\varepsilon_i \mid 1 \le i \le n\}$.

Note that the restricted root system of $\mathfrak{g} = \mathfrak{sp}(n, \mathbb{R})$ with respect to $\mathfrak{a}$ coincides with Σ, which is of type C_n. In fact, by direct calculations, we have that $E_{i,j} - E_{j+n,i+n} \in \mathfrak{g}_{\varepsilon_i - \varepsilon_j}$, $E_{i,j+n} + E_{j,i+n} \in \mathfrak{g}_{\varepsilon_i + \varepsilon_j}$, $E_{k,k+n} \in \mathfrak{g}_{2\varepsilon_k}$ for each $1 \le k \le n$, $1 \le i < j \le n$. We thus see that

$$\dim_{\mathbb{R}} \left\{ \left(\bigoplus_{i<j} \mathfrak{g}_{\varepsilon_i - \varepsilon_j} \right) \oplus \left(\bigoplus_{i<j} \mathfrak{g}_{\varepsilon_i + \varepsilon_j} \right) \oplus \left(\bigoplus_{i} \mathfrak{g}_{2\varepsilon_i} \right) \right\} = n^2.$$

Let us put

$$\Lambda := \{\alpha_1 := \varepsilon_1 - \varepsilon_2, \ldots, \alpha_{n-1} := \varepsilon_{n-1} - \varepsilon_n, \alpha_n := 2\varepsilon_n\}.$$

Λ is a set of simple roots of Σ. Then the set of positive roots associated with Λ is given by

$$\Sigma^+ := \{\varepsilon_i \pm \varepsilon_j \mid 1 \le i < j \le n\} \cup \{2\varepsilon_i \mid 1 \le i \le n\}.$$

We thus obtain the Iwasawa decomposition of $\mathfrak{g}$. In particular, the solvable part of the Iwasawa decomposition is as follows:

$$\mathfrak{a} \oplus \mathfrak{n} = \mathfrak{a} \oplus \left(\bigoplus_{i<j} \mathfrak{g}_{\varepsilon_i - \varepsilon_j} \right) \oplus \left(\bigoplus_{i<j} \mathfrak{g}_{\varepsilon_i + \varepsilon_j} \right) \oplus \left(\bigoplus_{i} \mathfrak{g}_{2\varepsilon_i} \right).$$

Put $c := \frac{1}{\sqrt{k}} > 0$. The inner product on $\mathfrak{a} \oplus \mathfrak{n}$ is given by

$$\langle X, Y \rangle = (1/c^2)\langle X_{\mathfrak{a}}, Y_{\mathfrak{a}} \rangle_{\mathfrak{g}} + (1/2c^2)\langle X_{\mathfrak{n}}, Y_{\mathfrak{n}} \rangle_{\mathfrak{g}}, \quad (X, Y \in \mathfrak{a} \oplus \mathfrak{n}).$$

We choose

$$Z := \frac{1}{2} \left(\begin{array}{c|c} 0 & -E_n \\ \hline E_n & 0 \end{array} \right) \in C(\mathfrak{k}),$$

where E_n is the identity matrix of size n. Then one can see that $(\mathrm{ad}(Z)|_{\mathfrak{p}})^2 = -\mathrm{id}_{\mathfrak{p}}$, and hence $J = (d\Phi)_e^{-1} \circ \mathrm{ad}(Z) \circ (d\Phi)_e$ is the complex structure on $\mathfrak{a} \oplus \mathfrak{n}$. We here take a basis of $\mathfrak{a} \oplus \mathfrak{n}$ as follows:

$$\begin{aligned}
A_i &:= (c/\sqrt{2})H_i \quad (1 \le i \le n) \\
X_{ij} &:= c(E_{i,j} - E_{j+n,i+n}) \quad (1 \le i < j \le n), \\
Y_{ij} &:= c(E_{i,j+n} + E_{j,i+n}) \quad (1 \le i < j \le n), \\
Z_i &:= \sqrt{2}c E_{i,i+n} \quad (1 \le i \le n).
\end{aligned}$$

Proposition 5.1. *Let* $(\mathfrak{a} \oplus \mathfrak{n}, \langle\,,\,\rangle, J)$ *is the solvable model of* $M = \mathrm{Sp}(n, \mathbb{R})/\mathrm{U}(n)$ *as above. Then we have*

(1) $\{A_i, Z_i, X_{jk}, Y_{jk} \mid 1 \leq i \leq n, 1 \leq j < k \leq n\}$ *is an orthonormal basis of* $\mathfrak{a} \oplus \mathfrak{n}$ *with respect to the inner product* $\langle\,,\,\rangle$,
(2) $J(A_i) = Z_i$, $J(Z_i) = -A_i$, $J(X_{ij}) = Y_{ij}$, $J(Y_{ij}) = -X_{ij}$.

Proof. (1) We first prove that $\{A_i, Z_i, X_{jk}, Y_{jk} \mid 1 \leq i \leq n, 1 \leq j < k \leq n\}$ is a basis of $\mathfrak{a} \oplus \mathfrak{n}$. In order to do this, we show that

$$\mathfrak{g}_{\varepsilon_i - \varepsilon_i} = \mathrm{span}_{\mathbb{R}}\{X_{ij}\},$$
$$\mathfrak{g}_{\varepsilon_i + \varepsilon_i} = \mathrm{span}_{\mathbb{R}}\{Y_{ij}\},$$
$$\mathfrak{g}_{2\varepsilon_k} = \mathrm{span}_{\mathbb{R}}\{Z_1, \ldots, Z_n\}.$$

For each inclusion $(\supset)$, one can show it by direct calculations. Hence we have

$$\mathfrak{n}' := \mathrm{span}_{\mathbb{R}}\{Z_i, X_{jk}, Y_{jk} \mid 1 \leq i \leq n, 1 \leq j < k \leq n\} \subset \mathfrak{n}.$$

We calculate the dimensions of both sides.

$$\dim_{\mathbb{R}} \mathfrak{n}' = n + \frac{1}{2}n(n-1) + \frac{1}{2}n(n-1) = n^2, \text{ and}$$

$$\dim_{\mathbb{R}} \mathfrak{n} = \dim_{\mathbb{R}} M - \mathrm{rank}M = n^2 + n - n = n^2.$$

By dimensional reason, we see $\mathfrak{n}' = \mathfrak{n}$, and which yields that each converse inclusion $(\subset)$ holds. By definition of the inner product on $\mathfrak{a} \oplus \mathfrak{n}$, we have

$$\langle A_i, A_j \rangle = \frac{1}{c^2} \left\langle \frac{c}{\sqrt{2}}(E_{i,i} - E_{i+n,i+n}), \frac{c}{\sqrt{2}}(E_{j,j} - E_{j+n,j+n}) \right\rangle$$

$$= \frac{1}{2}\mathrm{tr}(^t(E_{i,i} - E_{i+n,i+n})(E_{j,j} - E_{j+n,j+n}))$$

$$= \delta_{ij}.$$

Similarly, one can check that the basis $\{A_i, Z_i, X_{jk}, Y_{jk} \mid 1 \leq i \leq n, 1 \leq j < k \leq n\}$ is an orthonormal with respect to $\langle\,,\,\rangle$.
(2) We only show that $J(A_i) = Z_i$. By the definition of J, we have

$$J(A_i) = (d\varPhi)_e^{-1} \circ \mathrm{ad}(Z) \circ (d\varPhi)_e(A_i)$$

$$= (d\varPhi)_e^{-1} \circ \mathrm{ad}(Z)(A_i)$$

$$= (d\varPhi)_e^{-1}((c/\sqrt{2})(E_{i,i+n} + E_{i+n,i}))$$

$$= Z_i.$$

For the remaining equations, we omit the proof. $\square$

Based on the above arguments, we define the solvable model of $M = \mathrm{Sp}(n, \mathbb{R})/\mathrm{U}(n)$.

Definition 5.1. Let $c > 0$, $n \in \mathbb{N}$. We call $(\mathfrak{s}(c), \langle\,,\,\rangle, J)$ the *solvable model of* $\mathrm{Sp}(n, \mathbb{R})/\mathrm{U}(n)$ if this triplet satisfies the following conditions:
(1) $\mathfrak{s}(c) := \mathrm{span}_{\mathbb{R}}\{A_i, Z_i \mid 1 \le i \le n\} \oplus \mathrm{span}_{\mathbb{R}}\{X_{ij}, Y_{ij} \mid 1 \le i < j \le n\}$ is a $(n^2 + n)$-dimensional real Lie algebra whose nonzero bracket relations are defined by

- $[A_k, X_{kj}] = (c/\sqrt{2})X_{kj}$, $[A_k, X_{ik}] = -(c/\sqrt{2})X_{ik}$,
- $[A_k, Y_{kj}] = (c/\sqrt{2})Y_{kj}$, $[A_k, Y_{ik}] = (c/\sqrt{2})Y_{ik}$,
- $[A_i, Z_i] = \sqrt{2}cZ_i$,
- $[X_{ij}, X_{jk}] = cX_{ik}$,
- $[X_{ij}, Y_{jk}] = cY_{ik}$, $[X_{ij}, Y_{kj}] = \begin{cases} cY_{ik} \ (i < k) \\ cY_{ki} \ (i > k) \end{cases}$, $[X_{ij}, Y_{ij}] = \sqrt{2}cZ_i$,
- $[X_{ij}, Z_j] = \sqrt{2}cY_{ij}$;

(2) $\langle\,,\,\rangle$ is an inner product on $\mathfrak{s}(c)$ such that the above basis is orthonormal;
(3) J is a complex structure on $\mathfrak{s}(c)$ defined by

$$J(A_i) = Z_i, \ \ J(X_{ij}) = Y_{ij}.$$

Let $S(c)$ be the connected and simply-connected Lie group with Lie algebra $\mathfrak{s}(c)$. The solvable model $(S(c), \langle\,,\,\rangle, J)$ is isomorphic to $\mathrm{Sp}(n, \mathbb{R})/\mathrm{U}(n)$ with minimal sectional curvature $-c^2$.

Next, we study Lagrangian subalgebras in the solvable model $(\mathfrak{s}(c), \langle\,,\,\rangle, J)$ of $\mathrm{Sp}(n, \mathbb{R})/\mathrm{U}(n)$. We put

$$T_i := \cos\varphi_i A_i + \sin\varphi_i Z_i, \ \ V_{ij} := \cos\psi_{ij} X_{ij} + \sin\psi_{ij} Y_{ij},$$

and

$$s' := \mathrm{span}_{\mathbb{R}}\{T_i \mid 1 \le i \le n\} \oplus \mathrm{span}_{\mathbb{R}}\{V_{ij} \mid 1 \le i < j \le n\}.$$

Easy computations give that

$$[T_i, T_j] = 0 \in \mathfrak{s}',$$

$$[T_i, V_{ij}] = \frac{c}{\sqrt{2}}\cos\varphi_i V_{ij} \in \mathfrak{s}',$$

$$[T_j, V_{ij}] = -\frac{c}{2}\cos\varphi_j\cos\psi_{ij} X_{ij} + \frac{c}{\sqrt{2}}(\cos\varphi_j\sin\psi_{ij} - 2\sin\varphi_j\cos\psi_{ij})Y_{ij},$$

$$[V_{ij}, V_{jk}] = c\cos\psi_{ij}\{\cos\psi_{jk} X_{ik} + \sin\psi_{jk} Y_{ik}\} \ \ (n \ge 3),$$

$$[V_{ij}, V_{kj}] = c(\cos\psi_{ij}\sin\psi_{kj} - \sin\psi_{ij}\cos\psi_{kj})Y_{ik} \ \ (n \ge 3).$$

Lemma 5.1. *Suppose the subspace $\mathfrak{s}'$ satisfies the condition $[T_j, V_{ij}] \in \mathfrak{s}'$. Then the following hold:*

(1) *If $\varphi_j = \frac{\pi}{2}$, then $\psi_{ij} = \frac{\pi}{2}$.*

(2) *If $\varphi_j = 0$, then $\psi_{ij} \in \{0, \frac{\pi}{2}\}$,*

(3) *If $\varphi_j \in (0, \frac{\pi}{2})$, then $\psi_{ij} \in \{\varphi_j, \frac{\pi}{2}\}$.*

Proof. By similar arguments to the proof of Lemma 4.1, one can prove this Lemma. $\qquad\square$

Lemma 5.2. *Put $U = \displaystyle\sum_{j=1}^{n} a_j A_j + z_j Z_0 + \sum_{1 \leq i < j \leq n} x_{ij} X_{ij} + y_{ij} Y_{ij}$. Then, for each $j \in \{1, \ldots, n\}$ we have*

$$\mathrm{Ad}(\exp(tZ_j))(U) = U - ta_j c Z_j - \sqrt{2} c t \sum_{i=1}^{j-1} Y_{ij}.$$

This Lemma is proved directly by using Baker-Campbell-Hausdorff formula and bracket structures of the solvable model $\mathfrak{s}(c)$. We omit the proof.

When $n = 2$, $\mathfrak{s}'$ is a Lagrangian subalgebra if and only if $[T_i, V_{ij}] \in \mathfrak{s}'$ for each $1 \leq i < j \leq 2$. In this case, we can determine orbit equivalent classes of $S' \subset S(c)$ with Lie algebra $\mathfrak{s}'$ admitting a Lagrangian orbit.

Proposition 5.2. *Let $M = \mathrm{Sp}(2, \mathbb{R})/\mathrm{U}(2)$, S' be a connected closed subgroup of $S(c)$ with Lie algebra $\mathfrak{s}' = \mathrm{span}_{\mathbb{R}}\{T_1, T_2\} \oplus \mathrm{span}_{\mathbb{R}}\{V_{12}\}$. Suppose that $\mathfrak{s}'$ is a Lagrangian subalgebra. Then the action of S' is orbit equivalent to a connected closed subgroup of $S(c)$ whose Lie algebra is one of the following six:*

$$\mathrm{span}_{\mathbb{R}}\{A_1, A_2, X_{12}\}, \quad \mathrm{span}_{\mathbb{R}}\{A_1, A_2, Y_{12}\}, \quad \mathrm{span}_{\mathbb{R}}\{Z_1, A_2, X_{12}\},$$

$$\mathrm{span}_{\mathbb{R}}\{Z_1, A_2, Y_{12}\}, \quad \mathrm{span}_{\mathbb{R}}\{A_1, Z_2, Y_{12}\}, \quad \mathrm{span}_{\mathbb{R}}\{Z_1, Z_2, Y_{12}\}.$$

Proof. It follows from Lemma 5.1 and Lemma 5.2 that $\mathfrak{s}'$ is conjugate to one of the above Lie algebra, which complete the proof. $\qquad\square$

However, when $n \geq 3$, the conditions in Lemma 5.1 are not sufficient that $\mathfrak{s}'$ is a Lagrangian subalgebra. We need to check the conditions $[V_{ij}, V_{jk}], [V_{ij}, V_{kj}] \in \mathfrak{s}'$ for each $1 \leq i < j \leq n$ and $1 \leq k < l \leq n$. On the other hand, it follows from Lemma 5.1 and Lemma 5.2 that a subspace $\mathfrak{s}'$ satisfying the condition $[T_j, V_{ij}] \in \mathfrak{s}'$ is conjugate to

$$\bar{\mathfrak{s}}' := \mathrm{span}_{\mathbb{R}}\{\overline{T}_i \mid 1 \leq i \leq n\} \oplus \mathrm{span}_{\mathbb{R}}\{\overline{V}_{ij} \mid 1 \leq i < j \leq n\}.$$

Here $\overline{T}_i = A_i$ or Z_i, $\overline{V}_{ij} = X_{ij}$ or Y_{ij}, and $\overline{V}_{ij}$ coincide with Y_{ij} for each $i \in \{1, \ldots, j-1\}$ if $\overline{T}_j = Z_j$. Since $\mathfrak{s}'$ is a Lagrangian subalgebra if and only if so is $\bar{\mathfrak{s}}'$, it is sufficient to check the remaining subalgebra conditions for $\bar{\mathfrak{s}}'$. At the end of this paper, we describe the case of $n = 3$.

Proposition 5.3. *Let* $M = \mathrm{Sp}(3, \mathbb{R})/\mathrm{U}(3)$, *and* S' *be a connected closed subgroup of* $S(c)$ *with Lie algebra* $\bar{\mathfrak{s}}' = \mathrm{span}_{\mathbb{R}}\{\overline{T}_1, \overline{T}_2, \overline{T}_3\} \oplus \mathrm{span}_{\mathbb{R}}\{\overline{V}_{12}, \overline{V}_{13}, \overline{V}_{23}\}$. *Suppose that* $\bar{\mathfrak{s}}'$ *is a Lagrangian subalgebra. Then the action of* S' *is orbit equivalent to a connected closed subgroup of* $S(c)$ *whose Lie algebra is one of the following 26 algebras:*

$\mathrm{span}_{\mathbb{R}}\{A_1, A_2, A_3, X_{12}, X_{13}, X_{23}\}, \quad \mathrm{span}_{\mathbb{R}}\{Z_1, A_2, A_3, X_{12}, X_{13}, X_{23}\},$

$\mathrm{span}_{\mathbb{R}}\{A_1, A_2, A_3, X_{12}, Y_{13}, Y_{23}\}, \quad \mathrm{span}_{\mathbb{R}}\{Z_1, A_2, A_3, X_{12}, Y_{13}, Y_{23}\},$

$\mathrm{span}_{\mathbb{R}}\{A_1, A_2, A_3, Y_{12}, X_{13}, X_{23}\}, \quad \mathrm{span}_{\mathbb{R}}\{Z_1, A_2, A_3, Y_{12}, X_{13}, X_{23}\},$

$\mathrm{span}_{\mathbb{R}}\{A_1, A_2, A_3, Y_{12}, X_{13}, Y_{23}\}, \quad \mathrm{span}_{\mathbb{R}}\{Z_1, A_2, A_3, Y_{12}, X_{13}, Y_{23}\},$

$\mathrm{span}_{\mathbb{R}}\{A_1, A_2, A_3, Y_{12}, Y_{13}, X_{23}\}, \quad \mathrm{span}_{\mathbb{R}}\{Z_1, A_2, A_3, Y_{12}, Y_{13}, X_{23}\},$

$\mathrm{span}_{\mathbb{R}}\{A_1, A_2, A_3, Y_{12}, Y_{13}, Y_{23}\}, \quad \mathrm{span}_{\mathbb{R}}\{Z_1, A_2, A_3, Y_{12}, Y_{13}, Y_{23}\},$

$\mathrm{span}_{\mathbb{R}}\{A_1, Z_2, A_3, Y_{12}, X_{13}, X_{23}\}, \quad \mathrm{span}_{\mathbb{R}}\{Z_1, Z_2, A_3, Y_{12}, X_{13}, X_{23}\},$

$\mathrm{span}_{\mathbb{R}}\{A_1, Z_2, A_3, Y_{12}, X_{13}, Y_{23}\}, \quad \mathrm{span}_{\mathbb{R}}\{Z_1, Z_2, A_3, Y_{12}, X_{13}, Y_{23}\},$

$\mathrm{span}_{\mathbb{R}}\{A_1, Z_2, A_3, Y_{12}, Y_{13}, X_{23}\}, \quad \mathrm{span}_{\mathbb{R}}\{Z_1, Z_2, A_3, Y_{12}, Y_{13}, X_{23}\},$

$\mathrm{span}_{\mathbb{R}}\{A_1, Z_2, A_3, Y_{12}, Y_{13}, Y_{23}\}, \quad \mathrm{span}_{\mathbb{R}}\{Z_1, Z_2, A_3, Y_{12}, Y_{13}, Y_{23}\},$

$\mathrm{span}_{\mathbb{R}}\{A_1, A_2, Z_3, X_{12}, Y_{13}, Y_{23}\}, \quad \mathrm{span}_{\mathbb{R}}\{Z_1, A_2, Z_3, X_{12}, Y_{13}, Y_{23}\},$

$\mathrm{span}_{\mathbb{R}}\{A_1, A_2, Z_3, Y_{12}, Y_{13}, Y_{23}\}, \quad \mathrm{span}_{\mathbb{R}}\{Z_1, A_2, Z_3, Y_{12}, Y_{13}, Y_{23}\},$

$\mathrm{span}_{\mathbb{R}}\{A_1, Z_2, Z_3, Y_{12}, Y_{13}, Y_{23}\}, \quad \mathrm{span}_{\mathbb{R}}\{Z_1, Z_2, Z_3, Y_{12}, Y_{13}, Y_{23}\}.$

Proof. We only need to investigate the Lie subalgebra conditions of $\bar{\mathfrak{s}}'$ in $\mathfrak{s}(c)$. We assume that $\bar{\mathfrak{s}}'$ is a Lagrangian subalgebra. Then the derived ideal $[\mathfrak{s}', \mathfrak{s}'] = \mathrm{span}_{\mathbb{R}}\{\overline{V}_{12}, \overline{V}_{13}, \overline{V}_{23}\}$ must be a Lie subalgebra of $\mathfrak{s}'$. It follows from the bracket relations

$$[X_{12}, X_{23}] = cX_{13}, \quad [X_{12}, Y_{23}] = cY_{13} \quad and \quad [X_{13}, Y_{23}] = cY_{12}$$

that $(\overline{V}_{12}, \overline{V}_{13}, \overline{V}_{23}) \neq (X_{12}, Y_{13}, X_{23}), (X_{12}, X_{13}, X_{23})$. We recall that if $\overline{T}_j = Z_j$ for some j then $\overline{V}_{ij} = Y_{ij}$ for each $i < j$ since Lemma 5.1 (1). Taking the contraposition, we see that if $\overline{V}_{ij} = X_{ij}$ for some i, j then $\overline{T}_j = A_j$, which gives the desired conclusion. $\qquad\square$

Acknowledgments

The author would like to thank all organizers of ICDG 2018 for providing an opportunity to talk. He also thank to Hideya Hashimoto, Nobutaka

Boumuki, and Misa Ohashi for useful comments, and Hiroshi Tamaru and Toru Kajigaya for helpful discussions. The author was supported in part by JSPS KAKENHI(16K17603).

References

[1] A. Amarzaya and Y. Ohnita, Hamiltonian stability of certain minimal Lagrangian submanifolds in complex projective spaces. *Tohoku Math. J.* **55** (2003), 583–610.

[2] L. Bedulli and A. Gori, Homogeneous Lagrangian submanifolds, *Commun. Anal. Geom.* **16**(3) (2008), 591–615.

[3] G. Deltour, On a generalization of a theorem of McDuff. *J. Diff. Geom.* **93** (2013), 379–400.

[4] J. T. Cho, T. Hashinaga, A. Kubo, Y. Taketomi and H. Tamaru, Realizations of some contact metric manifolds as Ricci soliton real hypersurfaces. *J. Geom. Phys.* **123** (2018), 221–234.

[5] T. Hashinaga and T. Kajigaya, A class of non-compact homogeneous Lagrangian submanifolds in complex hyperbolic spaces. *Ann. Glob. Anal. Geom.* **51**(1) (2017), 21–33.

[6] S. Helgason, *Differential geometry, Lie groups, and symmetric spaces*, Graduate Studies in Mathematics **34**, American Mathematical Society, Providence, RI, 2001.

[7] H. Iriyeh and H. Ono, Almost all Lagrangian torus orbits in $\mathbb{C}P^n$ are not Hamiltonian volume minimizing. *Ann. Glob. Anal. Geom.* **50**(1) (2016), 85–96.

[8] H. Iriyeh, T. Sakai and H. Tasaki, Lagrangian Floer homology of a pair of real forms in Hermitian symmetric spaces of compact type. *J. Math. Soc. Japan* **65**(4) (2013), 1135–1151.

[9] H. Ma and Y. Ohnita, On Lagrangian submanifolds in complex hyperquadrics and isoparametric hypersurfaces in spheres. *Math. Z.* **261**(4) (2009), 749–785.

[10] H. Ma and Y. Ohnita, Hamiltonian stability of the Gauss images of homogeneous isoparametric hypersurfaces. *J. Diff. Geom.* **97**(2) (20014), 275–348.

[11] D. McDuff, The symplectic structure of Kähler manifolds of nonpositive curvature. *J. Diff. Geom.* **28**(3) (1988), 467–475.

[12] Y.G. Oh, Volume minimization of Lagrangian submanifolds under Hamiltonian deformations. *Math. Z.* **212**(2) (1993), 175–192.

[13] D. Petrecca and F. Podesta, Construction of homogeneous Lagrangian
      submanifolds in $\mathbb{C}P^n$ and Hamiltonian stability. *Tohoku Math. J.* **64**(2)
      (2012), 261–268.

Received December 22, 2018
Revised   February   2, 2019

COMPLEX CURVES AND ISOTROPIC MINIMAL SURFACES IN HYPERKÄHLER 4-MANIFOLDS

Naoya ANDO

Faculty of Advanced Science and Technology, Kumamoto University,
2–39–1 Kurokami, Kumamoto 860–8555 Japan
E-mail: andonaoya@kumamoto-u.ac.jp

We obtain a characterization of complex curves in Kähler surfaces. In addition, by a direct method, we see that in a hyperKähler 4-manifold, complex curves are just isotropic minimal surfaces compatible with the orientation of the space. We obtain characterizations of hyperKähler 4-manifolds among Kähler surfaces, in terms of isotropic minimal surfaces and deformations of the complex structures.

Keywords: Complex curve; isotropic minimal surface; hyperKähler manifold; deformation.

1. Introduction

A minimal surface in a 4-dimensional Riemannian manifold is said to be *isotropic* if at each point, principal curvatures do not depend on the choice of a unit normal vector, that is, if the ellipse of curvature is everywhere given by a circle. A complex curve in $E^4 = \mathbb{C}^2$ is an isotropic minimal surface, and a connected isotropic minimal surface in E^4 is congruent with a complex curve in $\mathbb{C}^2$ with respect to the standard complex structure I on $\mathbb{C}^2$. An isotropic minimal surface in E^4 is characterized in terms of a relation between the induced metric and a holomorphic cubic differential ([2]), and based on a rewrite of this characterization, an isotropic minimal surface in a 4-dimensional Riemannian space form is characterized ([4]). See [6, 15] for a characterization of such a surface by the curvature of the normal connection. See [5] for a characterization of an isotropic minimal surface in S^4 by the twistor space associated with S^4. Let N be an oriented Riemannian manifold of dimension 4 and $F : M \to N$ a conformal immersion of a connected Riemann surface M into N. Then F is an isotropic minimal immersion compatible with the orientation of N if and only if F has a horizontal lift into the twistor space $\tilde{N}$ associated with N ([9]).

The main purpose of the present paper is to study the relations between holomorphicity and isotropicity of minimal surfaces in Kähler and hyperKähler 4-manifolds. Let N be a Kähler surface. Then a complex curve in N has a horizontal lift into the twistor space associated with N and therefore we see from the above-mentioned result of [9] that a complex curve in N is an isotropic minimal surface compatible with the orientation of N. We will see that a connected complex curve in N is just a connected isotropic minimal surface in N with at least one complex point and compatible with the orientation of N (see § 4 below for the definition of complex points). We can find examples of totally geodesic surfaces with no complex points in $\mathbb{C}P^2$, $\mathbb{C}H^2$, $\mathbb{C}P^1 \times \mathbb{C}P^1$ and $\mathbb{C}H^1 \times \mathbb{C}H^1$ (see § 5 below).

Considering E^4 to be a hyperKähler 4-manifold with the standard complex structure I and additional two complex structures J, $K = IJ$, we see that a connected isotropic minimal surface in E^4 compatible with the orientation of E^4 is just a connected complex curve in E^4 with respect to the complex structure $aI + bJ + cK$ for an element (a, b, c) of S^2. The same result holds for the case where the space is a general hyperKähler 4-manifold. For a hyperKähler 4-manifold N, the twistor space $\tilde{N}$ associated with N is given by the product of N and S^2, and each element of S^2 gives a horizontal section of $\tilde{N}$. Therefore a connected complex curve with respect to the complex structure $aI + bJ + cK$ for an element (a, b, c) of S^2 is just a connected surface which has a horizontal lift into $\tilde{N} = N \times S^2$. Then we see from the result of [9] that a complex curve in N as above is just a connected isotropic minimal surface in N compatible with the orientation of N. We will supply a proof of it without using the twistor space.

A hyperKähler manifold N is a Ricci-flat Kähler manifold: The holonomy group of N is considered to be a subgroup of $\mathrm{Sp}(m)$ with $\dim N = 4m$ and therefore a subgroup of $\mathrm{SU}(2m)$. In addition, since $\mathrm{Sp}(1) = \mathrm{SU}(2)$, we see that a hyperKähler 4-manifold is just a Ricci-flat Kähler surface, if the manifold is simply connected. We can treat $K3$-surfaces as examples of hyperKähler 4-manifolds: They admit Ricci-flat Kähler metrics and are simply connected. We can refer to Chapters 6 and 7 of [10] for Ricci-flat Kähler and hyperKähler manifolds, respectively.

Let N be a Kähler surface and I its complex structure. We will obtain two characterizations of a hyperKähler 4-manifold. On a neighborhood V of each point of N, there exists an almost complex structure J with $IJ = -JI$ such that the metric h of N is Hermitian on (V, J) (see § 3 below). If $V = N$ and if J is parallel with respect to the Levi-Civita connection ∇ of h, then N is hyperKähler. We set $K := IJ$. In the present paper, a *deformation* of the complex structure I is defined to be an almost complex structure on N which is locally represented as $aI + bJ + cK$ for an S^2-valued function (a, b, c). Notice that a deformation of I is just a section of the twistor space associated with N and that there exists an almost complex structure on N which does not coincide with any deformations of I. We will obtain the first characterization of a hyperKähler 4-manifold as follows: N is hyperKähler if and only if there exists a deformation I' of I with $I' \neq \pm I$ such that the tangent bundle of N is locally represented as a direct sum of two I'-invariant involutive distributions of dimension 2 with isotropic minimal integral surfaces compatible with the orientation of the space. Therefore N is Ricci-flat if and only if the tangent bundle of N is locally represented as above for a deformation $I' \neq \pm I$ on a neighborhood of each point of N. We will obtain the second characterization of a hyperKähler 4-manifold as follows: N is hyperKähler if and only if there exists a deformation of I which is parallel with respect to ∇ and not identical with $\pm I$. Therefore N is Ricci-flat if and only if there exists a deformation of I on a neighborhood of each point of N as above.

2. Complex and Kähler manifolds

In this section, we will prepare basic notions related to complex and Kähler manifolds, referring to [11, Chap. 9] and [14].

Let N be a $2n$-dimensional almost complex manifold ($n \in \mathbb{N}$) and I its almost complex structure. It is known that the Nijenhuis tensor N_I of I vanishes if and only if N is a complex manifold such that I is its complex structure. If N_I vanishes, then for local complex coordinates $(z^1, \ldots, z^n)$ of N with $z^k = x^k + \sqrt{-1}y^k$, $I(\partial/\partial x^k)$ coincides with $\partial/\partial y^k$ ($k = 1, \ldots, n$) and we suppose that the orientation of N is given by an ordered basis $(\partial/\partial x^1, \partial/\partial y^1, \ldots, \partial/\partial x^n, \partial/\partial y^n)$.

For $q \in N$, let $T_q^{\mathbb{C}}(N)$ be the complexification of the tangent space $T_q(N)$ of N at q. For each element $Z = X + \sqrt{-1}Y$ of $T_q^{\mathbb{C}}(N)$ ($X, Y \in T_q(N)$), we denote by $\overline{Z}$ its conjugate: $\overline{Z} := X - \sqrt{-1}Y$. Let $T_q^{1,0}(N)$, $T_q^{0,1}(N)$ be subspaces of $T_q^{\mathbb{C}}(N)$ given by elements in the forms of $X - \sqrt{-1}IX$, $X + \sqrt{-1}IX$ ($X \in T_q(N)$), respectively. Then we obtain $T_q^{\mathbb{C}}(N) = T_q^{1,0}(N) \oplus T_q^{0,1}(N)$, and $Z \in T_q^{1,0}(N)$ is equivalent to $\overline{Z} \in T_q^{0,1}(N)$. We consider I to be a complex linear transformation of $T_q^{\mathbb{C}}(N)$. Then $T_q^{1,0}(N)$ and $T_q^{0,1}(N)$ are eigenspaces of I corresponding to the eigenvalues $\sqrt{-1}$, $-\sqrt{-1}$, respectively. In particular, in the case where (N, I) is a complex manifold, $\partial/\partial z^1, \ldots, \partial/\partial z^n$ (respectively, $\partial/\partial \overline{z}^1, \ldots, \partial/\partial \overline{z}^n$) form a basis of $T_q^{1,0}(N)$ (respectively, $T_q^{0,1}(N)$).

Let h be a Hermitian metric on an almost complex manifold (N, I). Then $h(IX, IY) = h(X, Y)$ for $X, Y \in T_q(N)$. We consider h a complex bilinear function on $T_q^{\mathbb{C}}(N)$. Then for Z, $W \in T_q^{\mathbb{C}}(N)$, h satisfies $h(\overline{Z}, \overline{W}) = \overline{h(Z, W)}$, and $h(Z, \overline{Z}) > 0$ if Z is nonzero. In addition, h satisfies $h(Z, W) = 0$ for Z, $W \in T_q^{1,0}(N)$. Let ∇ be the Levi-Civita connection of h. Then the Nijenhuis tensor N_I of I vanishes if and only if I and ∇ satisfy $(\nabla_{IX} I)Y = I(\nabla_X I)Y$ for arbitrary X, $Y \in T_q(N)$ and $q \in N$. Suppose that I is parallel with respect to ∇. Then N_I vanishes. We set

$$\Omega_I(X, Y) := h(IX, Y). \tag{1}$$

Then Ω_I is a 2-form on N and parallel with respect to ∇. Therefore Ω_I is closed and (N, h, I) is Kähler. Conversely, if (N, h, I) is Kähler, then we can show that I is parallel with respect to ∇.

Suppose that (N, h, I) is a Kähler manifold. Set $\partial_i := \partial/\partial z^i$, $\overline{\partial}_i = \partial_{\overline{i}} := \partial/\partial \overline{z}^i$, $h_{i\overline{j}} := h(\partial_i, \overline{\partial}_j)$. Then an $n \times n$ matrix $(h_{i\overline{j}})$ has the inverse matrix $(h^{i\overline{j}})$: $\sum_{k=1}^n h^{i\overline{k}} h_{j\overline{k}} = \delta_j^i$. Since I is parallel with respect to ∇, $\nabla_{X+\sqrt{-1}Y} \partial_i$ and $\nabla_{X+\sqrt{-1}Y} \overline{\partial}_i$ are eigenvectors of I corresponding to eigenvalues $\sqrt{-1}$, $-\sqrt{-1}$, respectively. Therefore $\nabla_{X+\sqrt{-1}Y} \partial_i$ (respectively, $\nabla_{X+\sqrt{-1}Y} \overline{\partial}_i$) is represented as a complex linear combination of $\partial_1, \ldots, \partial_n$ (respectively, $\overline{\partial}_1, \ldots, \overline{\partial}_n$) at each point. Since ∇ is torsion-free, we have $\nabla_{\partial_i} \overline{\partial}_j = \nabla_{\overline{\partial}_j} \partial_i = 0$. Christoffel symbols Γ_{ij}^k ($i, j, k = 1, \ldots, n$) with respect to local complex coordinates $(z^1, \ldots, z^n)$ are given by $\nabla_{\partial_i} \partial_j = \sum_{k=1}^n \Gamma_{ij}^k \partial_k$. We have

$$\Gamma_{ij}^k = \sum_{l=1}^n h^{k\overline{l}} \frac{\partial h_{j\overline{l}}}{\partial z^i}. \tag{2}$$

Let R be the curvature tensor field of ∇. We set

$$R(\partial_A, \partial_B)\partial_C = \sum_{l=1}^n R_{ABC}^l \partial_l + \sum_{l=1}^n R_{ABC}^{\overline{l}} \overline{\partial}_l$$

for $A, B, C \in \{1, \ldots n, \overline{1}, \ldots, \overline{n}\}$. Then we obtain $R_{ABk}^l = 0$ and $R_{AB\overline{k}}^l = 0$ for $k, l \in \{1, \ldots, n\}$, which imply $R_{ijA}^B = 0$ and $R_{\overline{ij}A}^B = 0$ for $i, j \in \{1, \ldots, n\}$. We obtain

$$R_{i\overline{j}k}^l = -\frac{\partial \Gamma_{ik}^l}{\partial \overline{z}^j}. \tag{3}$$

Let Ric be the Ricci tensor field of (N, h). Then Ric is symmetric. In addition, $\nabla I = 0$ implies $\text{Ric}\,(IX, IY) = \text{Ric}\,(X, Y)$. In particular, we have $\text{Ric}\,(\partial_i, \partial_j) = 0$. We set $\text{Ric}_{i\overline{j}} := \text{Ric}\,(\partial_i, \overline{\partial}_j)$. Then we obtain

$$\text{Ric}_{i\overline{j}} = -\sum_{k=1}^n \frac{\partial \Gamma_{ik}^k}{\partial \overline{z}^j} = -\frac{\partial^2 \log(\det(h_{i\overline{j}}))}{\partial z^i \partial \overline{z}^j}. \tag{4}$$

3. Another almost complex structure of a complex surface

Let (N, I) be a complex manifold. In the following, we suppose $n = 2$, so that N is of real dimension 4.

Lemma 3.1. *Let J be a $(1, 1)$-tensor field on N. Then J satisfies $IJ = -JI$ if and only if J is locally represented as*

$$(J\partial_1 \ J\partial_2) = (\bar{\partial}_1 \ \bar{\partial}_2) \begin{pmatrix} \alpha & \beta \\ \gamma & \delta \end{pmatrix}, \tag{5}$$

where α, β, γ, δ are complex-valued functions.

Proof. We set $\partial_{x^i} := \partial/\partial x^i$, $\partial_{y^i} := \partial/\partial y^i$ and

$$\left(J\partial_{x^1} \ J\partial_{y^1} \ J\partial_{x^2} \ J\partial_{y^2}\right) = \left(\partial_{x^1} \ \partial_{y^1} \ \partial_{x^2} \ \partial_{y^2}\right) \begin{pmatrix} a_1^1 & c_1^1 & a_2^1 & c_2^1 \\ b_1^1 & d_1^1 & b_2^1 & d_2^1 \\ a_1^2 & c_1^2 & a_2^2 & c_2^2 \\ b_1^2 & d_1^2 & b_2^2 & d_2^2 \end{pmatrix}.$$

Since $I\partial_k = \sqrt{-1}\partial_k$, if $IJ = -JI$, then $c_l^k = b_l^k$ and $d_l^k = -a_l^k$ for $k, l = 1, 2$. Therefore there exist complex-valued functions α, β, γ, δ satisfying (5). If J is locally represented as in (5), then we obtain $IJ = -JI$. $\qquad \square$

Lemma 3.2. *Let J be a $(1, 1)$-tensor field on N satisfying $IJ = -JI$. Then the following (a), (b), (c) are mutually equivalent:*

(a) *J is an almost complex structure on N;*
(b) *α, β, γ, δ as in Lemma 3.1 satisfy*

$$|\alpha| = |\delta|, \quad \beta \neq 0, \quad \gamma = -\frac{1 + |\alpha|^2}{|\beta|^2}\beta, \quad \alpha\delta = -\frac{|\alpha|^2}{|\beta|^2}\beta^2; \tag{6}$$

(c) *α, β, γ, δ as in Lemma 3.1 are represented as*

$$\begin{pmatrix} \alpha & \beta \\ \gamma & \delta \end{pmatrix} = \frac{\psi}{|\psi|^2} \begin{pmatrix} 0 & 1 \\ -1 & 0 \end{pmatrix} \begin{pmatrix} \rho_1 & \bar{\eta} \\ \eta & \rho_2 \end{pmatrix}, \tag{7}$$

where ρ_1, ρ_2 are positive-valued functions and η, ψ are complex-valued functions satisfying $\psi \neq 0$ and $\rho_1\rho_2 = |\psi|^2 + |\eta|^2$.

Proof. Let J be an almost complex structure on N. Then (5) implies

$$\alpha\bar{\alpha} + \bar{\beta}\gamma = -1, \quad \bar{\alpha}\beta + \bar{\beta}\delta = 0, \quad \alpha\bar{\gamma} + \gamma\bar{\delta} = 0, \quad \beta\bar{\gamma} + \delta\bar{\delta} = -1. \tag{8}$$

From the first and the fourth equations in (8), we obtain $|\alpha| = |\delta|$, $\beta \neq 0$ and $\gamma = -(1 + |\alpha|^2)/\bar{\beta}$. By the second or third equation in (8), we obtain $\alpha\delta = -(|\alpha|^2/|\beta|^2)\beta^2$. Therefore α, β, γ, δ satisfy (6). If α, β, γ, δ satisfy

(6), then they satisfy (8), and then J is an almost complex structure. Hence (a), (b) in Lemma 3.2 are equivalent to each other. We immediately see that (b), (c) in Lemma 3.2 are equivalent to each other.   □

Remark 3.1. Let h be a Hermitian metric on (N, I). Let J be an almost complex structure on N with $IJ = -JI$ such that h is a Hermitian metric on (N, J). Then J is locally represented as in (5) with (7) and

$$\rho_1 := h_{1\bar{1}}, \quad \rho_2 := h_{2\bar{2}}, \quad \eta := h_{1\bar{2}}, \quad \psi := \sqrt{\det(h_{i\bar{j}})}e^{\sqrt{-1}\theta} \qquad (9)$$

for a real-valued function θ. On a complex coordinate neighborhood V of each point of N, there exists an almost complex structure J with $IJ = -JI$ such that h is a Hermitian metric on (V, J). A 2-form Ω_J as in (1) for J is represented as $\Omega_J = -2\,\mathrm{Re}\,\Psi$, where Ψ is a complex 2-form of type $(2, 0)$ which is locally represented as $\psi\,dz^1 \wedge dz^2$. We see that Ψ is nowhere zero.

4. Complex curves and isotropic minimal surfaces in Kähler surfaces

Let (N, h, I) be a Kähler surface. Let M be a Riemann surface with complex structure I^M and $F : M \to N$ a conformal immersion of M into N. A point p of M is said to be *complex* with respect to F if $I \circ dF = dF \circ I^M$ on $T_p(M)$. Suppose that F is an isotropic minimal immersion. Then F is said to be *compatible with the orientation of N* if for a point p of M where principal curvatures of F are nonzero, $(T_1, T_2, \nabla_{T_1}T_1, \nabla_{T_1}T_2)$ is an ordered basis of the tangent space $T_{F(p)}(N)$ which gives the orientation of N, where $T_1 := dF(\partial/\partial u)$, $T_2 := dF(\partial/\partial v)$ for a local complex coordinate $w = u + \sqrt{-1}v$ on a neighborhood of p and ∇ is the Levi-Civita connection of h. If F is an isotropic minimal immersion compatible with the orientation of N and if p is a complex point with respect to F, then $I(\nabla_{T_1}T_1) = \nabla_{T_1}T_2$ at p. We will prove

Theorem 4.1. *Let (N, h, I) be a Kähler surface. Let M be a connected Riemann surface and $F : M \to N$ a conformal immersion. Then the following (a), (b) are equivalent to each other:*

(a) *F is a holomorphic immersion with respect to I;*
(b) *F is an isotropic minimal immersion with at least one complex point and compatible with the orientation of N.*

Proof. Let I^M be the complex structure on M and $F : M \to N$ a holomorphic immersion with respect to I. We have $dF \circ I^M = I \circ dF$. Let

$w = u + \sqrt{-1}v$ be a local complex coordinate on a neighborhood U of each $p \in M$. Let J be an almost complex structure on a neighborhood V of $F(p)$ with $IJ = -JI$ such that h is a Hermitian metric on (V, J). Set $K := IJ$. Then $T_1 := dF(\partial/\partial u)$, $T_2 := dF(\partial/\partial v) = IT_1$, $N_1 := JT_1$, $N_2 := KT_1$ have the same length, and N_1, N_2 are normal vector fields of F perpendicular to each other at any point. Let h^M be the induced metric by F. Then for a real-valued function α on U, h^M is locally represented as $h^M = e^{2\alpha}(du^2 + dv^2)$. Let μ_1, μ_2 be functions on U given by

$$\nabla_{T_1} T_1 = \alpha_u T_1 - \alpha_v T_2 + \mu_1 N_1 + \mu_2 N_2. \tag{10}$$

Then we have

$$\nabla_{T_1} T_2 = I \nabla_{T_1} T_1 = \alpha_v T_1 + \alpha_u T_2 - \mu_2 N_1 + \mu_1 N_2. \tag{11}$$

By (10) together with (11), we see that F is an isotropic minimal immersion compatible with the orientation of N. Let F be as in (b). Let N_1, N_2 be normal vector fields of F perpendicular to each other. Suppose that T_1, T_2, N_1, N_2 have the same length and that (T_1, T_2, N_1, N_2) gives the orientation of N. Then we have

$$(\nabla_{T_1} T_1 \ \nabla_{T_1} T_2 \ \nabla_{T_1} N_1 \ \nabla_{T_1} N_2)$$

$$= (T_1 \ T_2 \ N_1 \ N_2) \begin{pmatrix} \alpha_u & \alpha_v & -\mu_1 & -\mu_2 \\ -\alpha_v & \alpha_u & \mu_2 & -\mu_1 \\ \mu_1 & -\mu_2 & \alpha_u & -\beta \\ \mu_2 & \mu_1 & \beta & \alpha_u \end{pmatrix}, \tag{12}$$

where μ_1, μ_2 and β are functions and α is defined by $h^M = e^{2\alpha}(du^2 + dv^2)$. Therefore, since $\nabla I = 0$, we obtain

$$(-\nabla_{T_1} IT_2 \ \nabla_{T_1} IT_1 \ - \nabla_{T_1} IN_2 \ \nabla_{T_1} IN_1)$$

$$= (-IT_2 \ IT_1 \ - IN_2 \ IN_1) \begin{pmatrix} \alpha_u & \alpha_v & -\mu_1 & -\mu_2 \\ -\alpha_v & \alpha_u & \mu_2 & -\mu_1 \\ \mu_1 & -\mu_2 & \alpha_u & -\beta \\ \mu_2 & \mu_1 & \beta & \alpha_u \end{pmatrix}. \tag{13}$$

Let p be a complex point of M with respect to F. Then $IT_1 = T_2$ and $IN_1 = N_2$ at p. A system of ordinary differential equations with the Lipschitz condition has a unique solution for a given initial value. Therefore, comparing (12) with (13), we see that (T_1, T_2, N_1, N_2) and $(-IT_2, IT_1, -IN_2, IN_1)$ coincide with each other on the integral curve of $\partial/\partial u$ through p. Therefore, by analogous discussions on ∇_{T_2} instead of ∇_{T_1}, we see that (T_1, T_2, N_1, N_2) and $(-IT_2, IT_1, -IN_2, IN_1)$ coincide with each other on integral curves of $\partial/\partial v$. In particular, we obtain $IT_1 = T_2$ on a neighborhood of p. Since M is connected, F satisfies $I \circ dF = dF \circ I^M$ on $T_q(M)$ for any $q \in M$. This means that F is a holomorphic immersion with respect to I. Hence we obtain (a) from (b). $\square$

Remark 4.1. In (b) in Theorem 4.1, we can not remove the condition of the existence of a complex point (see Example 5.3 in the next section).

Remark 4.2. Let (N, h, I) be a Kähler surface and $F : M \to N$ a holomorphic immersion with respect to I. Let J be as in the above proof and T a 3-tensor field on a neighborhood U of each point p of M defined by

$$T(X, Y, Z) := h(J \circ dF(X),\ \sigma(Y, Z)), \tag{14}$$

where X, Y, Z are tangent vectors at each point of U and σ is the second fundamental form of F. We consider T to be a complex 3-linear function on the complexification of the tangent plane at each point of U. We can find local complex coordinates (z^1, z^2) on a neighborhood V of $F(p)$ such that $z^2 \circ F$ is constant on U. Then we can suppose $z^1 \circ F = w$ on U. Noticing that J is locally represented as in (5) with (7) and (9), we obtain

$$T\left(\frac{\partial}{\partial w}, \frac{\partial}{\partial w}, \frac{\partial}{\partial w} \right) = \frac{\psi}{|\psi|^2} h(\Gamma_{11}^1 \partial_1 + \Gamma_{11}^2 \partial_2,\ \eta \bar{\partial}_1 - \rho_1 \bar{\partial}_2) = -\psi \Gamma_{11}^2$$

and $T(\partial/\partial \overline{w}, \partial/\partial w, \partial/\partial w) = 0$. Therefore functions μ_1, μ_2 satisfying (10), (11) are given by $\mu_1 + \sqrt{-1}\mu_2 = -2\psi\Gamma_{11}^2/e^{2\alpha}$. We set $N := \mathbb{C}^2$. Let h be the standard metric of $\mathbb{C}^2$. We can choose a complex structure on $\mathbb{C}^2$ as the above J. Let $F : M \to \mathbb{C}^2$ be a holomorphic immersion with respect to I. Then a 3-tensor field T on M as in (14) is given by the real part of a holomorphic cubic differential Φ on M ([2]). We represent F as $F = (f^1, f^2)$ by two holomorphic functions f^1, f^2 on M. If we suppose $f^1 f_z^2 - f^2 f_z^1 \neq 0$ for any local complex coordinate z, then on a neighborhood of each point of M, there exists a local complex coordinate w such that f^1, f^2 satisfy $f_{ww}^i + \phi f^i = 0$ for $i = 1, 2$, where ϕ is given by $\Phi = \phi dw^3$ ([2]). This means that any complex curve in $\mathbb{C}^2$ is locally given by the composition of an affine Schwarz map and a parallel translation in $\mathbb{C}^2$. A characterization of a complex curve in $\mathbb{C}^2$ is given by a relation between the induced metric h^M and Φ which is represented as $\alpha_{w\overline{w}} = |\phi|^2/2e^{4\alpha}$ ([2]).

5. Examples of isotropic minimal surfaces

Example 5.1. Let N be the complex projective plane $\mathbb{C}P^2$ or the complex hyperbolic plane $\mathbb{C}H^2$. If $N = \mathbb{C}P^2$, then N has the Fubini-Study metric, which is represented for inhomogeneous coordinates (z^1, z^2) of N as

$$h = \frac{4}{\xi^2} \left(\xi \left(dz^1 d\bar{z}^1 + dz^2 d\bar{z}^2 \right) - \left(\bar{z}^1 dz^1 + \bar{z}^2 dz^2 \right) \left(z^1 d\bar{z}^1 + z^2 d\bar{z}^2 \right) \right),$$

where $\xi := 1 + z^1 \bar{z}^1 + z^2 \bar{z}^2$; if $N = \mathbb{C}H^2$, then N is given by

$$N = \{ (z^1, z^2) \in \mathbb{C}^2 \mid |z^1|^2 + |z^2|^2 < 1 \}$$

with the Bergman metric

$$h = \frac{4}{\xi^2} \left(\xi \left(dz^1 d\bar{z}^1 + dz^2 d\bar{z}^2 \right) + \left(\bar{z}^1 dz^1 + \bar{z}^2 dz^2 \right) \left(z^1 d\bar{z}^1 + z^2 d\bar{z}^2 \right) \right),$$

where $\xi := 1 - z^1 \bar{z}^1 - z^2 \bar{z}^2$. Based on these setting, we obtain

$$h_{1\bar{1}} = \frac{2}{\xi^2}(1 + \varepsilon z^2 \bar{z}^2), \quad h_{1\bar{2}} = \overline{h_{2\bar{1}}} = -\frac{2\varepsilon}{\xi^2}\bar{z}^1 z^2, \quad h_{2\bar{2}} = \frac{2}{\xi^2}(1 + \varepsilon z^1 \bar{z}^1), \quad (15)$$

where $\xi := 1 + \varepsilon(z^1 \bar{z}^1 + z^2 \bar{z}^2)$, and $\varepsilon = +1$ or -1 according to $N = \mathbb{C}P^2$ or $\mathbb{C}H^2$. By (2) together with (15), we obtain

$$\Gamma_{11}^1 = -\frac{2\varepsilon \bar{z}^1}{\xi}, \quad \Gamma_{12}^1 = -\frac{\varepsilon \bar{z}^2}{\xi}, \quad \Gamma_{12}^2 = -\frac{\varepsilon \bar{z}^1}{\xi}, \quad \Gamma_{22}^2 = -\frac{2\varepsilon \bar{z}^2}{\xi} \qquad (16)$$

and $\Gamma_{11}^2 = \Gamma_{22}^1 = 0$. Let I be the complex structure of N. Then by (3) together with (16), we see that holomorphic sectional curvatures of h are equal to ε for any two-dimensional I-invariant subspace of the tangent space of each point of N. Let J be an almost complex structure on an open set V of N with $IJ = -JI$ such that h is a Hermitian metric on (V, J). Then J is represented as in (5) with (7), (9) and (15). Let M be a Riemann surface and $F : M \to N$ a holomorphic immersion of M into N. Then we see from Theorem 4.1 that F is an isotropic minimal immersion compatible with the orientation of N. Let w be a local complex coordinate on an open set U of M with $F(U) \subset V$. We set $f^i := z^i \circ F$. Then by (16), we obtain

$$\nabla_{\frac{\partial}{\partial w}} \frac{\partial}{\partial w} = \sum_{k=1}^{2} \left(f_{ww}^k + \sum_{i,j=1}^{2} f_w^i f_w^j \Gamma_{ij}^k \right) \frac{\partial}{\partial z^k}$$

$$= \left(f_{ww}^1 - f_w^1 (\log \xi^2)_w \right) \frac{\partial}{\partial z^1} + \left(f_{ww}^2 - f_w^2 (\log \xi^2)_w \right) \frac{\partial}{\partial z^2} \qquad (17)$$

with $f_w^i := \partial f^i / \partial w$, $f_{ww}^i := \partial^2 f^i / \partial w^2$. Let h^M be the induced metric by F. We represent h^M as $h^M = e^{2\alpha} dw d\bar{w}$ on U. We can suppose that w satisfies $f^1 f_w^2 - f^2 f_w^1 \equiv 1$ (refer to [2]). Then we see that α is given by $e^{2\alpha} = 4\varepsilon(1 + \xi_{w\bar{w}})/\xi^2$. The following hold:

$$h \left(\nabla_{\frac{\partial}{\partial w}} \frac{\partial}{\partial w}, \frac{\partial}{\partial w} \right) = 0, \qquad h \left(\nabla_{\frac{\partial}{\partial w}} \frac{\partial}{\partial w}, \frac{\partial}{\partial \bar{w}} \right) = e^{2\alpha} \alpha_w. \qquad (18)$$

By (17) together with $f^1 f_w^2 - f^2 f_w^1 \equiv 1$, we obtain

$$h\left(\nabla_{\frac{\partial}{\partial \overline{w}}}\frac{\partial}{\partial w}, J\frac{\partial}{\partial w}\right) = T\left(\frac{\partial}{\partial w}, \frac{\partial}{\partial w}, \frac{\partial}{\partial w}\right) = \frac{\phi}{2} \qquad (19)$$

with

$$\phi := \frac{4e^{\sqrt{-1}\theta}}{\xi^{5/2}}(\xi_{ww} - (f_w^1 f_{ww}^2 - f_w^2 f_{ww}^1)). \qquad (20)$$

By (18) together with (19), we obtain

$$\nabla_{\frac{\partial}{\partial \overline{w}}}\frac{\partial}{\partial w} = 2\alpha_w \frac{\partial}{\partial w} + \frac{\phi}{e^{2\alpha}}J\frac{\partial}{\partial w}, \qquad (21)$$

which means (10), (11) with $\mu_1 + \sqrt{-1}\mu_2 = \phi/e^{2\alpha}$ and (20). We obtain

$$\nabla_{\frac{\partial}{\partial \overline{w}}}J\frac{\partial}{\partial w} = \left(\log \frac{e^{\sqrt{-1}\theta}}{\sqrt{\xi}}\right)_w J\frac{\partial}{\partial w} + \frac{e^{\sqrt{-1}\theta}}{\sqrt{\xi}}\left(f_{ww}^2 \frac{\partial}{\partial \overline{z}^1} - f_{ww}^1 \frac{\partial}{\partial \overline{z}^2}\right).$$

Therefore we obtain

$$h\left(\nabla_{\frac{\partial}{\partial \overline{w}}}J\frac{\partial}{\partial w}, J\frac{\partial}{\partial \overline{w}}\right) = \frac{\tau}{2},$$

$$\tau := \left(\log \frac{e^{\sqrt{-1}\theta}}{\sqrt{\xi}}\right)_w e^{2\alpha} + \frac{4\varepsilon}{\xi^3}(\xi_{ww\overline{w}} + \xi_{ww}\xi_{\overline{w}}). \qquad (22)$$

From (21) and (22), we obtain

$$\nabla_{\frac{\partial}{\partial \overline{w}}}J\frac{\partial}{\partial w} = -\frac{\phi}{e^{2\alpha}}\frac{\partial}{\partial \overline{w}} + \frac{\tau}{e^{2\alpha}}J\frac{\partial}{\partial w}. \qquad (23)$$

Using (17), (21) and (23), we obtain $\alpha_{w\overline{w}} = -\varepsilon e^{2\alpha}/4 + |\phi|^2/2e^{4\alpha}$. This can be considered to be an analogue of $\alpha_{w\overline{w}} = |\phi|^2/2e^{4\alpha}$ in Remark 4.2.

Example 5.2. Let N be given by one of the products $\mathbb{C}P^1 \times \mathbb{C}P^1$ and $\mathbb{C}H^1 \times \mathbb{C}H^1$. The metric of $\mathbb{C}P^1$ is represented as $(4/\xi^2)dzd\overline{z}$ by an inhomogeneous coordinate z of $\mathbb{C}P^1$ and the metric of $\mathbb{C}H^1 = \{z \in \mathbb{C} \mid |z| < 1\}$ is represented as in the same form, where $\xi := 1 + \varepsilon z\overline{z}$, and $\varepsilon = +1$ or -1 according to $\mathbb{C}P^1$ or $\mathbb{C}H^1$. Let h be the product metric of N: $h = 4/\xi_1^2 dz^1 d\overline{z}^1 + 4/\xi_2^2 dz^2 d\overline{z}^2$, where $\xi_k := 1 + \varepsilon z^k \overline{z}^k$. Then we obtain

$$h_{1\overline{1}} = \frac{2}{\xi_1^2}, \quad h_{1\overline{2}} = h_{2\overline{1}} = 0, \quad h_{2\overline{2}} = \frac{2}{\xi_2^2}. \qquad (24)$$

By (2) together with (24), we obtain

$$\Gamma_{11}^1 = -\frac{2\varepsilon \overline{z}^1}{\xi_1}, \quad \Gamma_{12}^1 = \Gamma_{12}^2 = \Gamma_{11}^2 = \Gamma_{22}^1 = 0, \quad \Gamma_{22}^2 = -\frac{2\varepsilon \overline{z}^2}{\xi_2}. \qquad (25)$$

By (3) together with (25), we see that holomorphic sectional curvatures of h are in either $[1/2, 1]$ or $[-1, -1/2]$, according to $\varepsilon = +1$ or -1. Let I be the natural complex structure of N. Let J be an almost complex structure on an open set V of N with $IJ = -JI$ such that h is a Hermitian metric on (V, J). Then J is represented as in (5) with (7), (9) and (24). Let M be a Riemann surface and $F : M \to N$ a holomorphic immersion of M into N. Then we see from Theorem 4.1 that F is an isotropic minimal immersion compatible with the orientation of N. Let w be a local complex coordinate on an open set U of M with $F(U) \subset V$ and set $f^i := z^i \circ F$. Then by (25), we obtain

$$\nabla_{\frac{\partial}{\partial w}} \frac{\partial}{\partial w} = \left(f^1_{ww} - f^1_w (\log \xi_1^2)_w\right) \frac{\partial}{\partial z^1} + \left(f^2_{ww} - f^2_w (\log \xi_2^2)_w\right) \frac{\partial}{\partial z^2}. \quad (26)$$

Let h^M be the induced metric by F. We represent h^M as $h^M = e^{2\alpha} dw d\overline{w}$ on U. We see that α is given by $e^{2\alpha} = 4\varepsilon(X_1 + X_2)$ with $X_i := (\xi_i)_{w\overline{w}}/\xi_i^2$. We have (18). By (26), we obtain (19) with

$$\phi := \frac{4e^{\sqrt{-1}\theta}}{\xi_1 \xi_2} \left(f^1_w f^2_w \left(\log \frac{\xi_2^2}{\xi_1^2}\right)_w - (f^1_w f^2_{ww} - f^2_w f^1_{ww})\right). \quad (27)$$

From (18) and (19), we obtain (21), which means (10) and (11) with $\mu_1 + \sqrt{-1}\mu_2 = \phi/e^{2\alpha}$ and (27). We obtain

$$\nabla_{\frac{\partial}{\partial w}} J \frac{\partial}{\partial w} = \sqrt{-1}\theta_w J \frac{\partial}{\partial w} + e^{\sqrt{-1}\theta} \left\{\left(\left(\frac{\xi_1}{\xi_2}\right)_w f^2_w + \frac{\xi_1}{\xi_2} f^2_{ww}\right) \frac{\partial}{\partial \overline{z}^1}\right.$$
$$\left. - \left(\left(\frac{\xi_2}{\xi_1}\right)_w f^1_w + \frac{\xi_2}{\xi_1} f^1_{ww}\right) \frac{\partial}{\partial \overline{z}^2}\right\}.$$

Therefore we obtain

$$h\left(\nabla_{\frac{\partial}{\partial w}} J \frac{\partial}{\partial w}, J \frac{\partial}{\partial \overline{w}}\right) = \frac{\tau}{2},$$
$$\tau := \sqrt{-1} e^{2\alpha} \theta_w + \frac{4\varepsilon}{\xi_1 \xi_2} \left((\xi_1 \xi_2 X_1)_w + (\xi_1 \xi_2 X_2)_w\right). \quad (28)$$

From (21) with (27), and (28), we obtain (23) with (27) and (28). Using (21), (23) with (27) and (28), and (26), we obtain

$$\alpha_{w\overline{w}} = -\frac{\varepsilon e^{2\alpha}}{4} \frac{X_1^2 + X_2^2}{(X_1 + X_2)^2} + \frac{|\phi|^2}{2e^{4\alpha}}.$$

Example 5.3. Let N, (z^1, z^2) and I be as in one of the above two examples. Then we see by (16) or (25) that a surface S in N given by $\operatorname{Im} z^1 = \operatorname{Im} z^2 = 0$ is totally geodesic. Therefore S is an isotropic minimal surface in N with zero principal curvatures. On the other hand, S can not be any complex curves in N. In addition, S has no complex points.

6. Complex curves and isotropic minimal surfaces in hyperKähler 4-manifolds

Let (N, h) be a 4-dimensional Riemannian manifold and ∇ the Levi-Civita connection of h. Let I, J, K be almost complex structures on N with $IJ = K$. Suppose that h is a Hermitian metric on each of almost complex manifolds (N, I), (N, J), (N, K) and that I, J, K are parallel with respect to ∇. Then N, h, I, J, K form a hyperKähler manifold. For each $(a, b, c) \in S^2$, $(N, h, aI + bJ + cK)$ is a Kähler manifold. The orientation of N does not depend on the choice of a complex structure $aI + bJ + cK$. As in § 1, we can obtain the following theorem by discussions in the twistor space associated with a hyperKähler 4-manifold together with the result of [9], while we will supply a proof of it without using the twistor space.

Theorem 6.1. *Let* (N, h, I, J, K) *be a* 4*-dimensional hyperKähler manifold. Let* M *be a connected Riemann surface and* $F : M \to N$ *a conformal immersion. Then the following* (a), (b) *are equivalent to each other*:

(a) *There exists a point* (a, b, c) *of* S^2 *such that* F *is a holomorphic immersion with respect to a complex structure* $aI + bJ + cK$;

(b) *The map* F *is an isotropic minimal immersion compatible with the orientation of* N.

Proof. We can obtain Theorem 6.1, referring to the proof of Theorem 4.1. We can obtain (a) from (b) more briefly. Let F be as in (b). Then there exists an S^2-valued function (a, b, c) on M satisfying $(aI + bJ + cK) \circ dF = dF \circ I^M$ on $T_p(M)$ for each $p \in M$. Set $I' := aI + bJ + cK$. Let $w = u + \sqrt{-1}v$ be a local complex coordinate of M and set $T_1 := dF(\partial/\partial u)$, $T_2 := dF(\partial/\partial v)$. Then we obtain

$$\nabla_{T_1} I' T_1 = (a_u I + b_u J + c_u K)T_1 + I'(\nabla_{T_1} T_1). \qquad (29)$$

Since F is as in (b), we have $\nabla_{T_1} I' T_1 = I'(\nabla_{T_1} T_1)$. Therefore we see by (29) that a_u, b_u, c_u are identically zero. In addition, since

$$\nabla_{I'T_1} I' T_1 = (a_v I + b_v J + c_v K)T_1 - \nabla_{T_1} T_1,$$

a_v, b_v, c_v are identically zero. Hence we obtain (a) from (b). $\qquad \square$

7. Characterizations of hyperKähler 4-manifolds

As in § 1, noticing holonomy groups, we see that a hyperKähler 4-manifold is just a Ricci-flat Kähler surface, if the manifold is simply connected. On the other hand, based on § 3, we can prove

Proposition 7.1. *Let* (N, h, I) *be a simply connected Kähler surface. Then the following conditions* (a), (b), (c), (d) *are mutually equivalent:*

(a) *There exists a complex structure* J *on* N *such that* N, h, I, J, $K := IJ$ *form a hyperKähler manifold;*

(b) *There exists a complex structure* J *on* N *with* $IJ = -JI$ *such that* (N, h, J) *is a Kähler manifold;*

(c) *There exists a nowhere zero holomorphic* 2-*form* $\Psi = \psi \, dz^1 \wedge dz^2$ *on* N *satisfying* $|\psi| = \sqrt{\det(h_{i\bar{j}})}$;

(d) (N, h) *is Ricci-flat.*

Proof. Conditions (a), (b) in Proposition 7.1 are equivalent to each other. Let J be as in (b) in Proposition 7.1. Then by $J\nabla_{\partial_i}\partial_j = \nabla_{\partial_i}J\partial_j$, we have

$$\frac{\partial}{\partial z^i}\begin{pmatrix} \alpha & \beta \\ \gamma & \delta \end{pmatrix} = \begin{pmatrix} \alpha & \beta \\ \gamma & \delta \end{pmatrix}\begin{pmatrix} \Gamma^1_{i1} & \Gamma^1_{i2} \\ \Gamma^2_{i1} & \Gamma^2_{i2} \end{pmatrix}, \tag{30}$$

where α, β, γ, δ are as in (5) for J. By (2), we obtain

$$\frac{\partial}{\partial z^i}\begin{pmatrix} h_{1\bar{1}} & h_{2\bar{1}} \\ h_{1\bar{2}} & h_{2\bar{2}} \end{pmatrix} = \begin{pmatrix} h_{1\bar{1}} & h_{2\bar{1}} \\ h_{1\bar{2}} & h_{2\bar{2}} \end{pmatrix}\begin{pmatrix} \Gamma^1_{i1} & \Gamma^1_{i2} \\ \Gamma^2_{i1} & \Gamma^2_{i2} \end{pmatrix}. \tag{31}$$

Using (c) in Lemma 3.2 and (31), we can rewrite (30) into $\partial\overline{\psi}/\partial z^i = 0$. Therefore ψ is holomorphic with respect to z^1, z^2. Then by $\Psi = \psi \, dz^1 \wedge dz^2$, we can define a holomorphic 2-form on N. If $\Psi = \psi \, dz^1 \wedge dz^2$ is as in (c) in Proposition 7.1, then we can find an almost complex structure J as in (5) with (7) and (9) so that J is parallel with respect to ∇. Hence (b), (c) in Proposition 7.1 are equivalent to each other. Suppose (c). Then $\log(\det(h_{i\bar{j}}))$ is the real part of a holomorphic function of z^1, z^2. Therefore $\partial\log(\det(h_{i\bar{j}}))/\partial z^1$, $\partial\log(\det(h_{i\bar{j}}))/\partial z^2$ are holomorphic with respect to z^1, z^2 and by this together with (4), we obtain (d) in Proposition 7.1. Suppose (d). Then $\partial\log(\det(h_{i\bar{j}}))/\partial z^1$, $\partial\log(\det(h_{i\bar{j}}))/\partial z^2$ are holomorphic with respect to z^1, z^2. Therefore $\log(\det(h_{i\bar{j}}))$ is represented as

$$\log(\det(h_{i\bar{j}})) = f(z^1, z^2) + g(\bar{z}^1, \bar{z}^2), \tag{32}$$

where f (respectively, g) is holomorphic with respect to z^1, z^2 (respectively, $\bar{z}^1$, $\bar{z}^2$). Since $\log(\det(h_{i\bar{j}}))$ is real-valued, we obtain

$$\log(\det(h_{i\bar{j}})) = \overline{f(z^1, z^2)} + \overline{g(\bar{z}^1, \bar{z}^2)}. \tag{33}$$

From (32) and (33), we obtain

$$f(z^1, z^2) - \overline{g(\bar{z}^1, \bar{z}^2)} = \overline{f(z^1, z^2)} - g(\bar{z}^1, \bar{z}^2). \tag{34}$$

The left side of (34) is holomorphic with respect to z^1, z^2, while the right side of (34) is holomorphic with respect to $\bar{z}^1$, $\bar{z}^2$. Therefore $\overline{f(z^1, z^2)} - g(\bar{z}^1, \bar{z}^2)$ is constant and this implies that $\log(\det(h_{i\bar{j}}))$ is the real part of a holomorphic function of z^1, z^2. Then we see that there exist a system of holomorphic coordinate neighborhoods $\{(V_\lambda, z_\lambda)\}_{\lambda \in \Lambda}$ of N and a family of holomorphic functions $\{\psi_\lambda\}_{\lambda \in \Lambda}$ satisfying the following: Each V_λ is homeomorphic to an open ball in $\mathbb{C}^2 = E^4$; if $V_\lambda \cap V_\mu \neq \emptyset$, then $V_\lambda \cap V_\mu$ is connected; the domain of ψ_λ is V_λ; $|\psi_\lambda| = \sqrt{\det(h_{i\bar{j}})}$ for $z^i := z_\lambda^i$. Since N is simply connected, we can choose $\{\psi_\lambda\}_{\lambda \in \Lambda}$ so that arguments θ_λ, θ_μ of ψ_λ, ψ_μ, respectively satisfy

$$\theta_\mu - \theta_\lambda + \arg \frac{\partial(z_\mu^1, z_\mu^2)}{\partial(z_\lambda^1, z_\lambda^2)} = 2n_{\lambda\mu}\pi$$

for an integer $n_{\lambda\mu}$, if $V_\lambda \cap V_\mu \neq \emptyset$. Then we can define a nowhere zero holomorphic 2-form Ψ on N by $\Psi := \psi_\lambda \, dz_\lambda^1 \wedge dz_\lambda^2$ on each V_λ. Hence we see that (c), (d) in Proposition 7.1 are equivalent to each other. $\qquad\square$

Remark 7.1. Although we do not suppose that N is simply connected, conditions (a), (b), (c) in Proposition 7.1 are mutually equivalent.

We will prove

Theorem 7.1. *Let (N, h, I) be a Kähler surface. Then the conditions* (a), (b), (c) *in Proposition 7.1 and the following conditions* (e), (f) *are mutually equivalent:*

(e) *There exists a deformation I' of I with $I' \neq \pm I$ such that the tangent bundle of N is locally represented as a direct sum of two I'-invariant involutive distributions $\mathcal{D}_1$, $\mathcal{D}_2$ of dimension 2 with isotropic minimal integral surfaces compatible with the orientation of N;*

(f) *There exists a deformation of I which is parallel with respect to the Levi-Civita connection ∇ of h and not identical with $\pm I$.*

Proof. The proof of Theorem 7.1 consists of three parts.

Part 1 Suppose (b) in Proposition 7.1. Then for the complex structure J in (b) in Proposition 7.1, two families of complex coordinate curves on a neighborhood of each point give J-invariant distributions $\mathcal{D}_1$, $\mathcal{D}_2$ as in (e) in Theorem 7.1 and therefore we obtain (e).

Part 2 Suppose (e). Let I' be a deformation of I as in (e) and $\mathcal{D}_1$, $\mathcal{D}_2$ as in (e) on a neighborhood V of each point of N. Let $F : M \to N$ be a conformal immersion of a Riemann surface M into N which gives an integral surface of $\mathcal{D}_1$. Let $w = u + \sqrt{-1}v$ be a local complex coordinate of M and set $T_1 := dF(\partial/\partial u)$, $T_2 := dF(\partial/\partial v)$. We can suppose $T_2 = I'T_1$. Since F is an isotropic minimal immersion compatible with the orientation of N, we have $I'\nabla_{T_1}T_1 = \nabla_{T_1}I'T_1$. Therefore using $\nabla_{T_1}T_1 + \nabla_{T_2}T_2 = 0$ and $\nabla_{T_1}T_2 = \nabla_{T_2}T_1$, we obtain $I'\nabla_{T_i}T_j = \nabla_{T_i}I'T_j$ for $i, j = 1, 2$. Let J' be a deformation of I on V satisfying $I'J' = -J'I'$. We set $N_1 := J'T_1$, $N_2 := I'N_1$. Then by $I'\nabla_{T_i}T_j = \nabla_{T_i}I'T_j$, we obtain $I'\nabla_{T_i}N_j = \nabla_{T_i}I'N_j$. In addition, considering integral surfaces of $\mathcal{D}_2$ similarly, we obtain $I'\nabla_X Y = \nabla_X I'Y$ for vector fields X, Y on V. This means that I' is parallel with respect to ∇. Hence we obtain (f) in Theorem 7.1.

Part 3 Suppose (f). There exists a deformation I' of I which is represented as $aI + bJ + cK$ on a neighborhood V of each point of N for an S^2-valued function (a, b, c) with $a \neq \pm 1$ and parallel with respect to ∇. We have

$$(\nabla_{\partial_i}(aI + bJ + cK))(\partial_j)$$

$$= \sqrt{-1}\frac{\partial a}{\partial z^i}\partial_j + \frac{\partial \xi}{\partial z^i}J\partial_j + \xi\left(\nabla_{\partial_i}(J\partial_j) - \sum_{k=1}^{2}\Gamma_{ij}^{k}J\partial_k\right), \tag{35}$$

where $\xi := b - \sqrt{-1}c$. Therefore by (31) together with (35), we obtain

$$((\nabla_{\partial_i}(aI + bJ + cK))(\partial_1) \quad (\nabla_{\partial_i}(aI + bJ + cK))(\partial_2))$$

$$= \sqrt{-1}\frac{\partial a}{\partial z^i}(\partial_1 \ \partial_2) + \left(\frac{\partial \xi}{\partial z^i} - \xi\frac{\partial \log \overline{\psi}}{\partial z^i}\right)(J\partial_1 \ J\partial_2). \tag{36}$$

From (36), we obtain $\partial a/\partial z^i \equiv 0$, $\partial(\overline{\xi}/\psi)/\partial \overline{z}^i \equiv 0$. Therefore a is constant and $J' := (1/\sqrt{1-a^2})(bJ + cK)$ is an almost complex structure parallel with respect to ∇. We see that J' can be considered as a complex structure on N with $IJ' = -J'I$ such that (N, h, J') is a Kähler manifold. Hence we obtain (b) in Proposition 7.1. $\qquad\square$

Remark 7.2. Since $\mathbb{C}P^2$, $\mathbb{C}H^2$, $\mathbb{C}P^1 \times \mathbb{C}P^1$ and $\mathbb{C}H^1 \times \mathbb{C}H^1$ are not Ricci-flat, Theorem 7.1 implies that there exist no non-trivial parallel deformations of I. Therefore a surface S in Example 5.3 in § 5 can not be any complex curves with respect to parallel deformations of I.

8. Related topics

Let n be a positive integer. Let N be a pseudo-Riemannian space form of dimension $n + 2$ which has signature $(p + 2, q)$, where p, q are non-negative integers satisfying $p + q = n$. Let F be a conformal space-like immersion of a Riemann surface M into N with zero mean curvature vector. Then F

defines a holomorphic quartic differential Q on M (see [3]). If $n = p = 1$, then Q is given by the Hopf differential of F. If $n = p = 2$, then $Q \equiv 0$ is equivalent to the isotropicity of a minimal immersion F. In the case where $N = S^4$, Q is just a differential given in [5].

We can find results on the relation between holomorphicity and stability of minimal surfaces. It follows from the Wirtinger inequality that a compact complex submanifold in a Kähler manifold minimizes the volume in its homology class ([7], [8, p. 653]). Let M be a parabolic Riemann surface (see [1]) and $F : M \to E^4$ a conformal minimal immersion. If F is stable and if the induced metric by F is complete, then the composition of F and an isometry of E^4 is holomorphic with respect to the standard complex structure I on $\mathbb{C}^2$ ([12]). In a general hyperKähler 4-manifold, a stable minimal surface is a complex curve if its normal bundle has a holomorphic section, while there exist a hyperKähler 4-manifold and a stable minimal surface in it which is not a complex curve with respect to any complex structures in the form of $aI + bJ + cK$ ([13]).

Acknowledgements

The author is grateful to the referee for his helpful comments. This work was partially supported by Grant-in-Aid for Scientific Research (17K05221), Japan Society for the Promotion of Science.

References

[1] L. V. Ahlfors and L. Sario, *Riemann surfaces*, Princeton Univ. Press, 1960.

[2] N. Ando, Local characterizations of complex curves in $\boldsymbol{C}^2$ and sphere Schwarz maps, *Intern. J. Math.* **27** (2016), 1650067.

[3] N. Ando, Surfaces in pseudo-Riemannian space forms with zero mean curvature vector, to appear in *Kodai Math. J.*

[4] N. Ando, Surfaces with zero mean curvature vector in 4-dimensional space forms, preprint.

[5] R. Bryant, Conformal and minimal immersions of compact surfaces into the 4-sphere, *J. Differential Geom.* **17** (1982), 455–473.

[6] E. Calabi, Minimal immersions of surfaces in Euclidean spheres, *J. Differential Geom.*, **1** (1967), 111–125.

[7] H. Federer, Some theorems on integral currents, *Trans. Amer. Math. Soc.* **117** (1965) 43–67.

[8] H. Federer, *Geometric measure theory*, Springer, 1969.

[9] T. Friedrich, On surfaces in four-spaces, *Ann. Glob. Anal. Geom.* **2** (1984). 257–287.

[10] D. D. Joyce, *Compact manifolds with special holonomy*, Oxford University Press, 2000.

[11] S. Kobayashi and K. Nomizu, *Foundations of differential geometry*, Vol. 2, John Wiley & Sons, 1969.

[12] M. J. Micallef, Stable minimal surfaces in Euclidean space, *J. Differential Geom.* **19** (1984), 57–84.

[13] M. J. Micallef and J. G. Wolfson, The second variation of area of minimal surfaces in four-manifolds, *Math. Ann.* **295** (1993), 245–267.

[14] A. Moroianu, *Lectures on Kähler geometry*, Cambridge University Press, 2007.

[15] R. A. Tribuzy and I. V. Guadalupe, Minimal immersions of surfaces into 4-dimensional space forms, *Rend. Semin. Mat. Univ. Padova* **73** (1985) 1–13.

Received October 8, 2018
Revised January 30, 2019

(α, ε)-STRUCTURES OF GENERAL NATURAL LIFT TYPE ON COTANGENT BUNDLES

Simona-Luiza DRUȚĂ-ROMANIUC

Department of Mathematics and Informatics,
"Gheorghe Asachi" Technical University of Iași,
Bd. Carol I, No. 1, 700506 Iași, Romania
E-mail: simonadruta@yahoo.com

Our purpose here is to give a characterization à la Etayo and Santamaria for the general natural (α, ε)-structures on the total space T^*M of the cotangent bundle of a Riemannian manifold (M, g). We first obtain the general natural α-structures on T^*M, and prove that they depend on four coefficients of the energy density t in a cotangent vector p. Moreover, we show that the α-structures obtained on T^*M are integrable if and only if (M, g) is of constant sectional curvature c, and three of the coefficients depend on the other coefficients, their derivatives, c, t and α. Then, we consider a general natural lifted metric G on T^*M, and by studying its ε-compatibility with the obtained α-structures, we characterize the two classes of general natural (α, ε)-structures on T^*M (according to the value 1 or -1 of the product $\alpha\varepsilon$). We show that for the class with $\alpha\varepsilon = -1$ there are two proportionality relations between the coefficients of the α-structure and those of the metric. We prove that in this case the fundamental tensor field Ω is a closed 2-form (i.e. the structure is of (almost) Kähler type) if and only if one of the proportionality factors is the derivative of the other one.

Keywords: Cotangent bundles; (α, ε)-structures; natural lifts; almost Kähler structures.

1. Introduction

Geometric structures on cotangent bundles have been of interest for many geometers in the last decades (see [4, 5, 7, 29, 32, 33, 35, 38] for example). Different from the situation of the tangent bundle (for which we quote [27, 28, 38]), the vertical lift of a vector field from the base manifold (M, g) to the total space T^*M of the cotangent bundle is not a vector field anymore, but a function on T^*M (see [37]), and the vertical lift of a 1-form from (M, g) to T^*M is a vector field on T^*M (see [38]). For these reasons, in the constructions of almost complex and almost product structures (generically called α-structures) on T^*M, musical isomorphisms are used (see [7] and the references therein, for example).

The four classes of manifolds with (α, ε)-structures, which are also called $(J^2 = \pm 1)$-metric manifolds, were studied in an unified manner by Etayo and Santamaria (see [12–14], for example). Some historical facts for each class of (α, ε)-structures were mentioned by the present author in [11] (see [17, 21] for almost Hermitian manifolds, [1–3, 6, 9, 15, 18] for almost para-Hermitian and in particular almost para-Kähler manifolds, [16, 34] for almost anti-Hermitian or Norden manifolds, and [9, 25, 26, 30, 36] for Riemannian almost product manifolds).

In the present paper we characterize in a unified way the (α, ε)-structures obtained as general natural lifts of the metric from the base manifold (M, g) to the total space of the cotangent bundle T^*M. In [11] we dealt with the same problem on the dual bundle, namely on the tangent bundle, where the constructions of the lifts were simpler. Now we first construct a $(1, 1)$-tensor field S on T^*M, naturally obtained from the metric g, the basic 1-form p, and eight coefficients which depend only on the energy density of p. The action of S on horizontal lifts of vector fields and on vertical lifts of 1-forms is expressed by using musical isomorphisms. We prove that S is an α-structure on T^*M, i.e. $S^2 = \alpha I$ with $\alpha = \pm 1$, if and only if four of its eight coefficients are expressed as certain functions of the other four. Moreover, we show that the obtained α-structure S is integrable if and only if the base manifold is a space form, and that only two of the involved coefficients are essential. Next, we consider the general natural lifted metric G constructed on T^*M by the author (see [7]). By studying the ε-compatibility with the general natural α-structures obtained on T^*M, we get two main classes of (α, ε)-structures according to the value of the product $\alpha\varepsilon$, which is ± 1. We can characterize the class with $\alpha\varepsilon = -1$ by the property that the coefficients of the α-structure are proportional to the coefficients of the metric. For this class of (α, ε)-structures, the fundamental tensor field Ω is a 2-form, hence we can obtain the general natural (α, ε)-structures of almost Kähler type on T^*M by studying the closedness of Ω. In the characterization of this special type of structures on T^*M, the essential parameters are four coefficients of the general natural α-structure S on T^*M and a proportionality factor λ (function of the energy density of p). Moreover, we prove that any general natural (α, ε)-structure (G, S) on T^*M is of Kähler type if and only if it depends on two coefficients of S and on λ.

The results provided in this paper generalize the previous ones which were obtained by the author for the general natural structures on T^*M, which are almost Hermitian (or $(-1, 1)$-structures) in [7], almost para-

Hermitian (or $(1, -1)$-structures) and Riemannian almost product (or $(1, 1)$-structures) in [10].

In this paper we assume that the manifolds, tensor fields and other geometric objects are differentiable of class C^∞ (i.e. smooth). The range of the indices h, i, j, k, l, m, r, is always $\{1, \ldots, n\}$ and the Einstein summation convention is used.

2. Adapted local frame fields on cotangent bundles

Let (M, g) be a smooth n-dimensional Riemannian manifold. We denote by $\pi : T^*M \to M$ the projection of its cotangent bundle. We recall from [38] that each coordinate neighborhood $(U, x^1, \ldots, x^n)$ of a point $P \in M$ induces a coordinate neighborhood $(\pi^{-1}(U), q^1, \ldots, q^n, p_1, \ldots, p_n)$ of a point $\widetilde{P} \in \pi^{-1}(U)$, where $q^i = x^i \circ \pi$ $(i = 1, \ldots, n)$, and $p_1, \ldots, p_n$ are the vector space coordinates of a 1-form p on $\pi^{-1}(U)$ with respect to the natural coframe $(dx^1_{\pi(p)}, \ldots, dx^n_{\pi(p)})$, that is, $p = p_i dx^i_{\pi(p)}$.

The tangent space $T_p T^*M$ of T^*M at p splits as

$$T_p T^*M = V_p T^*M \oplus H_p T^*M,$$

where $V_p T^*M = \operatorname{Ker} \pi_*$ is the vertical space and $H_p T^*M$ is the horizontal space, determined by the Levi Civita connection ∇^g of the metric g.

Identifying (by an abuse of notation) TT^*M with the set of all vector fields tangent to T^*M one has the direct sum decomposition

$$TT^*M = VT^*M \oplus HT^*M, \tag{1}$$

into the vertical and horizontal distributions, which are locally generated respectively by $\left\{ \frac{\partial}{\partial p_i} \right\}_{i=1}^n$ and $\left\{ \frac{\delta}{\delta q^j} \right\}_{j=1}^n$ on $\pi^{-1}(U)$, where according to formula given in [38, (5.1) p. 289] one has

$$\frac{\delta}{\delta q^j} = \frac{\partial}{\partial q^j} + \Gamma_{jh} \frac{\partial}{\partial p_h}, \quad \Gamma_{jh} = p_k \Gamma^k_{jh},$$

and $\Gamma^k_{jh}(\pi(p))$ are the Christoffel symbols of g. A local frame field on T^*M which is adapted to the direct sum decomposition (1) is given by the vector fields $\left\{ \frac{\partial}{\partial p_i}, \frac{\delta}{\delta q^j} \right\}_{i,j=1}^n$. In the sequel we denote this by $\{\partial^i, \delta_j\}_{i,j=1}^n$. The dual of the adapted local frame field $\{\partial^i, \delta_j\}_{i,j=1}^n$ on T^*M is the set $\{Dp_i, dq^j\}_{i,j=1}^n$ where Dp_i is the absolute differential of p_i given by

$$Dp_i = dp_i - \Gamma_{ih} dq^h.$$

We mention that in this paper the notations used on T^*M for the adapted local frame field and its dual are the same as in [7, 8, 10, 33], different from those on TM in [27, 28].

If X is a vector field and θ is a 1-form on M which are locally expressed on U as $X = X^i \frac{\partial}{\partial x^i}$ and $\theta = \theta_i dx^i$, then the horizontal lift of X and the vertical lift of θ to T^*M are denoted by X^H and θ^V respectively, and are locally expressed on $\pi^{-1}(U)$ as

$$X^H = X^i \delta_i, \quad \theta^V = \theta_i \partial^i.$$

We use the musical isomorphisms $\flat : \mathcal{T}_0^1(M) \to \mathcal{T}_1^0(M)$ and $\sharp : \mathcal{T}_1^0(M) \to \mathcal{T}_0^1(M)$ which are defined respectively by

$$X^\flat(Y) = g(X, Y) \quad \text{for} \quad X, Y \in \mathcal{T}_0^1(M),$$
$$g(\theta^\sharp, Y) = \theta(Y), \quad \text{for} \quad \theta \in \mathcal{T}_1^0(M), Y \in \mathcal{T}_0^1(M).$$

3. General natural α-structures on T^*M

We introduce here a $(1,1)$-tensor field which is obtained as a general natural lift of the metric g from the base manifold (M, g) to the total space T^*M of the cotangent bundle, and then we determine the conditions that this tensor field turns to be an α-structure on T^*M. To this aim, we recall first the definition of $(J^2 = \pm 1)$-manifolds or manifolds endowed with α-structures (see [12–14]).

Definition 3.1. A differentiable manifold M is said to be a $(J^2 = \pm 1)$-manifold or to have an α-structure J if J is a $(1,1)$-tensor field on M satisfying

$$J^2 = \alpha I, \quad \alpha \in \{-1, 1\}. \tag{2}$$

When $\alpha = -1$, the tensor field J is called an almost complex structure and when $\alpha = 1$, it is called an almost product structure. The couple (M, J) is called, respectively, an almost complex manifold and an almost product manifold. Moreover, (M, J) is called, respectively, a complex manifold and a locally product manifold if the structure J is integrable, that is, its Nijenhuis tensor field vanishes, i.e.

$$[JX, JY] - J[JX, Y] - J[X, JY] + \alpha[X, Y] = 0 \tag{3}$$

for all $X, Y \in \mathcal{T}_0^1(M)$.

The simplest almost complex structure J on T^*M and the almost product structures P and Q on T^*M mentioned in [5] are given respectively by

$$J(X^H) = (X^\flat)^V, \quad J(\theta^V) = -(\theta^\sharp)^H, \tag{4}$$

$$P(X^H) = -X^H, \qquad P(\theta^V) = \theta^V, \tag{5}$$

$$Q(X^H) = (X^\flat)^V, \qquad Q(\theta^V) = (\theta^\sharp)^H, \tag{6}$$

for every $X \in \mathcal{T}_0^1(M)$ and $\theta \in \mathcal{T}_1^0(M)$.

In this section we construct a natural lift $S \in \mathcal{T}_1^1(T^*M)$ of the metric g to T^*M, which generalizes the above mentioned structures and also generalizes the much more complicated structures of natural lift type on T^*M from [7, 8, 10].

A natural operator is briefly defined as a mapping between fibred manifolds which is invariant under the actions of local diffeomorphisms of the base manifolds (see [22]). The results in [20] and [23] which concern the natural lifts to the tangent bundle of a pseudo-Riemannian manifold lead us to the so called general natural lifts of the metric from the base manifold to the tangent bundle, which were introduced by Oproiu in [31]. Structures of general natural lift type on the cotangent bundle were studied by the author in [7, 10].

The main notion of this section is given as follows:

Definition 3.2. A $(1,1)$-tensor field S on T^*M is called a general natural lift of the metric g from the base manifold to T^*M if its action is given by

$$\begin{cases} SX_p^H = a_1(t)(X^\flat)_p^V + b_1(t)p(X)p_p^V - a_3(t)X_p^H - b_3(t)p(X)(p^\sharp)_p^H, \\ S\theta_p^V = a_4(t)\theta_p^V + b_4(t)g_{\pi(p)}^{-1}(p, \theta)p_p^V \\ \qquad\qquad + \alpha\big[a_2(t)(\theta^\sharp)_p^H + b_2(t)g_{\pi(p)}^{-1}(p, \theta)(p^\sharp)_p^H\big], \end{cases}$$

at every cotangent vector $p \in \pi^{-1}(U)$ on T^*M for arbitrary $X \in \mathcal{T}_0^1(M)$ and $\theta \in \mathcal{T}_1^0(M)$, where a_i, b_i $(i = 1, 2, 3, 4)$ are smooth functions of the energy density given by

$$t = \frac{1}{2}\|p\|^2 = \frac{1}{2}g_{\pi(p)}^{-1}(p, p) = \frac{1}{2}g^{ik}(x)p_i p_k \ (\geq 0),$$

the vector $p^\sharp$ is tangent to M in $\pi(p)$, $p^V = p_i \partial^i$ is the Liouville vector field on T^*M, and $(p^\sharp)^H = g^{0i}\delta_i$ is the geodesic spray on T^*M.

With respect to the adapted local frame field $\{\partial^i, \delta_j\}_{i,j=1}^n$ on T^*M, the expressions of S become

$$\begin{cases} S\delta_i = a_1(t)g_{ij}\partial^j + b_1(t)p_i p_j \partial^j - a_3(t)\delta_i - b_3(t)p_i g^{0j}\delta_j, \\ S\partial^i = a_4(t)\partial^i + b_4(t)g^{0i}p_j\partial^j + \alpha\big[a_2(t)g^{ij}\delta_j + b_2(t)g^{0i}g^{0j}\delta_j\big]. \end{cases} \tag{7}$$

By using the concept of M-tensor field on the cotangent bundle of a Riemannian manifold (see [7]), which can be defined in the same way as the M-tensor fields on the tangent bundle (see [28]), we can write

$$\begin{cases} S\delta_i = (S^{(1)})_{ij}\partial^j - (S^{(3)})_i^j\delta_j, \\ S\partial^i = (S^{(4)})_j^i\partial^j + \alpha(S^{(2)})^{ij}\delta_j, \end{cases} \tag{8}$$

where the M-tensor fields involved are

$$(S^{(1)}))_{ij} = a_1(t)g_{ij} + b_1(t)p_ip_j, \quad (S^{(2)})^{ij} = a_2(t)g^{ij} + b_2(t)g^{0i}g^{0j},$$
$$(S^{(3)})_j^i = a_3(t)\delta_j^i + b_3(t)g^{0i}p_j, \quad (S^{(4)})_i^j = a_4(t)\delta_i^j + b_4(t)g^{0j}p_i. \tag{9}$$

Definition 3.3. A general natural lift S of the metric g from the base manifold M to T^*M is called a general natural α-structure (or an α-structure of general natural lift type) on T^*M if it is an α-structure on T^*M, that is, it satisfies the identity

$$S^2 = \alpha I, \ \alpha \in \{-1, 1\}.$$

We recall a result from [7] which will be used throughout this paper.

Lemma 3.1. Let (M, g) be an n-dimensional Riemannian manifold with $n \geq 2$. If u, v are two smooth functions on T^*M such that either $ug_{ij} + vp_ip_j = 0$, $ug^{ij} + vg^{0i}g^{0j} = 0$ or $u\delta_j^i + vg^{0i}p_j = 0$ holds on the domain of an arbitrary induced local chart on T^*M, then we have $u = 0$, $v = 0$.

Now we can prove the following characterization:

Theorem 3.1. A general natural lift S of the metric g from the base manifold M to T^*M is a general natural α-structure on T^*M if and only if the following relations between the coefficients hold:

$$a_4 = a_3, \qquad b_4 = b_3,$$
$$a_1a_2 + \alpha a_3^2 = \tilde{a}_1\tilde{a}_2 + \alpha\tilde{a}_3^2 = 1, \tag{10}$$

where we set $\tilde{a}_i = a_i + 2tb_i$ $(i = 1, 2, 3)$.

Proof. By using (8) and the definition (2) of an α-structure, we obtain the following relations:

$$\left[(S^{(1)})_{ij}(S^{(4)})_l^j - (S^{(1)})_{jl}(S^{(3)})_i^j\right]\partial^l$$
$$+ \left[\alpha(S^{(1)})_{ij}(S^{(2)})^{jl} + (S^{(3)})_j^l(S^{(3)})_i^j\right]\delta_l = \alpha\delta_i,$$

$$\left[(S^{(4)})_j^i(S^{(4)})_l^j + \alpha(S^{(1)})_{jl}(S^{(2)})^{ij}\right]\partial^l$$
$$+ \alpha\left[(S^{(4)})_j^i(S^{(2)})^{jl} - (S^{(3)})_j^l(S^{(2)})^{ij}\right]\delta_l = \alpha\partial^i.$$

By identifying the horizontal and vertical components, we find that these yield the following system of equations:

$$\begin{cases} (S^{(1)})_{ij}(S^{(4)})_l^j - (S^{(1)})_{jl}(S^{(3)})_i^j = 0, \\ \alpha(S^{(1)})_{ij}(S^{(2)})^{jl} + (S^{(3)})_j^l(S^{(3)})_i^j = \alpha\delta_i^l, \\ (S^{(4)})_j^i(S^{(4)})_l^j + \alpha(S^{(1)})_{jl}(S^{(2)})^{ij} = \alpha\delta_l^i, \\ (S^{(4)})_j^i(S^{(2)})^{jl} - (S^{(3)})_j^l(S^{(2)})^{ij} = 0. \end{cases} \qquad (11)$$

By substituting (9) into (11) we obtain

$$\begin{cases} a_1(a_4 - a_3)g_{il} + [b_1(a_4 - a_3) + \widetilde{a}_1(b_4 - b_3)]p_i p_l = 0, \\ (\alpha a_1 a_2 + a_3^2 - \alpha)\delta_i^l + [\alpha\widetilde{a}_2 b_1 + \alpha a_1 b_2 + 2a_3 b_3 + 2t b_3^2]g^{0l}p_i = 0, \\ (a_4^2 + \alpha a_1 a_2 - \alpha)\delta_l^i + [\alpha\widetilde{a}_2 b_1 + \alpha a_1 b_2 + 2a_4 b_4 + 2t b_4^2]g^{0i}p_l = 0, \\ a_2(a_4 - a_3)g^{il} + [b_2(a_4 - a_3) + \widetilde{a}_2(b_4 - b_3)]g^{0i}g^{0l} = 0. \end{cases}$$

Thus, by using Lemma 3.1 we get

$$a_3 = a_4, \quad b_3 = b_4, \quad \alpha a_1 a_2 + a_3^2 = \alpha, \quad \alpha\widetilde{a}_2 b_1 + \alpha a_1 b_2 + a_3 b_3 + \widetilde{a}_3 b_3 = 0.$$

Multiplying the third equality in the above by $\alpha = \pm 1$, we have $a_1 a_2 + \alpha a_3^2 = 1$. Adding to this equality the forth one multiplied by $2\alpha t$, we obtain

$$a_1 a_2 + \alpha a_3^2 + 2t a_2 b_1 + 4t^2 b_1 b_2 + 2t a_1 b_2 + 2\alpha t a_3 b_3 + 2\alpha t a_3 b_3 + 4\alpha t^2 b_3^2 = 1,$$

which can be written in the shorter form $\widetilde{a}_1\widetilde{a}_2 + \alpha\widetilde{a}_3^2 = 1$. Thus, we obtain all the required relations. $\square$

We here give an example of general natural α-structure on T^*M.

Example 3.1. We take two nonzero real constants k_1, k_2 and an arbitrary smooth function u of the energy density t satisfying $k_2 + 2tu \neq 0$. If we set the coefficients so that

$$a_1 = \frac{1}{k_1}, \quad a_2 = k_1, \quad a_3 = 0, \quad b_1 = \frac{u}{k_1 k_2}, \quad b_2 = -\frac{uk_1}{k_2 + 2tu}, \quad b_3 = 0,$$

then the corresponding general natural lift of the metric from the base manifold M to T^*M is a general natural α-structure on T^*M. When we take $\alpha = 1$ in Definition 3.2, with the above coefficients, we obtain the almost product structure considered in [35].

When $\alpha = -1$, Theorem 3.1 yields the general natural almost complex structures on T^*M which were obtained by the author in [7, Theorem 3.1]. When $\alpha = 1$, Theorem 3.1 also yields the general natural almost product structures on T^*M which were obtained in [10, Theorem 3.3]. The complex and locally product structures of general natural lift type on T^*M in [7] and [10] can be obtained by particularizing the integrable α-structures. They will be characterized in the following theorem. We here point out a property on coefficients of α-structures. When $\alpha = -1$, the relation (10) guarantees that the coefficients a_1, a_2 of each α-structure do not vanish because $a_1 a_2 = 1 + a_3^2 > 0$.

Theorem 3.2. *Let (M, g) be a connected Riemannian manifold of dimension n (≥ 3), and let S be an α-structure of general natural lift type on T^*M. We suppose that its coefficients satisfy the condition*

$$a_1(t) - 2ta_1'(t) + 2t\,\frac{a_1(0)}{a_2^2(0)}\big(a_2'(0) - b_2(0)\big)\big(a_2(t) + 2ta_2'(t)\big) \neq 0, \qquad (12)$$

and suppose additionally $a_2(0) \neq 0$ when $\alpha = 1$. Then this α-structure S is integrable if and only if the base manifold is of constant sectional curvature $c = \alpha \frac{a_1(0)}{a_2^2(0)}\big(a_2'(0) - b_2(0)\big)$ and the coefficients b_1, b_2, b_3 are expressed as follows:

$$\begin{cases} b_1 = \dfrac{a_1 a_1' + 3a_3^2 c + 2ta_2^2 c^2 + \alpha c(1 - 2ta_1 a_2')}{a_1 - 2ta_1' + 2\alpha ct(a_2 + 2ta_2')}, \\[3mm] b_2 = \dfrac{a_2'(a_1 - 2ta_1') - 2\alpha t a_3'^2 - \alpha c a_2(a_2 + 2ta_2')}{a_1 - 2ta_1' + 2\alpha ct(a_2 + 2ta_2')}, \\[3mm] b_3 = \dfrac{a_1 a_3' - 2\alpha c(a_2 a_3 + 2ta_2' a_3 - ta_2 a_3')}{a_1 - 2ta_1' + 2\alpha ct(a_2 + 2ta_2')}. \end{cases} \qquad (13)$$

Proof. We define $N_S : \mathcal{T}_0^1(T^*M) \times \mathcal{T}_0^1(T^*M) \to \mathcal{T}_0^1(T^*M)$ by

$$N_S[X, Y] = [SX, SY] - S[SX, Y] - S[X, SY] + \alpha[X, Y].$$

By the relation (3), we see that S is integrable if and only if $N_S[X, Y] = 0$, for all $X, Y \in \mathcal{T}_0^1(T^*M)$. We compute N_S with respect to the adapted local frame field $\{\partial^i, \delta_j\}_{i,j=1}^n$, and study the vanishing conditions on its horizontal and vertical parts. The horizontal part $h[N_S(\partial^i, \partial^j)]$ of $N_J(\partial^i, \partial^j)$ is

$$h[N_S(\partial^i, \partial^j)]$$
$$= \alpha\big\{(S^{(3)})_l^i \partial^l (S^{(2)})^{jh} - (S^{(3)})_l^j \partial^l (S^{(2)})^{hi}$$
$$+ [\partial^j (S^{(3)})_l^i - \partial^i (S^{(3)})_l^j](S^{(2)})^{lh} + (S^{(3)})_l^h [\partial^i (S^{(2)})^{lj} - \partial^j (S^{(2)})^{li}]\big\}\delta_h.$$

By substituting (9) into this relation, we find that it turns to the form

$$h[N_S(\partial^i, \partial^j)] = \alpha\big[2a_2'(a_3 + b_3 t) - a_2(a_3' - b_3) - 2a_3 b_2\big]\big(g^{hj}g^{0i} - g^{hi}g^{0j}\big)\delta_h,$$

hence obtain that $h[N_S(\partial^i, \partial^j)]$ vanishes if and only if

$$a_2' = \frac{a_2(a_3' - b_3) + 2a_3 b_2}{2(a_3 + b_3 t)}. \tag{14}$$

The vertical part $v[N_S(\partial^i, \partial^j)]$ of $N_S(\partial^i, \partial^j)$, is given as

$$v[N_S(\partial^i, \partial^j)]$$
$$= \{(S^{(3)})_m^i \partial^m (S^{(3)})_h^j - (S^{(3)})_m^j \partial^m (S^{(3)})_h^i$$
$$+ [\partial^j (S^{(3)})_m^i - \partial^i (S^{(3)})_m^j](S^{(3)})_h^m$$
$$+ \alpha(S^{(1)})_{mh}[\partial^i (S^{(2)})^{jm} - \partial^j (S^{(2)})^{im}] + p_s R_{hml}^s (S^{(2)})^{mi}(S^{(2)})^{lj}\}\partial^h.$$

By substituting (9) into this relation, we find that this becomes to the form

$$v[N_S(\partial^i, \partial^j)]$$
$$= \{[\alpha(a_1 a_2' - a_1 b_2) - 2a_3' b_3 t](\delta_h^i g^{0j} - \delta_h^j g^{0i})$$
$$+ a_2^2 p_s R_{hml}^s g^{mi} g^{lj} + a_2 b_2 p_s R_{hml}^s (g^{mi} g^{0l} g^{0j} + g^{lj} g^{0m} g^{0i})\}\partial^h,$$

where R is the curvature tensor field of the Levi-Civita connection ∇^g associated to the metric g on the base manifold. Trivially, R does not depend on the cotangent vector p. Thus, by computing the derivative of the coefficient of ∂^h in $v[N_S(\partial^i, \partial^j)]$ with respect to the cotangential coordinates p_s at $p = 0$, we have

$$\alpha[a_1(0)a_2'(0) - a_1(0)b_2(0)](\delta_h^i g^{sj} - \delta_h^j g^{si}) + a_2^2(0)R_{hml}^s g^{mi} g^{lj}.$$

Since $a_2 \neq 0$ when $\alpha = -1$ and since we supposed $a_2(0) \neq 0$ when $\alpha = 1$, we find that this derivative vanishes if and only if

$$R_{hml}^s = \alpha \frac{a_1(0)}{a_2^2(0)}(a_2'(0) - b_2(0))(\delta_m^s g_{lh} - \delta_l^s g_{mh}), \tag{15}$$

where the coefficient depends only on $q^1, ..., q^n$. If the base manifold (M, g) is connected and of dimension $n > 2$, by Schur's theorem we obtain that M is a space form, and that its constant sectional curvature is

$$c = \alpha \frac{a_1(0)}{a_2^2(0)}(a_2'(0) - b_2(0)). \tag{16}$$

The vertical and horizontal components of $N_S(\delta_i, \delta_j)$ are given as

$$v[N_S(\delta_i, \delta_j)] = \{(S^{(1)})_{im} \partial^m (S^{(1)})_{jh} - (S^{(1)})_{jm} \partial^m (S^{(1)})_{ih}$$
$$+ [p_s R_{hml}^s (S^{(3)})_j^l + p_s R_{lmj}^s (S^{(3)})_h^l](S^{(3)})_i^m$$
$$+ p_s R_{lim}^s (S^{(3)})_j^m (S^{(3)})_h^l + \alpha p_s R_{hij}^s\}\partial^h,$$

$$h[N_S(\delta_i, \delta_j)] = \big\{(S^{(1)})_{jm}\partial^m(S^{(3)})_i^h - (S^{(1)})_{im}\partial^m(S^{(3)})_j^h$$
$$+ \alpha(S^{(2)})^{lh}[p_s R_{mlj}^s(S^{(3)})_i^m + p_s R_{mil}^s(S^{(3)})_j^m]\big\}\delta_h.$$

When the equality (15) holds, that is, when the base manifold is of constant sectional curvature c of the form (16), in particular when S is integrable, by using (9), the above components become of the forms

$$v[N_S(\delta_i, \delta_j)] = \big[a_1'\tilde{a}_1 - a_1 b_1 + c(\alpha + 3a_3^2 + 4ta_3 b_3)\big](g_{hj}p_i - g_{hi}p_j)\partial^h$$
$$h[N_S(\delta_i, \delta_j)] = \big[a_3'\tilde{a}_1 - a_1 b_3 - 2\alpha c a_2(a_3 + tb_3)\big](\delta_i^h p_j - \delta_j^h p_i)\delta_h,$$

and they vanish simultaneously if and only if

$$\begin{cases} a_1' = [a_1 b_1 - c(\alpha + 3a_3^2 + 4ta_3 b_3)]/\tilde{a}_1, \\ a_3' = [a_1 b_3 + 2c\alpha a_2(a_3 + tb_3)]/\tilde{a}_1. \end{cases} \tag{17}$$

By differentiating with respect to t the relation $a_1 a_2 + \alpha a_3^2 = 1$ which was obtained in Theorem 3.1, we obtain

$$a_1 a_2' + a_1' a_2 + 2\alpha a_3 a_3' = 0. \tag{18}$$

We solve the system of linear equations given by (14) and (17) with respect to b_1, b_2, b_3 by taking into account (18) and the first relation in (10). Then we obtain the expressions (13) as solutions of this system. Thus, if S is integrable, we get the conditions.

On the other hand, we suppose that the conditions hold. Then we see in the above that the equalities (14) and (17) hold. Thus, we find that $h[N_S(\partial^i, \partial^j)]$, $v[N_S(\delta_i, \delta_j)]$ and $h[N_S(\delta_i, \delta_j)]$ vanish. Since the equality (15) holds and the base manifold is of constant sectional curvature (16), we find that the vertical part $v[N_S(\partial^i, \partial^j)]$ is expressed as

$$v[N_S(\partial^i, \partial^j)] = \big[\alpha(a_1 a_2' - a_1 b_2) - 2a_3' b_3 t - a_2 c(a_2 + 2tb_2)\big](\delta_h^i g^{0j} - \delta_h^j g^{0i})\partial^h.$$

By substituting b_2, b_3 of our conditions into this expression and by using the relation (10), we can conclude that this part also vanishes. Finally, we compute the mixed component of the Nijenhuis tensor field. We have

$$N_S(\partial^i, \delta_j)$$
$$= \big\{(S^{(3)})_m^i\partial^m(S^{(1)})_{jh} - (S^{(1)})_{jm}\partial^m(S^{(3)})_h^i + (S^{(1)})_{mh}\partial^i(S^{(3)})_j^m$$
$$- (S^{(3)})_h^m\partial^i(S^{(1)})_{jm} - \alpha[p_s R_{hml}^s(S^{(3)})_j^l + p_s R_{lmj}^s(S^{(3)})_h^l](S^{(2)})^{im}\big\}\partial^h$$
$$- \big[(S^{(3)})_m^i\partial^m(S^{(3)})_j^h + (S^{(3)})_m^h\partial^i(S^{(3)})_j^m + \alpha(S^{(1)})_{jm}\partial^m(S^{(2)})^{ih}$$
$$+ \alpha\partial^i(S^{(1)})_{jm}(S^{(2)})^{mh} + p_s R_{lmj}^s(S^{(2)})^{lh}(S^{(2)})^{mi}\big]\delta_h.$$

By substituting (9) into this expression, and by taking into account that the base manifold has constant sectional curvature c, we obtain that the vertical and the horizontal components of $N_S(\partial^i, \delta_j)$ become of the forms

$$v[N_S(\partial^i, \delta_j)] = \left\{ \left[a_1(a_3' - b_3) - 2\alpha a_3 c(a_2 + 2tb_2) + 2ta_1'b_3 \right] g_{hj} g^{0i} \right.$$
$$- \left[a_1(a_3' - b_3) - 2\alpha a_2 c(a_3 + tb_3) + 2ta_3'b_1 \right] \delta_h^i p_j$$
$$\left. + \left[(a_3'b_1 - a_1'b_3 + \alpha c(2a_3b_2 - a_2b_3) \right] p_h p_j g^{0i} \right\} \partial^h,$$

$$h[N_S(\partial^i, \delta_j)] = -\left\{ \left[\alpha a_2'(a_1 + 2tb_1) + \alpha a_2 b_1 + 2a_3 b_3 - a_2^2 c \right] g^{hi} p_j \right.$$
$$+ \left[\alpha b_1(a_2 + 2tb_2) + \alpha a_1 b_2 + 2b_3(a_3 + tb_3) \right] \delta_j^i g^{0h}$$
$$+ \left[\alpha a_1'a_2 + 2a_3'(a_3 + tb_3) + \alpha a_1 b_2 + a_2 c(a_2 + 2tb_2) \right] \delta_j^h g^{0i}$$
$$+ \left[\alpha b_1'(a_2 + 2tb_2) + \alpha b_2'(a_1 + 2tb_1) + 2b_3'(a_3 + 2tb_3) \right.$$
$$\left. + \alpha a_1'b_2 + 3\alpha b_1 b_2 + a_3'b_3 + 3b_3^2 - a_2 b_2 c \right] p_j g^{0h} g^{0i} \right\} \delta_h.$$

By substituting b_1, b_2, b_3 of our conditions and by using the relation (10), we find that these components vanish. Thus, we find that all the components of the Nijenhuis tensor field vanish under our conditions. Therefore, we see that S is integrable, and get the conclusion. $\qquad\square$

Corollary 3.1. *Let (M, g) be a connected Riemannian manifold of dimension n (≥ 3). A general natural α-structure given is Example 3.1 is integrable if and only if the base manifold has constant sectional curvature c and $u = \alpha c k_1^2 k_2$.*

Proof. The structure given in Example 3.1 has $a_2(0) = k_1 \neq 0$. It satisfies the condition (12) if and only if $2tu(0) \neq -k_2$. Thus, we can apply Theorem 3.2. Since the relations in (13) are reduced to $u = \alpha c k_1^2 k_2$, we get the conclusion. $\qquad\square$

Remark 3.1. When $\alpha = 1$, Corollary 3.1 becomes [35, Theorem 4.2]. It gives the necessary and sufficient conditions that the almost product structure considered in [35] is integrable.

By taking into account Corollary 3.1, we have the following:

Example 3.2. Let (M, g) be an n (> 2)-dimensional Riemannian manifold of constant sectional curvature c. Given nonzero real constants k_1, k_2 satisfying $1 + 2\alpha c t k_1^2 \neq 0$, we set the coefficients as $a_1 = \frac{1}{k_1}$, $a_2 = k_1$, $a_3 = 0$, $b_1 = \alpha c k_1$, $b_2 = -\alpha c k_1^3/(1 + 2\alpha c t k_1^2)$, $b_3 = 0$. Then the corresponding general natural lift of g to T^*M is an integrable general natural α-structure on T^*M.

4. Cotangent bundles with general natural (α, ε)-structures

In this section we study in an unified way the compatibility and anti-compatibility relations between α-structures of general natural lift type on T^*M which were characterized in Theorem 3.1 and general natural lifted metrics on T^*M which were constructed in [7]. Thus we shall obtain the conditions under which (T^*M, G, S) is an (α, ε)-manifold. In the sequel, we recall the notion of (α, ε)-manifold, which was studied in [12–14].

Definition 4.1. Let M be a $2m$-dimensional manifold and let ε be ± 1. When $\varepsilon = -1$, we take a semi-Riemannian metric g of signature (m, m), and when $\varepsilon = 1$, we take a Riemannian metric g. We say that an α-structure J on (M, g) satisfies ε-compatibility with the metric g if

$$g(JX, JY) = \varepsilon g(X, Y) \tag{19}$$

holds for all $X, Y \in \mathcal{T}_0^1(M)$. In this case, we call the pair (g, J) an (α, ε)-structure on M. We call (M, g) an (α, ε)-manifold or a $(J^2 = \pm 1)$-metric manifold, if it is endowed with an α-structure satisfying ε-compatibility with the metric g.

Now we consider a general natural lifted metric on T^*M which is defined in [7] by

$$\begin{cases} G_p(X^H, Y^H) = c_1(t)g_{\pi(p)}(X, Y) + d_1(t)p(X)p(Y), \\ G_p(\theta^V, \omega^V) \ = c_2(t)g_{\pi(p)}^{-1}(\theta, \omega) + d_2(t)g_{\pi(p)}^{-1}(p, \theta)g_{\pi(p)}^{-1}(p, \omega), \\ G_p(X^H, \theta^V) = G_p(\theta^V, X^H) = c_3(t)\theta(X) + d_3(t)p(X)g_{\pi(p)}^{-1}(p, \theta), \end{cases} \tag{20}$$

for $X, Y \in \mathcal{T}_0^1(M)$ and $\theta, \omega \in \mathcal{T}_1^0(M)$ at an arbitrary $p \in T^*M$, where the coefficients c_i, d_i $(i = 1, 2, 3)$ are smooth functions of the energy density of p. By setting $\tilde{c}_i = c_i + 2td_i$, $(i = 1, 2, 3)$, we find that this metric G is nondegenerate if $c_1c_2 - c_3^2 \neq 0$ and $\tilde{c}_1\tilde{c}_2 - \tilde{c}_3^2 \neq 0$ hold, and is positive definite if

$$\tilde{c}_1 > 0, \quad \tilde{c}_2 > 0, \quad \tilde{c}_1\tilde{c}_2 - \tilde{c}_3^{\ 2} > 0, \tag{21}$$

hold. With respect to the adapted local frame field $\{\partial^i, \delta_j\}_{i,j=1}^n$ on T^*M, (20) is expressed as

$$\begin{cases} G(\delta_i, \delta_j) = (G^{(1)})_{ij} = c_1(t)g_{ij} + d_1(t)p_ip_j, \\ G(\partial^i, \partial^j) = (G^{(2)})^{ij} = c_2(t)g^{ij} + d_2(t)g^{0i}g^{0j}, \\ G(\partial^i, \delta_j) = G(\delta_j, \partial^i) = (G^{(3)})_j^i = c_3(t)\delta_j^i + d_3(t)g^{0i}p_j. \end{cases} \tag{22}$$

The study of the ε-compatibility between G and S reveals two classes of (α, ε)-structures of general natural lift type on T^*M. These structures are classified according to the signature of the product $\alpha\varepsilon$ $(= \pm 1)$.

Theorem 4.1. *Let* (M, g) *be a Riemannian manifold. When the total space* T^*M *of the cotangent bundle is endowed with a general natural lifted metric* G *and with a general natural α-structure S, one has the following:*

(I) (G, S) *is an (α, ε)-structure with $\alpha\cdot\varepsilon = 1$ if and only if the coefficients of G and S satisfy the relations*

$$a_1 c_2 - \alpha a_2 c_1 = 2a_3 c_3, \qquad \widetilde{a}_1 \widetilde{c}_2 - \alpha \widetilde{a}_2 \widetilde{c}_1 = 2\widetilde{a}_3 \widetilde{c}_3; \qquad (23)$$

(II) (G, S) *is an (α, ε)-structure with $\alpha\cdot\varepsilon = -1$ if and only if the coefficients of G and S satisfy the proportionality relations*

$$c_2 = \varepsilon\lambda a_2, \ \ c_i = \lambda a_i, \ \ \widetilde{c}_2 = \varepsilon\widetilde{\lambda}\widetilde{a}_2, \ \ \widetilde{c}_i = \widetilde{\lambda}\widetilde{a}_i, \ \ (i = 1, 3), \qquad (24)$$

with functions λ, μ of t satisfying $\lambda \neq 0$ and $\widetilde{\lambda} = \lambda + 2t\mu \neq 0$.

Proof. By using the adapted local frame field $\{\partial^i, \delta_j\}_{i,j=1}^n$, we find that the ε-compatibility relation $G(SX, SY) = \varepsilon G(X, Y)$ for all $X, Y \in \mathcal{T}_0^1(T^*M)$ turns to

$$G(S\delta_i, S\delta_j) = \varepsilon(G^{(1)})_{ij}, \ \ G(S\partial^i, S\partial^j) = \varepsilon(G^{(2)})^{ij}, \ \ G(S\partial^i, S\delta_j) = \varepsilon(G^{(3)})^i_j.$$

By using (8), we see that these three relations become

$$(S^{(1)})_{il}(S^{(1)})_{jm}(G^{(2)})^{lm} - (S^{(1)})_{il}(S^{(3)})^m_j(G^{(3)})^l_m$$
$$- (S^{(3)})^l_i(S^{(1)})_{jm}(G^{(3)})^m_l + (S^{(3)})^l_i(S^{(3)})^m_j(G^{(1)})_{lm} - \varepsilon G^{(1)}_{ij} = 0$$

$$(S^{(3)})^i_l(S^{(3)})^j_m(G^{(2)})^{lm} + \alpha(S^{(3)})^i_l(S^{(2)})^{jm}(G^{(3)})^l_m$$
$$+ \alpha(S^{(2)})^{il}(S^{(3)})^j_m(G^{(3)})^m_l + (S^{(2)})^{il}(S^{(2)})^{jm}(G^{(1)})_{lm} - \varepsilon(G^{(2)})^{ij} = 0$$

$$(S^{(3)})^i_l(S^{(1)})_{jm}(G^{(2)})^{lm} - (S^{(3)})^i_l(S^{(3)})^m_j(G^{(3)})^l_m$$
$$+ \alpha(S^{(2)})^{il}(S^{(1)})_{jm}(G^{(3)})^m_l - \alpha(S^{(2)})^{il}(S^{(3)})^m_j(G^{(1)})_{lm} - \varepsilon(G^{(3)})^i_j = 0.$$

By substituting (9) and (22) into these relations, they are written as

$$T_{11}g_{ij} + T_{12}p_ip_j = 0, \ \ T_{21}g^{ij} + T_{22}g^{0i}g^{0j} = 0, \ \ T_{31}\delta^i_j + T_{32}g^{0i}p_j = 0, \ \ (25)$$

where T_{kl} $(k = 1, 2, 3,\ l = 1, 2)$ are functions of the coefficients a_i, b_i, c_i, d_i $(i = 1, 2, 3)$. By using Lemma 3.1, it follows from (25) that all the coefficients T_{kl} vanish.

By the conditions $T_{k1} = 0$ $(k = 1, 2, 3)$, we obtain a systems of linear equations with the unknowns c_i $(i = 1, 2, 3)$:

$$\begin{cases} (a_3^2 - \varepsilon)c_1 + a_1^2 c_2 - 2a_1 a_3 c_3 = 0, \\ a_2^2 c_1 + (a_3^2 - \varepsilon)c_2 + 2\alpha a_2 a_3 c_3 = 0, \\ -\alpha a_2 a_3 c_1 + a_1 a_3 c_2 + (\alpha a_1 a_2 - a_3^2 - \varepsilon)c_3 = 0. \end{cases} \tag{26}$$

By the conditions $T_{k2} = 0$ $(k = 1, 2, 3)$, we obtain another system,

$$\begin{cases} (a_3^2 - \varepsilon)d_1 + 2a_3(b_3 c_1 - b_1 c_3) + 2a_1(b_1 c_2 - b_3 c_3) + a_1^2 d_2 \\ \quad - 2a_1 a_3 d_3 + 2b_3^2 c_1 t + 2b_1^2 c_2 t - 4tb_1 b_3(c_3 + 2td_3) \\ \quad + 4tb_3(a_3 + tb_3)d_1 + 4tb_1(a_1 + tb_1)d_2 - 4t(a_3 b_1 + a_1 b_3)d_3 = 0, \\[4pt] a_2^2 d_1 + (a_3^2 - \varepsilon)d_2 + 2(a_2 + tb_2)b_2(c_1 + 2td_1) \\ \quad + 2(a_3 + tb_3)b_3(c_2 + 2td_2) - 2\alpha[a_3 b_2 + (a_2 + 2tb_2)b_3]c_3 \\ \quad + 2\alpha(a_2 + 2tb_2)(a_3 + 2tb_3)d_3 = 0, \\[4pt] -\alpha[a_3 b_2 + (a_2 + 2tb_2)b_3]c_1 - \alpha(a_2 + 2tb_2)(a_3 + 2tb_3)d_1 \\ \quad + [a_3 b_1 + (a_1 + 2tb_1)b_3]c_2 + (a_1 + 2tb_1)(a_3 + 2tb_3)d_2 \\ \quad + (\alpha a_1 a_2 - a_3^2 - \varepsilon)d_3 + \alpha[a_2 b_1 + (a_1 + 2tb_1)b_2](c_3 + 2td_3) \\ \quad - 2b_3(a_3 + tb_3)(c_3 + 2td_3) = 0. \end{cases} \tag{27}$$

Multiplying each equation in (27) by $2t$ and by adding to it the corresponding equation in (26), we obtain the following linear system on $\tilde{c}_i$ $(i = 1, 2, 3)$:

$$\begin{cases} (\tilde{a}_3^2 - \varepsilon)\tilde{c}_1 + \tilde{a}_1^2 \tilde{c}_2 - 2\tilde{a}_1 \tilde{a}_3 \tilde{c}_3 = 0, \\ \tilde{a}_2^2 \tilde{c}_1 + (\tilde{a}_3^2 - \varepsilon)\tilde{c}_2 + 2\alpha \tilde{a}_2 \tilde{a}_3 \tilde{c}_3 = 0, \\ -\alpha \tilde{a}_2 \tilde{a}_3 \tilde{c}_1 + \tilde{a}_1 \tilde{a}_3 \tilde{c}_2 + (\alpha \tilde{a}_1 \tilde{a}_2 - \tilde{a}_3^2 - \varepsilon)\tilde{c}_3 = 0. \end{cases} \tag{28}$$

The two systems (26) and (28) coincide with those obtained in the study of the general natural (α, ε)-structures on the total space TM of the tangent bundle, except for the fact that the coefficients and the unknowns are now functions of the energy density t of the cotangent vector p. These systems were solved in the proof of [11, Theorem 4.2] for both classes of structures presented in the statements. The obtained solutions satisfy the relations in (I) and (II). Thus, we obtain our result on the background of the cotangent bundle as well. $\qquad \square$

Remark 4.1. The general natural Riemannian almost product structures on T^*M (see [10, Theorem 4.1]) can be obtained from Theorem 4.1 (I) by taking $\alpha = \varepsilon = 1$. Moreover, as corollaries of Theorem 4.1 (II), we can characterize general natural almost Hermitian structures on T^*M (see [7, Theorem 4.1] and almost para-Hermitian structures on T^*M [10, Theorem 4.3]).

Example 4.1. On T^*M we take the general natural α-structure S given in Example 3.1 and a general natural lifted metric G whose coefficients are $c_1 = \alpha a_1$, $d_1 = \alpha b_1$, $c_2 = a_2$, $d_2 = b_2$, $c_3 = d_3 = 0$. Then the pair (G, S) is a general natural (α, ε)-structure on T^*M of class (I) in Theorem 4.1.

Example 4.2. On T^*M we take a general natural α-structure S given in Example 3.1 and a general natural lifted metric G whose coefficients are $c_1 = \lambda a_1$, $d_1 = \lambda b_1$, $c_2 = \varepsilon \lambda a_2$, $d_2 = \varepsilon \lambda b_2$, $c_3 = d_3 = 0$ with an arbitrary nonzero function λ of t. Then the pair (G, S) is an example of general natural (α, ε)-structure on T^*M of class (II) in Theorem 4.1.

5. General natural (α, ε)-structures of Kähler type on T^*M

In this section we investigate general natural (α, ε)-structures on T^*M with $\alpha \cdot \varepsilon = -1$. In this case, the fundamental tensor field Ω of type $(0, 2)$ defined by

$$\Omega(X, Y) = G(X, SY), \quad X, Y \in \mathcal{T}_0^1(T^*M)$$

is a 2-form. We here study its closedness. When it is closed (i.e. $d\Omega = 0$), we say that (G, S) is of almost-Kähler type, and say that it is of Kähler type if additionally satisfies that S is integrable.

Theorem 5.1. *A general natural (α, ε)-structure (G, S) with $\alpha \cdot \varepsilon = -1$ on T^*M is of almost-Kähler type if and only if the functions which appeared in Theorem 4.1 (II) satisfy $\mu = \lambda'$. Moreover, if the α-structure S is integrable, then (G, S) is an (α, ε)-structure of Kähler type on T^*M.*

Proof. By use of the local adapted frame field $\{\partial^i, \delta_j\}_{i,j=1}^n$ on T^*M, the fundamental tensor field associated to (G, S) has the expressions

$$\Omega(\partial^i, \partial^j) = (S^{(3)})_h^i (G^{(2)})^{jh} + \alpha (S^{(2)})^{ih} (G^{(3)})_h^j,$$

$$\Omega(\delta_i, \delta_j) = (S^{(1)})_{ih} (G^{(3)})_j^h - (S^{(3)})_i^h (G^{(1)})_{jh},$$

$$\Omega(\delta_i, \partial^j) = (S^{(3)})_h^j (G^{(3)})_i^h + \alpha (S^{(2)})^{jh} (G^{(1)})_{hi},$$

$$\Omega(\partial^i, \delta_j) = (S^{(1)})_{jh} (G^{(2)})^{hi} - (S^{(3)})_j^h (G^{(3)})_h^i.$$

By substituting (22) and (9) into these equalities and by using Theorem 4.1
(II), the above expressions yield

$$\Omega(\partial^i, \partial^j) = \Omega(\delta_i, \delta_j) = 0, \;\; \Omega(\delta_i, \partial^j) = -\Omega(\partial^j, \delta_i) = \alpha(\lambda\delta_i^j + \mu p_i g^{0j}),$$

or equivalently

$$\Omega = \alpha(\lambda\delta_i^j + \mu p_i g^{0j})dq^i \wedge Dp_j. \tag{29}$$

Computing the differential of Ω in the same way as that in [10], we get

$$d\Omega = \frac{\alpha}{2}(\mu - \lambda')(g^{0h}\delta_i^j - g^{0j}\delta_i^h)Dp_h \wedge Dp_j \wedge dq^i,$$

hence find that Ω is closed if and only if $\mu = \lambda'$.   □

Remark 5.1. The family of general natural (α, ε)-structures from Class
II is of almost Kähler (respectively Kähler) type on T^*M if and only if
it depends on the coefficients a_1, a_3, b_1, b_3, λ (respectively on a_1, a_3, λ,
only), provided that on $\mathbb{R} \setminus \{0\}$ the function λ is not of the form $\kappa/\sqrt{t}$
with a constant $\kappa \in \mathbb{R}$. In the particular case of $(-1, 1)$-manifolds, the
supplementary conditions $a_1 \cdot \lambda$, $\widetilde{a_1} \cdot \widetilde{\lambda} > 0$ must be satisfied.

Example 5.1. If we take a constant function λ in Example 4.2, then (G, S)
is a general natural (α, ε)-structure of almost-Kähler type on T^*M. Moreover, if we take an integrable general natural α-structure given in Example 3.2, we obtain a general natural (α, ε)-structure of Kähler type on T^*M.

Acknowledgement

The author would like to thank the anonymous referee for the valuable
suggestions, and the editor, professor Adachi, for his scientific support and
for the detailed correspondence, which led to the improvement of the paper.

References

[1] P. Alegre, A. Carriazo, Slant submanifolds of para-Hermitian manifolds, *Mediterr. J. Math.* **14**, 5 (2017), Art. 214, 14 pp.

[2] D.V. Alekseevsky, C. Medori and A. Tomassini, Homogeneous para-
Kahler Einstein manifolds, *Russ. Math. Surv.* **64**, 1 (2009), 1–43.

[3] C. Bejan, A classification of the almost parahermitian manifolds, *Proc.
Conference on Diff. Geom. and Appl., Dubrovnik*, 1988, 23–27.

[4] M. Boucetta and Z. Saassai, A class of Poisson structures compatible
with the canonical Poisson structure on the cotangent bundle, *C. R.
Acad. Sci. Paris, Ser. I* **349** (2011), 331–335.

[5] V. Cruceanu, *Selected Papers*, Editura PIM, Iaşi, 2006.

[6] V. Cruceanu and F. Etayo, On almost para-Hermitian manifolds, *Algebras Groups Geom.* **16**, 1 (1999), 47–61.

[7] S. L. Druţă, Cotangent bundles with general natural Kähler structures, *Rév. Rou. Math. Pures Appl.*, **54** (2009), 13–23.

[8] S. L. Druţă-Romaniuc, Natural diagonal Riemannian almost product and para-Hermitian cotangent bundles, *Czech. Math. J.*, **62** (137) (2012), 937–949.

[9] ———, General natural Riemannian almost product and para-Hermitian structures on tangent bundles, *Taiwanese J. Math.* **16**, 2 (2012), 497–510.

[10] ———, Riemannian almost product and para-Hermitian cotangent bundles of general natural lift type, *Acta Math. Hung.* **139**, 3 (2013), 228–244.

[11] ———, General natural (α, ε)-structures, *Mediterr. J. Math.* **15**, 6 (2018), Art. 228, 13 pp.

[12] F. Etayo and R. Santamaria, $(J^2 = \pm 1)$-metric manifolds, *Publ. Math. Debrecen* **57** (2000), 435–444.

[13] ———, Distinguished connections on $(J^2 = \pm 1)$-metric manifolds, *Archivum mathematicum (Brno)* **52** (2016), 159–203.

[14] ———, The well adapted connection of a $(J^2 = \pm 1)$-metric manifold, *Rev. R. Acad. Cienc. Exactas Fs. Nat. Ser. A Math.* **111**, 2 (2017), 355–375.

[15] P. M. Gadea and J. Muñoz Masqué, Classification of almost para-Hermitian manifolds, *Rend. Mat. Appl.* (7) **11** (1991), 377–396.

[16] G. Ganchev and A. V. Borisov, Note on the almost complex manifolds with a Norden metric, *C. R. Acad. Bulgare Sci.* **39**, 5 (1986), 31–34.

[17] A. Gray and L. M. Hervella, The sixteen classes of almost Hermitian manifolds and their linear invariants, *Ann. Mat. Pura Appl.* **123**, 1 (1980), 35–58.

[18] C. Ida and A. Manea, On the Integrability of generalized almost para-Norden and para-Hermitian Structures, *Mediterr. J. Math.* **14**, 4 (2017), Art. 173, 21 pp.

[19] S. Ivanov, S. Zamkovoy, Parahermitian and paraquaternionic manifolds, *Diff. Geom. Appl.* **23**, 2 (2005), 205–234.

[20] J. Janyška, Natural 2-forms on the tangent bundle of a Riemannian manifold, *Rend. Circ. Mat. Palermo, Serie II* **32** (1993), 165–174, Supplemento, The Proceedings of the Winter School Geometry and Topology Srní-January 1992.

[21] S. Kobayashi and K. Nomizu, *Foundations of Differential Geometry,* vol. I and II, Interscience, N. York, (1963, 1969).

[22] I. Kolář, P. Michor and J. Slovak, *Natural Operations in Differential Geometry,* Springer-Verlag, Berlin, 1993.

[23] O. Kowalski and M. Sekizawa, Natural transformations of Riemannian metrics on manifolds to metrics on tangent bundles - a classification, *Bull. Tokyo Gakugei Univ.* (4) **40** (1988), 1–29.

[24] D. Luczyszyn and Z. Olszak, On paraholomorphically pseudosymmetric para-Kählerian manifolds, *J. Korean Math. Soc.* **45**, 4 (2008), 953–963.

[25] D. Mekerov and M. Manev, Natural connection with totally skew-symmetric torsion on Riemann almost product manifolds, *Int. J. Geom. Methods Mod. Phys.* **9**, 1 (2012), 14.

[26] I. Mihai and C. Nicolau, Almost product structures on the tangent bundle of an almost paracontact manifold, *Demonstratio Math.* **15** (1982), 1045–1058.

[27] K. P. Mok, Infinitesimal automorphisms on the tangent bundle, *J. Math. Kyoto Univ.* **17**, 2 (1977), 399–412.

[28] K. P. Mok, E. M. Patterson and Y. C. Wong, Structure of symmetric tensors of type (0,2) and tensors of type (1,1) on the tangent bundle, *Trans. Amer. Math. Soc.* **234** (1977), 253–278.

[29] M. I. Munteanu, CR-structures on the unit cotangent bundle and Bochner type tensor, *An. Șt. Univ. Al. I. Cuza Iași, Math.* **44**, 1 (1998), 125–136.

[30] A. M. Naveira, A classification of Riemannian almost product manifolds, *Rend. Math.* **3** (1983), 577–592.

[31] V. Oproiu, A generalization of natural almost Hermitian structures on the tangent bundles, *Math. J. Toyama Univ.* **22** (1999), 1–14.

[32] V. Oproiu, N. Papaghiuc and G.Mitric, Some classes of parahermitian structures on cotangent bundles, *An. Șt. Univ. "Al. I. Cuza", Iași, Mat. N.S.* **43** (1997), 7–22.

[33] V. Oproiu and D. D. Poroșniuc, A class of Kaehler Einstein structures on the cotangent bundle of a space form, *Publ. Math. Debrecen* **66** (2005), 457–478.

[34] N. Papaghiuc, Some almost complex structures with Norden metric on the tangent bundle of a space form, *An. Științ. Univ. "Al. I. Cuza", Iași, Mat. N.S.* **46**, 1 (2000), 99–110.

[35] E. Peyghan and A. Heydari, A class of locally symmetric para-Kähler Einstein structures on the cotangent bundle, *Int. Math. Forum* **5**, 3 (2010), 145–153.

[36] M. Staikova and K. Gribachev, Cannonical connections and conformal invariants on Riemannian almost product manifolds, *Serdica Math. J.* **18** (1992), 150–161.

[37] K. Yano and E. M. Patterson, Vertical and complete lifts from a manifold to its cotangent bundle, *J. Math. Soc. Japan* **19**, 1 (1967), 289–311.

[38] K. Yano and S. Ishihara, *Tangent and Cotangent Bundles*, M. Dekker Inc., New York, 1973.

Received January   7, 2019
Revised   February 21, 2019

EIGENVALUES OF REGULAR KÄHLER GRAPHS
HAVING COMMUTATIVE ADJACENCY OPERATORS

Toshiaki ADACHI*

*Department of Mathematics, Nagoya Institute of Technology,
Nagoya 466-8555, Japan
E-mail: adachi@nitech.ac.jp*

A Kähler graph consists of two graphs having the same set of vertices. When
these two graphs are regular and their adjacency operators are commutative,
the probabilistic adjacency operators of derived graphs are selfadjoint. In this
paper we study maximum eigenvalues of derived graphs of such a regular Kähler
graph whose constituent graphs are connected, and investigate whether derived
graphs are connected or not.

Keywords: Kähler graphs; bicolored paths; derived graphs; adjacency operators; regular; bipartite; vertex-transitive; circuits.

1. Introduction

A non-directed graph consists of a set of vertices and a set of edges, and
is regarded as a 1-dimensional CW-complex. It is commonly said that
graphs are discrete models of Riemanian manifolds. The author introduced
the notion of Kähler graphs in [3] to give discrete models of manifolds
admitting magnetic fields. In Riemannian geometry, closed 2-forms, such as
Kähler forms on Kähler manifolds, are said to be magnetic fields (see [11]).
Under the influence of Lorentz forces of these magnetic fields, we can define
trajectories of charged particles on manifolds. Since they are generalizations
of geodesics, they help us to study manifolds from Riemannian geometric
point of view (see [1, 6, 9], for example). A Kähler graph is a compound of
two graphs having the common set of vertices. Since paths on a graph are
considered to correspond to geodesics, the author defined "curved paths"
on a Kähler graph by giving different roles to these two graphs. Since he
did not give correspondences of magnetic fields on graphs, we may say that
he gave discrete models of manifolds admitting F-geodesics (see [7]).

In this paper we study maximum eigenvalues of regular finite Kähler
graphs having commutative adjacency operators. Generally, probabilistic

---

*The author is partially supported by Grant-in-Aid for Scientific Research (C)
(No. 16K05126), Japan Society for the Promotion of Science.

adjacency operators of Kähler graphs are not selfadjoint with respect to the ordinary inner product on the space of functions of the set of vertices. This property corresponds to the property that magnetic mean operators, which are generators of random walks formed by trajectory-segments, are not selfadjoint with respect to Riemannian metrics (cf. [2]). But, on Kähler manifolds of constant holomorphic sectional curvature, these mean operators for Kähler magnetic fields are selfadjoint. Here, a Kähler magnetic field is a constant multiple of the Kähler form (see [1]). Therefore, it is natural to study Kähler graphs having selfadjoint probabilistic adjacency operators by considering them as correspondences of homogeneous manifolds. Since regular graphs are considered to be corresponded to manifolds of constant sectional curvature, we take Kähler graphs whose ingredient graphs are regular. We are interested in their derived graphs which are formed by "bicolored paths" corresponding to trajectory-segments. For an ordinary regular graph, its maximum eigenvalue is given by its degree, and its multiplicity is given by the number of connected components because we can apply Perron-Frobenius theorem. By making use of this property, we investigate maximum eigenvalues of probabilistic adjacency operators of derived graphs of Kähler graphs and study their connectivity.

2. Kähler graphs

We shall start by introducing some notations and terminologies on graphs. A (non-directed) graph $G = (V, E)$ is a pair of sets of vertices and edges, and is represented as a 1-dimensional CW-complex. We say two vertices $v, w \in V$ are *adjacent* to each other if there is an edge joining them. In this case we denote as $v \sim w$. For a vertex $v \in V$ we set $d(v)$ as the cardinality of the set $\{w \in V \mid w \sim v\}$, and call it the *degree* at v. When the degree does not depend on the choice of vertices, we call this graph *regular*. When both the sets of vertices and edges are finite we call this graph finite, and when degree at each vertex is finite we call this graph locally finite. A graph is said to be simple if it does not contain loops, hairs and multiple edges. Here, a loop is an edge joining a vertex and itself, a hair is an edge one of whose end-vertices has degree 1, and multiple edges are edges joining the same pair of vertices.

A simple non-directed graph $G = (V, E)$ is said to be *Kähler* if the set E of edges is divided into a disjoint union $E^{(p)} + E^{(a)}$ of two subsets and satisfies that both the *principal* graph $G^{(p)} = (V, E^{(a)})$ and the *auxiliary* graph $G^{(a)} = (V, E^{(a)})$ do not have hairs. That is, for each vertex $v \in V$, its principal degree $d^{(p)}(v)$ and its auxiliary degree $d^{(a)}(v)$ at v, which are degrees at v in $G^{(p)}$ and $G^{(a)}$ respectively, are greater than 1. For two vertices v, w, we denote as $v \sim_p w$ if they are adjacent in $G^{(p)}$, and denote as $v \sim_a w$ if they are adjacent in $G^{(a)}$. Given a relatively prime positive integers p, q, we say a $(p+q)$-step path $\gamma = (v_0, v_1, \ldots, v_{p+q}) \in V \times \cdots \times V$ to be a (p, q)-primitive bicolored path if it satisfies the following properties:

i) $v_{i+1} \neq v_{i-1}$ for $1 \leq i \leq p+q-1$,
ii) $v_i \sim_p v_{i+1}$ for $0 \leq i \leq p-1$,
iii) $v_i \sim_a v_{i+1}$ for $p \leq i \leq p+q-1$.

The first condition shows that this path does not have backtrackings, the second tells that its first p-step $(v_0, \ldots, v_p)$ is a path in $G^{(p)}$, and the third tells that the last q-step $(v_p, \ldots, v_{p+q})$ is a path in $G^{(a)}$. Here, a path on a graph means a chain of edges, or equivalently a sequence of adjacent vertices. For a (p, q)-primitive bicolored path $\gamma = (v_0, \ldots, v_{p+q})$, we put $o(\gamma) = v_0$ and $t(\gamma) = v_{p+q}$, and call them its origin and terminus. An $m(p+q)$-step path $(v_0, \ldots, v_{m(p+q)})$ is called a (p, q)-bicolored path if each path $\gamma_j = (v_{(j-1)(p+q)}, \ldots, v_{j(p+q)})$ is a (p, q)-primitive bicolored path for $j = 1, \ldots, m$. Since paths on a graph are considered as correspondences of geodesics on a Riemannian manifold, we consider (p, q)-bicolored paths as "bended" curves under the action of a magnetic field of strength q/p.

For a (p, q)-primitive bicolored path $\gamma = (v_0, v_1 \ldots, v_{p+q})$, we define its *probabilistic weight* by

$$\omega(\gamma) = \frac{1}{d^{(a)}(v_p) \prod_{i=1}^{q-1} \{d^{(a)}(v_{p+i}) - 1\}}.$$

For a (p, q)-bicolored path $\gamma = \gamma_1 \cdots \gamma_m$ we set its probabilistic weight as $\omega(\gamma) = \prod_{j=1}^{m} \omega(\gamma_j)$. Since we can not show the direction where the Lorentz force of a magnetic field acts to, we treat terminuses of "bended" paths probabilistically.

3. Adjacency operators

Let $G = (V, E^{(p)} + E^{(a)})$ be a locally finite Kähler graph. We denote by $C(V)$ the set of all complex valued functions of V whose supports are finite sets. Given a pair (p, q) of relatively prime positive numbers, we denote by $\mathfrak{P}_{p,q}$ the set of all (p, q)-primitive bicolored paths. We define the (p, q)-*adjacency operator* $\mathcal{A}_{p,q} : C(V) \to C(V)$ by

$$\mathcal{A}_{p,q} f(v) = \sum_{\gamma \in \mathfrak{P}_{p,q}, \, o(\gamma)=v} \omega(\gamma) f(t(\gamma)).$$

It is the generating operator for the random walk formed by (p, q)-bicolored paths. In order to make clear that this operator is different from ordinary adjacency operators, we sometimes call it a probabilistic adjacency operator.

We denote by $\mathfrak{P}_{p,0}$ the set of all p-step paths without backtrackings on $G^{(p)}$, and by $\mathfrak{P}_{0,q}$ the set of all q-step paths without backtrackings on $G^{(a)}$. We define two operators $\mathcal{A}_p^{(p)}, \mathcal{Q}_q^{(a)} : C(V) \to C(V)$ by

$$\mathcal{A}_p^{(p)} f(v) = \sum_{\sigma \in \mathfrak{P}_{p,0}, \, o(\sigma)=v} f\big(t(\sigma)\big), \quad \mathcal{Q}_q^{(a)} f(v) = \sum_{\tau \in \mathfrak{P}_{0,q}, \, o(\tau)=v} \omega(\tau) f\big(t(\tau)\big),$$

where we denote by $\omega(\tau)$ the probabilistic weight of τ by regarding it as $(0, q)$-primitive bicolored path. Clearly, $\mathcal{A}_p^{(p)}$ coincides with the adjacency operator $\mathcal{A}^{(p)}$ of the principal graph, and $\mathcal{Q}_1^{(a)}$ coincides with the transition operator $\mathcal{P}^{(a)}$ of the auxiliary graph. Here, the adjacency operator of the principal graph and the transition operator of the auxiliary graph which act on $C(V)$ are defined by

$$\mathcal{A}^{(p)} f(v) = \sum_{w:w \sim_p v} f(w), \quad \mathcal{P}^{(a)} f(v) = \frac{1}{d^{(a)}(v)} \sum_{w:w \sim_a v} f(w),$$

respectively. By using the p-step adjacency operator $\mathcal{A}_p^{(p)}$ and the q-step probabilistic transition operator $\mathcal{Q}_q^{(a)}$, we have

Lemma 3.1 ([10]). *The (p, q)-adjacency operator for an arbitrary Kähler graph $(V, E^{(p)}+E^{(a)})$ is decomposed into a composition of two operators as $\mathcal{A}_{p,q} = \mathcal{A}_p^{(p)} \circ \mathcal{Q}_q^{(a)}$. In particular, it is not selfadjoint with respect to the ordinary inner product on $C(V)$, in general.*

Considering (p, q)-primitive bicolored paths as directed edges, we call the directed graph $\overrightarrow{G}_{p,q} = (V, \mathfrak{P}_{p,q})$ the (p, q)-derived graph of a Kähler graph G. This may have multiple edges and loops. A directed graph is said to be irreducible if for an arbitrary pair (v, w) of vertices there is a directed path, a chain of directed edges, whose origin is v and whose terminus is w, and is said to be strongly irreducible if we can take a directed edge instead of a directed path. By applying Perron-Frobenius theorem, we have the following.

Proposition 3.1. *Let G be a finite Kähler graph. If its derived graph $\overrightarrow{G}_{p,q}$ is irreducible, then there exist positive integer $k_{p,q}$ and a positive constant $\lambda_{p,q}$ satisfying the following properties:*

  i) $e^{2\sqrt{-1}\pi j/k_{p,q}} \lambda_{p,q}$ $(j = 0, 1, \ldots, k_{p,q} - 1)$ *are simple eigenvalues of $\mathcal{A}_{p,q}$;*

EIGENVALUES OF REGULAR KÄHLER GRAPHS

ii) *The absolute values of other eigenvalues of* $\mathcal{A}_{p,q}$ *are less than* $\lambda_{p,q}$.

The constant $\lambda_{p,q}$ *satisfies the estimate*

$$\min_{v \in V} d^{(p)}(v) \left\{ \min_{v \in V} d^{(p)}(v) - 1 \right\}^{p-1} \leq \lambda_{p,q} \leq \max_{v \in V} d^{(p)}(v) \left\{ \max_{v \in V} d^{(p)}(v) - 1 \right\}^{p-1},$$

and when $G_{p,q}$ *is strongly irreducible then we have* $k_{p,q} = 1$.

4. Kähler graphs having commutative adjacency operators

Since adjacency operators for ordinary graphs are selfadjoint, it seems unnatural that adjacency operators for Kähler graphs are not selfadjoint. But we introduced Kähler graphs as discrete models of Kähler manifolds admitting Kähler magnetic fields. Magnetic mean operators which are generators of random walks formed by trajectory-segments for Kähler magnetic fields are not selfadjoint in general, but are selfadjoint when the underlying manifolds are of constant holomorphic sectional curvature (see [2, 3]). We therefore interested in studying Kähler graphs which corresponds to complex space forms.

An ordinary graph is said to be regular if its degree function is constant. We call a Kähler graph *regular* if both of its principal and auxiliary graphs are regular. When the auxiliary graph of a Kähler graph is regular the probabilistic weight of each (p,q)-primitive bicolored path γ is given as $\omega(\gamma) = \{d^{(a)}(d^{(a)} - 1)^{q-1}\}^{-1}$. Thus we have

$$\mathcal{A}_{p,q} = \frac{1}{d^{(a)}(d^{(a)} - 1)^{q-1}} \, \mathcal{A}_p^{(p)} \circ \mathcal{A}_q^{(a)}$$

in this case, where $\mathcal{A}_q^{(a)} : C(V) \to C(V)$ is defined by

$$\mathcal{A}_q^{(a)} f(v) = \sum_{\tau \in \mathfrak{P}_{0,q}, \, o(\tau) = v} f(t(\tau)).$$

Moreover, for ordinary regular graphs their multistep adjacency operators are expressed by adjacency operators. We inductively define monic polynomials $F_k(\cdot; d)$ $(k = 1, 2, \dots)$ by the following relations:

$$\begin{cases} F_{k+2}(t; d) = t F_{k+1}(t; d) - (d-1) F_k(t; d), \\ F_1(t; d) = t, \quad F_2(t; d) = t^2 - d. \end{cases} \quad (4.1)$$

When a Kähler graph is regular, we have $\mathcal{A}_p^{(p)} = F_p(\mathcal{A}^{(p)}; d^{(p)})$ and $\mathcal{A}_q^{(a)} = F_q(\mathcal{A}^{(a)}; d^{(a)})$ (see [10]). Thus, if the adjacency operators $\mathcal{A}^{(p)}, \mathcal{A}^{(a)}$ of the principal and the auxiliary graphs of a Kähler graph are commutative, then we find for an arbitrary (p, q) that $\mathcal{A}_p^{(p)}$ and $\mathcal{A}_q^{(a)}$ are commutative, and hence that the (p,q)-adjacency operator $\mathcal{A}_{p,q}$ is selfadjoint with respect to the ordinary inner product on $C(V)$.

For a Kähler graph $G = (V, E^{(p)} + E^{(a)})$, we define its dual graph G^* by $(V, F^{(p)} + F^{(a)})$ with $F^{(p)} = E^{(a)}$ and $F^{(a)} = E^{(p)}$. The adjacency operator of a Kähler graph G are commutative if and only if for an arbitrary vertex v, the set of terminuses of $(1,1)$-primitive bicolored paths in G of origin v coincides with the set of terminuses of $(1,1)$-primitive bicolored paths in G^* of origin v by considering their multiplicities. When G is a regular Kähler graph, if its adjacency operators are commutative, then $\mathcal{A}_p^{(p)}$ and $\mathcal{A}_q^{(a)}$ are commutative. Hence, if we denote by $\mathfrak{P}_{p,q}(v,w)$ the set of all (p,q)-primitive bicolored paths of origin v and terminus w for a pair (v,w) of vertices including the case $v = w$, the cardinalities of $\mathfrak{P}_{p,q}(v,w)$ and of $\mathfrak{P}_{p,q}(w,v)$ coincide with each other. We can therefore define a bijection $\phi_{v,w} : \mathfrak{P}_{p,q}(v,w) \to \mathfrak{P}_{p,q}(w,v)$ satisfying $\phi_{w,v} = (\phi_{v,w})^{-1}$ in this case. By identifying (p,q)-primitive bicolored paths through these bijections, we obtain a non-directed graph $G_{p,q} = (V, \mathfrak{P}_{p,q}/\sim)$, where $\mathfrak{P}_{p,q}/\sim$ denotes the set of quotients under the identification. We call this a non-directed derived graph of G. This graph may have loops and multiple edges. Clearly, $\overrightarrow{G}_{p,q}$ is irreducible if and only if $G_{p,q}$ is connected.

We should note that there are many Kähler graphs having commutative adjacency operators. For example, we take an ordinary finite graph $G = (V, E)$ satisfying $2 \leq \min\{d(v) \mid v \in V\} \leq \max\{d(v) \mid v \in V\} \leq n_G - 2$, where n_G denotes the cardinality of the set of V. We define the set E^c so that a pair (v,w) of distinct vertices is an element of E^c if and only if v, w are not adjacent to each other. Then the complete graph $G^K = (V, E + E^c)$ is a Kähler graph having commutative adjacency operators (see [10]). In [5] Chen and the author show a way of constructing regular Kähler graphs having commutative adjacency operators. Given Kähler graphs $G = (V, E^{(p)} + E^{(a)})$, $H = (W, F^{(p)} + F^{(a)})$, we call a map $\varphi : V \to W$ homomorphism if it maps principal and auxiliary edges to principal and auxiliary edges, respectively. That is, it is a homomorphism if the following holds: If $v \sim_p v'$ in G then $\varphi(v) \sim_p \varphi(v')$ in H, and if $v \sim_a v'$ in G then $\varphi(v) \sim_a \varphi(v')$ in H. When φ is bijective and both φ and φ^{-1} are homomorphisms, it is called an isomorphism. We call a Kähler graph G *vertex-transitive* if for an arbitrary pair (v, v') of distinct vertices there is an isomorphism φ with $\varphi(v) = v'$. Clearly, every vertex-transitive Kähler graph is regular.

Lemma 4.1 ([5]). *Let n, d_1, d_2 be positive numbers satisfying $n \geq 5$, $d_1 \geq$*

2, $d_2 \geq 2$ and $d_1 + d_2 \leq n - 1$. *If one of the following conditions holds, then we can construct a vertex-transitive Kähler graph* G *having commutative adjacency operators such that* $d^{(p)} = d_1$, $d^{(a)} = d_2$, *the cardinality* n_G *of the set of vertices is* n, *and both* $G^{(p)}$ *and* $G^{(a)}$ *are connected*:

  i) n *is odd and both* d_1, d_2 *are even*,
  ii) n *is even and at least one of* d_1, d_2 *is even, but* $(n, d_1, d_2) \neq (6, 2, 2)$,
  iii) $n \equiv 0 \ (mod \ 4)$ *and* $d_1 + d_2 = n - 2$,
  iv) $n \equiv 0 \ (mod \ 4)$, $d_1 + d_2 < n - 2$ *and* $d_1 \equiv d_2 \ (mod \ 4)$,
  v) $n = 4\alpha\beta \geq 12$, $d_1 \geq \alpha$, $d_2 \geq \alpha$ *with some positive integers* α, β *and* $d_1 + d_2 < n - 2$.

We now study maximum eigenvalues of probabilistic adjacency operators for finite Kähler graphs having commutative adjacency operators. An ordinary graph is said to be a *circuit* if it is homeomorphic to a circle S^1 as a CW-complex. This is equivalent that it is a connected regular graph of degree 2. We call a graph (V, E) *bipartite* if the set of vertices is decomposed into two disjoint subsets as $V = V_+ + V_-$ so that every edge joins a vertex in V_+ and a vertex in V_-. This means that there are no edges joining two vertices in V_+ and no edges joining two vertices in V_-. We shall call this decomposition of the set of vertices the vertex-decomposition of a bipartite graph. For every finite regular bipartite graph, the cardinalities of V_+ and V_- coincide, hence the cardinality n_G of the set of vertices is even. It is also clear that a finite circuit graph G is bipartite if and only if n_G is even.

Theorem 4.1. *Let* $G = (V, E^{(p)} + E^{(a)})$ *be a finite regular Kähler graph. Suppose that its principal and auxiliary graphs are connected, and that its adjacency operators are commutative. If we additionally suppose that one of the following conditions*

  i) *at least one of* $G^{(p)}$, $G^{(a)}$ *is not bipartite, and at least one of them is not a circuit,*
  ii) *both* $G^{(p)}, G^{(a)}$ *are bipartite and their vertex-decompositions do not coincide, and they are not circuits,*

holds, then for an arbitrary pair (p, q) *of relatively prime positive integers, the non-directed derived graph* $G_{p,q}$ *is connected and is not bipartite. Hence* $\lambda_{p,q} = d^{(p)}(d^{(p)} - 1)^{p-1}$ *is a simple eigenvalue of* $\mathcal{A}_{p,q}$ *and absolute values of other eigenvalues are less than* $\lambda_{p,q}$.

Theorem 4.2. *Let* $G = (V, E^{(p)} + E^{(a)})$ *be a finite regular Kähler graph. Suppose that its principal and auxiliary graphs are connected bipartite* graphs having the same vertex-decompositions, and that at least one of these graphs is not a circuit. Moreover we suppose that the adjacency operators are commutative.

(1) When $p + q$ is even, then $G_{p,q}$ has two connected components corresponding to the vertex-decomposition of $G^{(p)}$ and is not bipartite. The multiplicity of the eigenvalue $\lambda_{p,q} = d^{(p)}(d^{(p)} - 1)^{p-1}$ of $\mathcal{A}_{p,q}$ is 2, and absolute values of other eigenvalues are less than $\lambda_{p,q}$.

(2) When $p + q$ is odd, then $G_{p,q}$ is connected, bipartite and has the same vertex-decomposition as of $G^{(p)}$. The eigenvalues $\pm\lambda_{p,q}$ of $\mathcal{A}_{p,q}$ are simple, and absolute values of other eigenvalues are less than $\lambda_{p,q}$.

We now need to study the case that both of the principal and the auxiliary graphs are circuits. To do this we suppose that our Kähler graphs are vertex-transitive.

Theorem 4.3. *Let $G = (V, E^{(p)} + E^{(a)})$ be a finite vertex-transitive Kähler graph with odd n_G. Suppose that its principal and auxiliary graphs are circuits and that the adjacency operators are commutative. Then the same assertions hold as in Theorem 4.1 with $\lambda_{p,q} = 2$.*

Theorem 4.4. *Let $G = (V, E^{(p)} + E^{(a)})$ be a finite vertex-transitive Kähler graph. Suppose that its principal and auxiliary graphs are connected bipartite circuits having the same vertex-decompositions and that the adjacency operators are commutative. Then the same assertions hold as in Theorem 4.2 with $\lambda_{p,q} = 2$.*

5. Eigenvalues of derived graphs for regular graphs

In this section, we recall basic properties on eigenvalues of the adjacency operators of ordinary regular graphs. The following result is well-known with the aid of Perron-Frobenius theorem.

Proposition 5.1. *Let G be a connected regular graph of degree d.*

(1) *The value d is a simple eigenvalue of its adjacency operator $\mathcal{A}$, and its eigenfunction is a constant function.*

(2) *The value $-d$ is an eigenvalue of $\mathcal{A}$ if and only if it is bipartite. In this case this eigenvalue is simple. Its eigenfunction is constant on each V_+ and V_-, and has opposite signatures, where $V = V_+ + V_-$ is the vertex-decomposition.*

(3) *The absolute values of other eigenvalues are less than d.*

This proposition guarantees that for a finite regular graph of degree d the multiplicity of the eigenvalue d of its adjacency operator shows the number of its connected components.

For an ordinary simple graph $G = (V, E)$ we denote the set of all p-step paths without backtrackings by $\mathfrak{P}_p$. For $\sigma = (v_0, \ldots, v_p) \in \mathfrak{P}_p$ we denote its reversed path $(v_p, v_{p-1}, \ldots, v_0) \in \mathfrak{P}_p$ by σ^{-1}. We identify a p-step path σ and its reverse σ^{-1}, and regard the set $\mathfrak{P}_p/\sim$ of quotients under this identification as a set of edges, then we can obtain a non-directed graph $G_p = (V, \mathfrak{P}_p/\sim)$. We call this graph the p-step graph of G.

Proposition 5.2. *Let G be a connected regular graph of degree $d \geq 3$. For an arbitrary positive integer p, its p-step graph G_p satisfies the following:*

(1) *If G is not bipartite, it is connected and is not bipartite;*
(2) *If G is bipartite, it has two connected components and is not bipartite when p is even, and it is connected and is bipartite when p is odd.*

Proof. Since G_p is a regular graph of degree $d(d-1)^{p-1}$, the multiplicity of the eigenvalue $\lambda_p = d(d-1)^{p-1}$ of the adjacency operator $\mathcal{A}_p$ of G_p is equal to the number of connected components, and $-\lambda_p$ is an eigenvalue of $\mathcal{A}_p$ if and only if G_p is bipartite.

Since G is connected, by Proposition 5.1 and by the following Lemma 5.1, for an eigenvalue λ ($\neq \pm d$) we have $|F_k(\lambda; d)| < F_k(d; d) = \lambda_k$. Since $F_{2j}(-d; d) = \lambda_{2j}$ and $F_{2j-1}(-d; d) = -\lambda_{2j-1}$, we get the assertion. $\qquad\square$

Lemma 5.1. *Let d be an integer with $d \geq 3$. If $|t| < d$ then $|F_k(t; d)| < d(d-1)^{k-1}$ for all k.*

Proof. Since we can show that the polynomial $F_k(t; d)$ contains only terms of odd degree when k is odd and contains only terms of even degree when k is even, we are enough to consider the case $0 \leq t < d$. We note that $|F_1(t; d)| = |t| < d$ and $F_2(t; d) = t^2 - d$ satisfies $-d \leq F_2(t; d) < d(d-1)$, hence the assertion is true for $k = 1, 2$.

When $0 \leq t < d - 2$, the relation (4.1) on our polynomials shows that $|F_{k+2}(t; d)| \leq t|F_{k+1}(t; d)| + (d-1)|F_k(t; d)|$. Thus we can show the assertion by induction in this case.

Next we study the case $2\sqrt{d-1} \leq t < d$. If we set $\alpha = \{t - \sqrt{t^2 - 4d + 4}\}/2$, $\beta = \{t + \sqrt{t^2 - 4d + 4}\}/2$, the relation (4.1) shows that

$$F_k(t; d) = \frac{1}{\beta - \alpha}\{\beta^{k-1}(t^2 - \alpha t - d) - \alpha^{k-1}(t^2 - \beta t - d)\}.$$

Since we have

$$t^2 - 2d \geq 4(d-1) - 2d = 2(d-2) > 0,$$
$$(t^2 - 2d)^2 - t^2(t^2 - 4d + 4) = 4(d^2 - t^2) > 0,$$

and $\beta \geq \alpha > 0$, we see $t^2 - \alpha t - d \geq t^2 - \beta t - d > 0$. Thus we find $F_k(t;d) \geq (\beta^{k-1} - \alpha^{k-1})(t^2 - \alpha t - d)/(\beta - \alpha) \geq 0$ for every k. We now show that $F_{k+1}(t;d) \leq (d-1)F_k(t;d)$ by induction. We note

$$(d-1)F_1(t;d) - F_2(t;d) = -t^2 + (d-1)t + d > 0 \quad \text{for } 0 \leq t < d.$$

If we suppose $F_{k+1}(t;d) < (d-1)F_k(t;d)$, then we have

$$F_{k+2}(t;d) = tF_{k+1}(t;d) - (d-1)F_k(t;d)$$
$$< (t-1)F_{k+1}(t;d) < (d-1)F_{k+1}(t;d)$$

for $0 \leq t < d$ because $F_k(t;d)$ and $F_{k+1}(t;d)$ are not negative. Thus we find

$$F_k(t;d) < (d-1)^{k-2}F_2(t;d) < t(d-1)^{k-1} < d(d-1)^{k-1}$$

for $0 \leq t < d$, and get the assertion in this case.

Thirdly, we study the case $d - 2 \leq t < 2\sqrt{d-1}$. This inequality shows that $d = 3, 4, 5, 6$. In this case α, β are complex numbers which satisfy $|\alpha| = |\beta| = \sqrt{d-1}$. As $\alpha\beta = d-1$, we have

$$|F_k(t;d)| \leq \left| \frac{\beta^{k-1} - \alpha^{k-1}}{\beta - \alpha} \right| |t^2 - d| + (d-1)t \left| \frac{\beta^{k-2} - \alpha^{k-2}}{\beta - \alpha} \right|$$
$$= \left| \sum_{j=0}^{k-2} \beta^{k-2-j}\alpha^j \right| |t^2 - d| + (d-1)t \left| \sum_{j=0}^{k-3} \beta^{k-3-j}\alpha^j \right|$$
$$\leq (k-1)(d-1)^{(k-2)/2} |t^2 - d| + (k-2)(d-1)^{(k-1)/2} t$$

for $k \geq 3$. Since $d - 2 \leq t < 2\sqrt{d-1}$, we have $|t^2 - d| < 3(d-1)$. Thus we find $|F_k(t;d)| < (5k-7)(d-1)^{k/2}$. Hence, if $5k - 7 \leq d(d-1)^{(k-2)/2}$ holds, we can get the assertion. When $d \geq 5$ this inequality holds for every $k \geq 3$, when $d = 4$ this holds for every $k \geq 5$, and when $d = 3$ this holds for every $k \geq 10$.

Finally, we study the remaining cases by direct computation under the condition $d - 2 \leq t < 2\sqrt{d-1}$. We have $F_3(t;d) = t(t^2 - 2d + 1)$, hence

$$-6 = F_3(2;4) \leq F_3(t;4) \leq F_3(2\sqrt{3};4) = 10\sqrt{3},$$
$$-2(5/3)^{3/2} = F_3(\sqrt{5/3};3) \leq F_3(t;3) \leq F_3(2\sqrt{2};3) = 6\sqrt{2},$$

and get $|F_3(t;d)| < d(d-1)^2$. As we have $F_4(t;d) = t^4 - (3d-2)t^2 + d(d-1)$, we find

$$-25/4 = F_4(\sqrt{7/2};3) \le F_4(t;3) < F_4(2\sqrt{2};3) = 14,$$
$$-13 = F_4(\sqrt{5};4) \le F_4(t;4) \le F_4(1;4) = 3,$$

and get $|F_4(t;d)| < d(d-1)^3$. We have $F_5(t;3) = t(t^4 - 9t^2 + 16)$, hence

$$F_5(t;3) < F_5(2\sqrt{2};3) = 32\sqrt{2}, \quad F_5(t;3) > 2\sqrt{2} \times (-17/4),$$

and get $|F_5(t;3)| < 3 \cdot 2^4$. We have $F_6(t;3) = t^6 - 11t^4 + 30t^2 - 12$, hence $F_6(t;3) > 8 \cdot (-11/4) - 12 = -56$. As $F_6(t;3) < f(t) := t^6 - (21/2)t^4 + 30t^2 - 12$, we have $F_6(t;3) < f(2\sqrt{2}) = 68$, and get $|F_6(t;3)| < 3 \cdot 2^5$. We have $F_7(t;3) = t(t^6 - 13t^4 + 48t^2 - 44)$, which is monotone increasing in $1 \le t < 2\sqrt{2}$. We hence have $-44 \le F_7(t;3) < 40\sqrt{2}$, and get $|F_7(t;3)| < 3 \cdot 2^6$. We have

$$F_8(t;3) = t^8 - 15t^6 + 70t^4 - 104t^2 + 24,$$
$$F_9(t;3) = t(t^8 - 17t^6 + 96t^4 - 200t^2 + 112).$$

With the aid of computer, we find

$$-30 < F_8(t;3) < F_8(2\sqrt{2};3) = 88, \quad -50 < F_9(t;3) < F_9(2\sqrt{2};3) = 96\sqrt{2},$$

and get $|F_8(t;3)| < 3 \cdot 2^7$ and $|F_9(t;3)| < 3 \cdot 2^8$. We obtain our assertion. □

We note that such a property in Lemma 5.1 does not holds for circuit graphs.

Lemma 5.2. *Let G be a circuit graph having n_G vertices.*

(1) *When p and n_G are relatively prime, its p-step graph G_p is a circuit.*

(2) *When p and n_G are not relatively prime, we denote by $r(p, n_G)$ the greatest common divisor of p and n_G. Then G is a union of $r(p, n_G)$ circuits having $n_G/r(p, n_G)$ vertices.*

We have $F_k(2;2) = 2$ and $F_k(-2;2) = (-1)^k 2$. For t with $|t| < 2$, we have the following on $F_k(t;2)$.

Corollary 5.1. *Let λ ($|\lambda| < 2$) be an eigenvalue of the adjacency operator of a circuit graph G having n_G vertices.*

(1) *When p and n_G are relatively prime, we have $|F_p(\lambda;2)| < 2$.*

(2) *We suppose p and n_G are not relatively prime.*

   a) *If λ corresponds to an eigenfunction which is constant on each component of G_p, then we have $F_p(\lambda;2) = 2$.*

    b) *When $n_G/r(p, n_G)$ is even, we denote by $V_+ + V_-$ the vertex-decomposition of G_p. If λ corresponds to an eigenfunction f such that for each component $\mathcal{C}$ of G_p its restrictions $f|_{\mathcal{C} \cap V_+}$ and $f|_{\mathcal{C} \cap V_-}$ on $\mathcal{C} \cap V_+$ and on $\mathcal{C} \cap V_-$ are constant and satisfy $f|_{\mathcal{C} \cap V_-} = -f|_{\mathcal{C} \cap V_+}$, then we have $F_p(\lambda; 2) = -2$.*

    c) *Otherwise, we have $|F_p(\lambda; 2)| < 2$.*

Example 5.1. The eigenvalues of a circuit graph G having 12 vertices are $0, 0, 1, 1, -1 - 1, \sqrt{3}, \sqrt{3}, -\sqrt{3}, -\sqrt{3}, 2, -2$. The eigenvalues of G_3 are $0, 0, 0, 0, 0, 0, 2, 2, 2, -2, -2, -2$. We note that the 3-step graph G_3 has 3 bipartite components and $F_3(\pm 1; 2) = \mp 2$, $F_3(\pm\sqrt{3}; 2) = F_3(0; 2) = 0$.

Example 5.2. The eigenvalues of a circuit graph G having 16 vertices are $0, 0, \pm\sqrt{2-\sqrt{2}}, \pm\sqrt{2-\sqrt{2}}, \pm\sqrt{2}, \pm\sqrt{2}, \pm\sqrt{2+\sqrt{2}}, \pm\sqrt{2+\sqrt{2}}, \pm 2$. The eigenvalues of G_4 are $0, 0, 0, 0, 0, 0, 0, 0, 0, \pm 2, \pm 2, \pm 2$. We note that the 4-step graph G_4 has 4 bipartite components and $F_4(\pm\sqrt{2}; 2) = -2$, $F_4(0; 2) = 2$, $F_4(\pm\sqrt{2-\sqrt{2}}; 2) = F_4(\pm\sqrt{2+\sqrt{2}}; 2) = 0$.

6. Proofs of theorems and some examples

We are now in the position to prove our main results. We study eigenvalues of the adjacency operator of the regular graph $G_{p,q}$ of degree $d^{(p)}(d^{(p)} - 1)^{p-1}d^{(a)}(d^{(a)} - 1)^{q-1}$ which is given by $\mathcal{A}_p^{(p)}\mathcal{A}_q^{(a)}$. The commutativity $\mathcal{A}^{(p)}\mathcal{A}^{(a)} = \mathcal{A}^{(a)}\mathcal{A}^{(p)}$ is equivalent to the condition that $\mathcal{A}^{(p)}$ and $\mathcal{A}^{(a)}$ are simultaneously diagonalizable. Thus, if $g \in C(V)$ is a common eigenfunction satisfying $\mathcal{A}^{(p)}g = \lambda g$ and $\mathcal{A}^{(a)}g = \mu g$, we have $\mathcal{A}_p^{(a)}\mathcal{A}_q^{(a)}g = F_p(\lambda; d^{(p)})F_q(\mu; d^{(a)})g$. In particular, when g is constant, we have $\mathcal{A}_p^{(a)}\mathcal{A}_q^{(a)}g = d^{(p)}(d^{(p)} - 1)^{p-1}d^{(a)}(d^{(a)} - 1)^{q-1}g$,

Proof of Theorem 4.1. (1) First we study the case that the condition i) holds. We take an eigenfunction $g \in C(V)$ for $\mathcal{A}^{(p)}$ which is not constant. Let λ and μ be its eigenvalues. That is, we have $\mathcal{A}^{(p)}g = \lambda g$ and $\mathcal{A}^{(a)}g = \mu g$.

    We suppose that $G^{(p)}$ is not bipartite. If $G^{(p)}$ is not a circuit, as $|\lambda| < d^{(p)}$, by Lemma 5.1 we have $|F_p(\lambda; d^{(p)})| < d^{(p)}(d^{(p)} - 1)^{p-1}$, hence find $|F_p(\lambda; d^{(p)})F_q(\mu; d^{(a)})| < d^{(p)}(d^{(p)} - 1)^{p-1}d^{(a)}(d^{(a)} - 1)^{q-1}$. If $G^{(p)}$ is a circuit, as we supposed that $G^{(p)}$ is not bipartite, we see that the cardinality of V is odd, hence $G^{(a)}$ is not bipartite. We then have $|\mu| < d^{(a)}$. Since $G^{(a)}$ is not a circuit in this case by the assumption, we have $|F_q(\mu; d^{(a)})| < d^{(a)}(d^{(a)} - 1)^{q-1}$, and have $|F_p(\lambda; d^{(p)})F_q(\mu; d^{(a)})| < d^{(p)}(d^{(p)} - 1)^{p-1}d^{(a)}(d^{(a)} - 1)^{q-1}$.

When $G^{(a)}$ is not bipartite, we get the same inequality along the same lines by changing the roles of $G^{(p)}$ and $G^{(a)}$.

(2) Next we study the case that the condition ii) holds. Let $V = V_+^{(p)} + V_-^{(p)}$ and $V = V_+^{(a)} + V_-^{(a)}$ denote the vertex-decompositions of $G^{(p)}$ and $G^{(a)}$, respectively. We take $h^{(p)}, h^{(a)} \in C(V)$ which are defined by

$$h^{(p)}(v) = \begin{cases} 1, & \text{if } v \in V_+^{(p)}, \\ -1, & \text{if } v \in V_-^{(p)}, \end{cases} \qquad h^{(a)}(v) = \begin{cases} 1, & \text{if } v \in V_+^{(a)}, \\ -1, & \text{if } v \in V_-^{(a)}. \end{cases}$$

We have $\mathcal{A}^{(p)} h^{(p)} = -d^{(p)} h^{(p)}$ and $\mathcal{A}^{(a)} h^{(a)} = -d^{(a)} h^{(a)}$. Since the vertex-decompositions of $G^{(p)}$ and $G^{(a)}$ do not coincide, we have $\mathcal{A}^{(a)} h^{(p)} = \mu h^{(p)}$ with $|\mu| < d^{(a)}$ and $\mathcal{A}^{(p)} h^{(a)} = \lambda h^{(a)}$ with $|\lambda| < d^{(p)}$. Since $G^{(p)}$, $G^{(a)}$ are not circuits, if a common eigenfunction g is not constant, we have $\mathcal{A}_p^{(p)} \mathcal{A}_q^{(a)} g = \nu g$ with $|\nu| < d^{(p)}(d^{(p)} - 1)^{p-1} d^{(a)}(d^{(a)} - 1)^{q-1}$ by Lemma 5.1. We hence get the assertion by Proposition 5.1. $\square$

Proof of Theorem 4.2. We use the same notations as in the proof of Theorem 4.1. Since we may consider $V_+^{(p)} = V_+^{(a)}$ and $V_-^{(p)} = V_-^{(a)}$, when $p + q$ is odd, the graph $G_{p,q}$ is bipartite having vertex-decomposition $V = V_+^{(p)} + V_-^{(p)}$, and when $p + q$ is even, for an arbitrary pair (v, v') of vertices in $V_+^{(p)}$ and $V_-^{(p)}$, it does not have edges joining them.

We now study eigenvalues of $\mathcal{A}_p^{(p)} \mathcal{A}_q^{(a)}$. As we have $h^{(p)} = h^{(a)}$, we find

$$\mathcal{A}_p^{(p)} \mathcal{A}_q^{(a)} h^{(p)} = F_p(-d^{(p)}; d^{(p)}) F_q(-d^{(a)}; d^{(a)}) h^{(p)}$$

$$= \begin{cases} d^{(p)}(d^{(p)} - 1)^{p-1} d^{(a)}(d^{(a)} - 1)^{q-1} h^{(p)}, & \text{when } p + q \text{ is even}, \\ -d^{(p)}(d^{(p)} - 1)^{p-1} d^{(a)}(d^{(a)} - 1)^{q-1} h^{(p)}, & \text{when } p + q \text{ is odd}. \end{cases}$$

When the common eigenfunction g is neither constant nor constant multiple of $h^{(p)}$, we have $|\lambda| < d^{(p)}$ and $|\mu| < d^{(a)}$. Since either $G^{(p)}$ or $G^{(a)}$ is not a circuit, by Lemma 5.1, we have either $|F_p(\lambda; d^{(p)})| < d^{(p)}(d^{(p)} - 1)^{p-1}$ or $|F_q(\mu; d^{(a)})| < d^{(a)}(d^{(a)} - 1)^{q-1}$ holds. Hence we get $|F_p(\lambda; d^{(p)}) F_q(\mu; d^{(a)})| < d^{(p)}(d^{(p)} - 1)^{p-1} d^{(a)}(d^{(a)} - 1)^{q-1}$. Thus we obtain that $G_{p,q}$ is connected and bipartite when $p + q$ is odd, and that it has two connected components when $p + q$ is even, by Proposition 5.1. $\square$

Proof of Theorem 4.4. We denote V as $\{0, 1, \dots, n_G - 1\}$ and consider numbers modulo n_G. We may consider $i \sim_p i \pm 1$. Then the vertex-decomposition of $G^{(p)}$ is given as the set $V_+ = \{0, 2, \dots, n_G - 2\}$ of even numbers and the set $V_- = \{1, 3, \dots, n_G - 1\}$ of odd numbers. Since G is vertex-transitive, the expression of $G^{(p)}$ guarantees that the isomorphism which maps j to k is either a rotation $\varphi_{j,k} : V \to V$ given by $\varphi_{j,k}(i) = i + k - j$ or a composition $\psi_{j,k} : V \to V$ of a rotation and a reflection given by $\psi_{j,k}(i) = -i + j + k$.

We have $0 \sim_a s_1$ and $0 \sim_a -s_2$ with $2 \le s_1, s_2 \le n_G - 2$ and $s_1 + s_2 \ne n_G$. Since the vertex-decompositions of $G^{(p)}$ and $G^{(a)}$ coincide, we see that s_1, s_2 are odd. If we suppose $s_1 \ne s_2$, the circuit $G^{(a)}$ is constructed as
$$0 \sim_a s_1 \sim_a s_1 + s_2 \sim_a 2s_1 + s_2 \sim_a 2(s_1 + s_2) \sim_a \cdots \sim_a -s_2 \sim_a 0.$$
This means that $i \sim_a i + s_1, i - s_2$ when i is even and that $i \sim_a i + s_2, i - s_1$ when i is odd. Therefore $(1,1)$-primitive bicolored paths of origin 0 in G are $(0, \pm 1, \pm 1 - s_1)$, $(0, \pm 1, \pm 1 + s_2)$, and those in G^* are $(0, s_1, s_1 \pm 1)$, $(0, -s_2, -s_2 \pm 1)$. Since the adjacency operators of G are commutative, the sets of their terminuses coincide. As $s_1 \ne s_2$, we have three cases:

  i) $1 + s_1 \equiv 1 - s_1 \pmod{n_G}$,   ii) $1 + s_1 \equiv -1 - s_1 \pmod{n_G}$,

  iii) $1 + s_1 \equiv -1 + s_2 \pmod{n_G}$.

In the case i), we have $s_1 = n_G/2$ hence $n_G \equiv 2 \pmod 4$. We therefore have $1 - s_2 \equiv 1 + s_2 \pmod{n_G}$, which shows $s_2 = n_G/2$. In the case iii), we have $s_1 - s_2 + 2 = 0$. If $-1 + s_1 \equiv 1 - s_2 \pmod{n_G}$, we need $-1 - s_2 \equiv 1 + s_2 \pmod{n_G}$. These show that $s_1 = s_2 = (n_G/2) - 1$. If $-1 + s_1 \equiv 1 + s_2 \pmod{n_G}$, we have $s_1 - s_2 - 2 = 0$. In the case ii), we have $s_1 + 1 = n_G/2$. We then have $-1 + s_1 \equiv 1 + s_2$, $1 - s_2 \equiv 1 + s_2 \pmod{n_G}$, hence have $s_2 - 1 = n_G/2$ which means $s_1 + s_2 = n_G$. In each case we have a contradiction. Thus we have $s_1 = s_2$, hence have $i \sim_a \pm s$ with some odd integer s with $0 < s < n_G/2$ for all i. As $G^{(a)}$ is a circuit, we see that s and n_G are relatively prime. Also, we find that rotations give vertex-transitivity of G.

Now we study the regular graph $G_{p,q}$ of degree 4. By (p, q)-primitive bicolored paths, we have $i \sim i \pm (p + qs)$ and $i \sim i \pm (p - qs)$. Hence we can join i and $i + 2qs$ by a (p, q)-bicolored path because $2qs = (p + qs) - (p - qs)$. Since s and n_G are relatively prime, we have integers α, β satisfying $\alpha s + \beta n_G = 1$. We hence have $2q = 2\alpha qs + 2\beta q n_G$ and get $2q \equiv 2\alpha qs \pmod{n_G}$. Thus we can join i and $i + 2q$ by a (p, q)-bicolored path.

(1) We study the case that $p + q$ is odd. As we see in the proof of Theorem 4.2, $G_{p,q}$ is bipartite. Hence we are enough to show that it is connected. To do this, we shall show that $2q$ and $p + qs$ are relatively prime. Since s is odd, we find $p + qs$ is also odd. Therefore, the condition that $2q$ and $p + qs$ are relatively prime is equivalent to the condition that q and $p + qs$ are relatively prime. If we suppose $q = \ell a$ and $p + qs = \ell' a$ with some integers ℓ, ℓ', a satisfying $a \ge 3$, then we have $p = \ell' a - qs = (\ell' - \ell s)a$. Since p and q are relatively prime, this is a contradiction. Hence we obtain that $2q$ and $p + qs$ are relatively prime. This means that i and $i + 1$ are joined by a (p, q)-bicolored path, hence $G_{p,q}$ is connected.

(2) We next study the case $p + q$ is even. As we see in the proof of Theorem 4.2, there are no (p, q)-primitive bicolored paths joining an arbitrary pair of vertices in V_+ and V_-. Hence we are enough to show that arbitrary pairs of vertices in V_+ are joined by a (p, q)-bicolored path and so are for an arbitrary pair of vertices in V_-. We note that $p + qs$ and $p - qs$ are even in this case. By the same argument as in the case $p + q$ is odd, we find that q and $(p + qs)/2$ are relatively prime. This means that i and $i + 2$ are joined by a (p, q)-bicolored path. Thus we obtain that $G_{p,q}$ has two components corresponding to the vertex-decomposition of $G^{(p)}$.

We here suppose $G_{p,q}$ is bipartite. This means that each component forms a bipartite graph. Therefore we have $n_G \equiv 0 \pmod{4}$ and $p + qs \equiv 2$, $p - qs \equiv 2 \pmod{4}$. If we denote as $p + qs = 4b + 2$, $p - qs = 4c + 2$ with some integers b, c, we have $p = 2(b + c + 1)$ and $qs = 2(b - c)$. Since s is odd, we find both p and q are even, which is a contradiction. Thus we find that $G_{p,q}$ is not bipartite. $\qquad\square$

Proof of Theorem 4.3. We denote as $V = \{0, 1, \ldots, n_G - 1\}$ and consider numbers modulo n_G and $i \sim_p i \pm 1$. We use the same notations $\varphi_{j,k}$ and $\psi_{j,k}$ for isomorphisms of G as in the proof of Theorem 4.4.

Since n_G is odd, we have odd s_1, s_2 satisfying $s_1 \neq s_2$, $2 \le |s_1|, |s_2| \le n_G - 2$ and $0 \sim_a s_1$, $0 \sim_a s_2$. If we suppose $s_1 + s_2 \neq \pm n_G$, that is $s_2 \not\equiv -s_1 \pmod{n_G}$, then by the same argument as in the proof of Theorem 4.4, we have a contradiction. Thus there is odd s with $2 \le |s| \le n_G - 2$ which satisfies $i \sim_a i \pm s$.

Since s is odd, we can apply the same arguments as in the proof of Theorem 4.4. When $p+q$ is odd, we can join i and $i+1$ by a (p, q)-bicolored path for all i, hence $G_{p,q}$ is connected. When $p+q$ is even, we can join i and $i+2$ by a (p, q)-bicolored path for all i. As n_G is odd, this shows i and $i+1$ are joined by a (p, q)-bicolored path. Hence $G_{p,q}$ is also connected. $\qquad\square$

We should consider the remaining case, which is the case that both of principal and auxiliary graphs are bipartite, their vertex-decompositions are different and at least one of them is a circuit.

Lemma 6.1. Let $G = (V, E^{(p)} + E^{(a)})$ be a finite vertex-transitive Kähler graph. Suppose that

i) its principal graph $G^{(p)}$ is connected and bipartite,
ii) its auxiliary graph $G^{(a)}$ is a circuit,
iii) the vertex-decompositions of $G^{(p)}$ and $G^{(a)}$ do not coincide.

We denote by $V = V_+^{(p)} + V_-^{(p)}$ the vertex-decomposition of $G^{(p)}$. Then, each component of $V_+^{(p)}$ in $G^{(a)}$ consists of two vertices, and so does for $V_-^{(p)}$. In particular, the cardinality n_G of V is factored by 4.

Proof. We denote V as $\{0, 1, \ldots, n_G - 1\}$ and consider numbers modulo n_G. We may consider $i \sim_a i \pm 1$ for every i. Then the vertex-decomposition of $G^{(a)}$ is given by $V_+^{(a)} = \{0, 2, \ldots, n_G - 2\}$ and $V_-^{(a)} = \{1, 3, \ldots, n_G - 1\}$. Since the vertex-decompositions of $G^{(p)}$ and $G^{(a)}$ do not coincide, we may suppose $0, -1(\equiv n_G - 1) \in V_+^{(p)}$ and $1 \in V_-^{(p)}$ by changing numbers of vertices if we need (see Fig. 1).

Since $G^{(a)}$ is a circuit, the isomorphism which maps a vertex j to a vertex k is either a rotation $\varphi_{j,k} : V \to V$ given by $\varphi_{j,k}(i) = i + k - j$ or a composition $\psi_{j,k} : V \to V$ of a rotation and a reflection which is given by $\psi_{j,k}(i) = -i + j + k$. As $G^{(p)}$ is connected, every isomorphism φ of G preserves the vertex-decomposition of $G^{(p)}$, that is, it satisfies either $\varphi(V_\pm^{(p)}) = V_\pm^{(p)}$ or $\varphi(V_\pm^{(p)}) = V_\mp^{(p)}$.

We shall show that $2 \in V_-^{(p)}$. The rotation $\varphi_{0,1}$ maps $0 \in V_+^{(p)}$ to $1 \in V_-^{(p)}$ and $-1(\equiv n_G - 1) \in V_+^{(p)}$ to $0 \in V_+^{(p)}$, hence it is not an isomorphism of G. Therefore $\psi_{0,1}$ should be an isomorphism. As it maps $0 \in V_+^{(p)}$ to $1 \in V_-^{(p)}$ and $-1 \in V_+^{(p)}$, we have $2 = \psi_{0,1}(-1) \in V_-^{(p)}$ (see Fig. 2).

Next we suppose that $V_+^{(p)}$ in $G^{(a)}$ has a component having 3 or more vertices. We may consider $-2\,(\equiv n_G - 2) \in V_+^{(p)}$ (see Fig. 3). The rotation $\varphi_{-1,0}$ maps $-1 \in V_+^{(p)}$ to $0 \in V_+^{(p)}$ and $0 \in V_+^{(p)}$ to $1 \in V_-^{(p)}$. The composition $\psi_{-1,0}$ maps $-1 \in V_+^{(p)}$ to $0 \in V_+^{(p)}$ and $-2 \in V_+^{(p)}$ to $1 \in V_-^{(p)}$ (see Fig. 4). Thus, $\varphi_{-1,0}$ and $\psi_{-1,0}$ are not isomorphisms. This is a contradiction. Hence each component of $V_+^{(p)}$ in $G^{(a)}$ consists of 2 vertices. By the same argument we can show that each component of $V_-^{(p)}$ in $G^{(a)}$ consists of 2 vertices.   □

Fig. 1.   Fig. 2.   images of $\varphi_{0,1}, \psi_{0,1}$ Fig. 3.   Fig. 4.

Lemma 6.2. Let $G = (V, E^{(p)} + E^{(a)})$ be a finite vertex-transitive Kähler graph. We pose the same assumption as in Lemma 6.1. We represent as $V = \{0, 1, \ldots, n_G - 1\}$ and $i \sim_a i \pm 1$ for every $i \in V$. We then have $i \sim_p i + 2a$ *if and only if* $i \sim_p i - 2a$ *for every* $i \in V$. *Moreover* $G^{(p)}$ *is not a circuit and the adjacency operators of* $G^{(p)}$ *and* $G^{(a)}$ *are not commutative. There are integers* b, c *with* $0 \le b < n_G/8$, $0 < c < n_G/4$ *such that* $i \sim_p i \pm (4b + 2)$ *and that either* $i \sim_p i + 4c + 1$ *and* $i \not\sim_p i - 4c - 1$ *or* $i \sim_p i - 4c - 1$ *and* $i \not\sim_p i + 4c + 1$ *holds.*

Proof. We use the same notations as in the proof of Lemma 6.1. By Lemma 6.1, we may suppose
$$V_+^{(p)} = \{j \in V \mid j \equiv 0, 3 \ (\mathrm{mod}\ 4)\}, \quad V_-^{(p)} = \{j \in V \mid j \equiv 1, 2 \ (\mathrm{mod}\ 4)\}$$
by relabeling numbers of vertices. We then find that the isomorphism which maps j to $j + \ell$ is $\psi_{j,j+\ell}$ if ℓ is odd, and is $\varphi_{j,j+\ell}$ if ℓ is even (see Fig. 5). Thus, if $i \sim_p i + 2a$, then we have $i - 2a = \varphi_{i,i+2a}^{-1}(i) \sim_p \varphi_{i,i+2a}^{-1}(i + 2a) = i$.

We suppose that i_0 is joined by principal edges only to $i_0 \pm 2a_1, \ldots, i_0 \pm 2a_s$ with some numbers $a_1, \ldots, a_s$. Since $G^{(p)}$ is bipartite, these numbers are odd. The vertex-transitivity shows that the same property holds for all i. This means that $G^{(p)}$ is not connected, which is a contradiction. Next we suppose that $0 \in V$ is joined by principal edges only to vertices of odd number. As $0 \in V_+^{(p)}$, these odd numbers are $4c_1 + 1, \ldots, 4c_t + 1$. Since the isomorphism which maps 0 to $4c_k + 1$ is $\psi_{0,4c_k+1}$, we find that the vertex $4c_k + 1$ is joined by principal edges to $4(c_k - c_1), \ldots, 4(c_k - c_t)$ which are 0 modulo 4. This means that we can join 0 by principal paths only to vertices of $0, 1$ modulo 4. This shows that $G^{(p)}$ is not connected, which is a contradiction. In order to show the assertion, we study the case that 0 is joined by principal edges only to $n_G/2$ and a vertex $4c + 1$. This may occurs when $n_G/4$ is odd. By vertex-transitivity, we have $4c + 1 \sim_p 0, 4c+1+(n_G/2)$ and $n_G/2 \sim_p 0, 4c+1+(n_G/2)$ because $n_G/2$ is even. This shows that $\{0, 4c + 1, n_G/2, 4c+1 + (n_G/2)\}$ forms a connected component of $G^{(p)}$. Since $n_G \ge 8$ because we can not construct Kähler graphs with $n_G \le 4$, we see $G^{(p)}$ is not connected, which is a contradiction.

By our argument and by the property that $G^{(p)}$ is bipartite, the vertex-transitivity guarantees that there are positive integers $a_1, \ldots, a_s, c_1, \ldots, c_t$ with $a_k \le n_G/4$ and $c_k < n_G/4$ such that there is k_0 with $a_{k_0} < n_G/4$ and the adjacency of $G^{(p)}$ is expressed as follows:

- $i \sim_p i \pm 2a_k$ $(k = 1, \ldots, s)$ for every i;
- $i \sim_p i + 4c_\ell + 1$ and $i \not\sim_p i - 4c_\ell - 1$ for $\ell = 1, \ldots, t$ when $i \equiv 0, 2 \ (\mathrm{mod}\ 4)$;
- $i \sim_p i - 4c_\ell - 1$ and $i \not\sim_p i + 4c_\ell + 1$ for $\ell = 1, \ldots, t$ when $i \equiv 1, 3 \ (\mathrm{mod}\ 4)$.

Thus we find $d^{(p)} \ge 3$. Since a_{k_0} is odd, by setting $a_{k_0} = 2b + 1$, we have $i \sim_p i \pm (4b + 2)$ and $0 \le b < n_G/8$.

To show that the adjacency operators are not commutative, we study

$(1,1)$-primitive bicolored paths on G and on G^* of origin 0. On G they are
$$(0,1,1 \pm 2a_k),\ (0,-1,-1 \pm 2a_k),\ (0,1,-4c_\ell),\ (0,-1,-4c_\ell - 2),$$
and on G^* they are
$$(0, \pm 2a_k, \pm 2a_k+1),\ (0, \pm 2a_k, \pm 2a_k-1),\ (0, 4c_\ell+1, 4c_\ell+2),\ (0, 4c_\ell+1, 4c_\ell).$$
Since $n_G \equiv 0 \pmod 4$ and $0 < c_\ell < n_G/4$, we find that the terminuses of $(1,1)$-primitive bicolored paths in G and G^* do not coincide. Hence the adjacency operators are not commutative. This completes the proof. $\square$

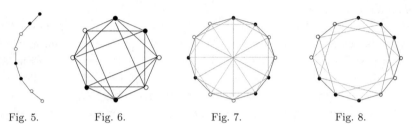

Fig. 5.   Fig. 6.   Fig. 7.   Fig. 8.

Remark 6.1. The assertions in Lemmas 6.1, 6.2 hold if we change the roles of the principal and auxiliary graphs.

Corollary 6.1. *Let G be a finite vertex-transitive Kähler graph. If $G^{(p)}$ and $G^{(a)}$ are bipartite circuits, then their vertex-decompositions coincide.*

Corollary 6.2. *Let G be a finite vertex-transitive Kähler graph having commutative adjacency operators. If $G^{(p)}$ and $G^{(a)}$ are connected bipartite graphs and if at least one of them is a circuit, then their vertex-decompositions coincide.*

Principal and auxiliary graphs of the Kähler graphs in Figs. 6, 7, 8 are bipartite graph having different vertex-decompositions. Here edges on circumferences are auxiliary and interior edges are principal. The adjacency operators of the graph in Fig. 6 are not commutative. Principal graphs in Figs. 7, 8 have 2 and 3 components, respectively. Fig. 8 shows that the assertion of Lemma 6.1 does not hold when $G^{(p)}$ is not connected.

We here study a bit more on Kähler graphs whose principal and auxiliary bipartite graphs have different vertex-decompositions.

Lemma 6.3. *Let $G = (V, E^{(p)} + E^{(a)})$ be a finite vertex-transitive Kähler graph. We pose the same assumption as in Lemma 6.1 except that $G^{(p)}$ is connected. Also, we suppose that the adjacency operators are commutative. Under the same representation of $G^{(a)}$ as in Lemma 6.2, there is an integer α with $2 \le \alpha < n_G/2$ satisfying that the components of $G^{(p)}$ is formed by*

$W_k \cup W_{\alpha+k}$ $(k = 0, \ldots, \alpha-1)$ *and the vertex-decomposition of* $G^{(p)}$ *is given by* $\left(\bigcup_{k=0}^{\alpha-1} W_k\right) + \left(\bigcup_{k=\alpha}^{2\alpha-1} W_k\right)$, *where* $W_k = \{i \in V \mid i \equiv k \ (\mathrm{mod}\ 2\alpha)\}$.

Proof. We use the same notations of isomorphisms as in the proof of Lemma 6.1. We suppose that we have s satisfying $2 \le s \le n_G-1$, $s \ne n_G/2$ and $0 \sim_p s$, $0 \not\sim_p -s$. Since the isomorphism which maps 0 to s is $\psi_{0,s}$, we have i_0 $(1 \le i_0 \le s)$ such that the triplet of isomorphisms which maps 0 to i_0-1, i_0, i_0+1 is either $(\varphi_{0,i_0-1}, \psi_{0.i_0}, \varphi_{0,i_0+1})$ or $(\varphi_{0,i_0-1}, \psi_{0,i_0}, \psi_{0,i_0+1})$. Here, we regard the identity as $\varphi_{0,0}$. Considering $(1,1)$-primitive bicolored paths of origin i_0 on G and those on G^*, we find a contradiction in the following manner. Let $u_1, \ldots, u_K$ $(2 \le u_1 < \cdots < u_K \le n_G-2)$ be vertices with $0 \sim_p u_k$. Hence we have $s = u_{k_0}$ with some k_0. When the triplet of isomorphisms is the former, the terminuses of $(1,1)$-primitive bicolored paths of origin i_0 on G are $\{i_0+n_G-u_k \pm 1 \mid k = 1, \ldots, K\}$, and those on G^* are $\{i_0+u_k \pm 1 \mid k = 1, \ldots, K\}$. As the adjacency operators are commutative, $U = \{n_G-u_k \pm 1 \mid k = 1, \ldots, K\}$ and $U^* = \{u_k \pm 1 \mid k = 1, \ldots, K\}$ coincide with each other by taking acount of their multiplicities. Considering the smallest numbers among U and among U^*, we find $n_G - u_1 - 1 = u_1 - 1$, which is equivalent to $u_1 = n_G - u_1$. We next study $U \setminus \{n_G-u_1 \pm 1\}$ and $U^* \setminus \{u_1 \pm 1\}$. Considering the smallest numbers we have $n_G-u_2 = u_2$. Continuing this argument we come to find $0 \sim_p -s$, which is a contradiction. When the triplet is the latter, the terminuses of $(1,1)$-primitive bicolored paths of origin i_0 on G^* are $\{i_0+u_k-1, i_0+n_G-u_k+1 \mid k = 1, \ldots, K\}$. By the same argument, we have a contradiction. Hence we find that $i \sim_p i+s$ if and only if $i \sim_p i-s$. By Lemma 6.2 we have $G^{(p)}$ is not connected, hence if $i \sim_p i+s$ then s and n_G are not relatively prime. Moreover, as $G^{(p)}$ is bipartite, $\ell s \not\equiv 0 \ (\mathrm{mod}\ n_G)$ for every odd integer ℓ. Therefore, there are such integers $s_1, \ldots, s_t$ with $2 \le s_k \le n_G/2$ satisfying that two vertices i, j are adjacent in $G^{(p)}$ if and only if $j = i \pm s_k$ with some k. Hence if we put α the greatest common devisor of $s_1, \ldots, s_t$ and n_G, then $G^{(p)}$ consists of α connected components $W_0 \cup W_\alpha, W_1 \cup W_{\alpha+1}, \ldots, W_{\alpha-1} \cup W_{2\alpha-1}$. Since $G^{(p)}$ is bipartite, we see n_G/α is even. Considering α-step graph $G_\alpha^{(a)}$ of $G^{(a)}$, we have a Kähler graph G_k satisfying the following:

- the set of vertices is $W_k \cup W_{\alpha+k}$,
- its auxiliary graph is a circuit (i.e. $i \sim_a j$ if and only if $|i-j| = \alpha$),
- its principal graph is connected and bipartite,
- its adjacency operators are commutative.

Hence, by Corollary 6.2, we find that the common vertex-decomposition of its principal and auxiliary graphs is $W_k + W_{\alpha+k}$. Thus, we get the conclusion. $\qquad \square$

We study some special derived graphs for Kähler graphs whose principal and auxiliary graphs are bipartite and have different vertex-decompositions.

Proposition 6.1. *Let $G = (V, E^{(p)} + E^{(a)})$ be a vertex-transitive Kähler graph having commutative adjacency operators. We suppose that $G^{(p)}$ and $G^{(a)}$ are bipartite graphs whose vertex-decompositions do not coincide and that $G^{(a)}$ is a circuit. We moreover suppose that $G^{(p)}$ consists of α connected components.*

(1) *When $q \equiv 0 \pmod{2\alpha}$, or when α is odd, $q \equiv \alpha \pmod{2\alpha}$ and p is even, the graph $G_{p,q}$ is bipartite having the same vertex-decomposition as of $G^{(p)}$, and has α components corresponding to those of $G^{(p)}$.*
(2) *When α is odd, $q \equiv \alpha \pmod{2\alpha}$ and p is odd, or when α is even and $q \equiv \alpha \pmod{2\alpha}$, the graph $G_{p,q}$ is not bipartite, and has 2α connected components corresponding to the vertex-decompositions and components of $G^{(p)}$.*

Proof. We represent as $V = \{0, 1, \ldots, n_G - 1\}$ and $i \sim_a i \pm 1$ for every $i \in V$. According to the proof of Lemma 6.3, when we consider $\ell\alpha$-step paths in $G^{(a)}$ with $\ell = 1, 2, \ldots$, we may consider Kähler graphs G_k ($k = 0, \ldots, \alpha-1$) whose adjacency operators are commutative. Here, we use the same notations as in the proof of Lemma 6.3. Considering q-step auxiliary paths on G is equivalent to considering q/α-step auxiliary paths on G_k. In the case (1), we have $p+(q/\alpha)$ is odd because p, q are relatively prime, and in the case of (2) we have $p+(q/\alpha)$ is even. Since the auxiliary graph of G_k is a circuit, by applying Theorems 4.2, 4.4 to G_k, we get the conclusion.   □

Example 6.1. The regular Kähler graph G given in Figs. 9, 10 has $n_G = 8$, $d^{(p)} = d^{(a)} = 3$ and has commutative adjacency operators. Its $G^{(p)}$ and $G^{(a)}$ are connected non-bipartite graphs and are isomorphic. The eigenvalues of $\mathcal{A}^{(p)}, \mathcal{A}^{(a)}$ and $\mathcal{A}_{p,q}$ with some (p, q) are given as follows:

$$\mathrm{Ev}(\mathcal{A}^{(p)}) = \{3, \ \sqrt{5}, \ 1, -1, -1, \ -1, -1, \ -\sqrt{5}\},$$
$$\mathrm{Ev}(\mathcal{A}^{(a)}) = \{3, -\sqrt{5}, \ -1, \ 1, -1, -1, -1, \ \sqrt{5}\},$$
$$\mathrm{Ev}(\mathcal{A}_{1,1}) = \Big\{3, \ -\frac{5}{3}, -\frac{1}{3}, -\frac{1}{3}, \ \frac{1}{3}, \ \frac{1}{3}, \ \frac{1}{3}, \ -\frac{5}{3}\Big\},$$
$$\mathrm{Ev}(\mathcal{A}_{2,1}) = \Big\{6, -\frac{2\sqrt{5}}{3}, \frac{2}{3}, -\frac{2}{3}, \frac{2}{3}, \ \frac{2}{3}, \ \frac{2}{3}, \frac{2\sqrt{5}}{3}\Big\}.$$

Fig. 9. $G^{(p)}$ Fig. 10. $G^{(a)}$

Here, eigenvalues are listed according to the order of common eigenfunctions.

Example 6.2. The regular Kähler graph G given in Figs. 11, 12 has $n_G = 8$, $d^{(p)} = 4$, $d^{(a)} = 3$ and has commutative adjacency operators. Its $G^{(p)}$ is a connected non-bipartite graph and its $G^{(a)}$ is a connected bipartite graph. The eigenvalues of $\mathcal{A}^{(p)}, \mathcal{A}^{(a)}$ and $\mathcal{A}_{p,q}$ with some (p,q) are as follows:

$$\mathrm{Ev}(\mathcal{A}^{(p)}) = \{4, \quad 2, \quad 0, \quad 0, \quad 0, -2, \quad -2, \quad -2\},$$

$$\mathrm{Ev}(\mathcal{A}^{(a)}) = \{3, -3, -1, -1, -1, \quad 1, \quad 1, \quad 1\},$$

$$\mathrm{Ev}(\mathcal{A}_{1,1}) = \left\{4, -2, \quad 0, \quad 0, \quad 0, -\frac{2}{3}, -\frac{2}{3}, -\frac{2}{3}\right\},$$

$$\mathrm{Ev}(\mathcal{A}_{1,2}) = \left\{4, \quad 2, \quad 0, \quad 0, \quad 0, \quad \frac{2}{3}, \quad \frac{2}{3}, \quad \frac{2}{3}\right\},$$

$$\mathrm{Ev}(\mathcal{A}_{1,4}) = \{4, \quad 2, \quad 0, \quad 0, \quad 0, \quad 0, \quad 0, \quad 0\},$$

$$\mathrm{Ev}(\mathcal{A}_{2,1}) = \left\{12, \quad 0, \quad \frac{4}{3}, \quad \frac{4}{3}, \quad \frac{4}{3}, \quad 0, \quad 0, \quad 0\right\}.$$

Fig. 11. $G^{(p)}$ Fig. 12. $G^{(a)}$

Example 6.3. The regular Kähler graph G given in Fig. 13 has $n_G = 8$, $d^{(p)} = 3$, $d^{(a)} = 2$ and has commutative adjacency operators. Here, edges on the circumference are auxiliary and interior edges are principal. Its $G^{(p)}$ is a connected non-bipartite graph and its $G^{(a)}$ is a bipartite circuit. The eigenvalues of $\mathcal{A}^{(p)}, \mathcal{A}^{(a)}$ and $\mathcal{A}_{p,q}$ with some (p,q) are given as follows:

$$\mathrm{Ev}(\mathcal{A}^{(p)}) = \{3, \quad 1, \quad 1, \quad \sqrt{2}-1, \sqrt{2}-1, \quad -1, -\sqrt{2}-1, \quad -\sqrt{2}-1\},$$

$$\mathrm{Ev}(\mathcal{A}^{(a)}) = \{2, \quad 0, \quad 0, \quad -\sqrt{2}, \quad -\sqrt{2}, \quad -2, \quad \sqrt{2}, \quad \sqrt{2}\},$$

$$\mathrm{Ev}(\mathcal{A}_{1,1}) = \left\{3, \quad 0, \quad 0, \frac{\sqrt{2}-2}{3}, \frac{\sqrt{2}-2}{3}, \frac{2}{3}, -\frac{2+\sqrt{2}}{3}, -\frac{2+\sqrt{2}}{3}\right\},$$

$$\mathrm{Ev}(\mathcal{A}_{1,2}) = \left\{3, -\frac{2}{3}, -\frac{2}{3}, \quad 0, \quad 0, \quad -\frac{2}{3}, \quad 0, \quad 0\right\}.$$

Fig. 13.

Example 6.4. The regular Kähler graph G given in Figs. 14, 15 has $n_G = 8$, $d^{(p)} = d^{(a)} = 3$ and has commutative adjacency operators. Its $G^{(p)}$ and $G^{(a)}$ are connected bipartite graphs having different vertex-decompositions. They are isomorphic. The eigenvalues of $\mathcal{A}^{(p)}, \mathcal{A}^{(a)}$ and $\mathcal{A}_{p,q}$ with some (p,q) are given as follows:

$$\mathrm{Ev}(\mathcal{A}^{(p)}) = \{3, \quad 1, \quad 1, \quad 1, -1, \quad -1, \quad -1, \quad -3\},$$

$$\mathrm{Ev}(\mathcal{A}^{(a)}) = \{3, \quad -1, \quad -1, \quad -3, \quad 1, \quad 1, \quad 1, \quad 1\},$$

$$\mathrm{Ev}(\mathcal{A}_{1,1}) = \left\{3, -\frac{1}{3}, -\frac{1}{3}, -1, \quad \frac{1}{3}, -\frac{1}{3}, -\frac{1}{3}, -1\right\},$$

$$\mathrm{Ev}(\mathcal{A}_{2,1}) = \left\{6, -\frac{2}{3}, -\frac{2}{3}, \quad 2, \quad \frac{2}{3}, \quad \frac{2}{3}, \quad \frac{2}{3}, \quad 2\right\},$$

$$\mathrm{Ev}(\mathcal{A}_{4,1}) = \{24, \quad 0, \quad 0, \quad 0, \quad 0, \quad 0, \quad 0, \quad 8\}.$$

Fig. 14. $G^{(p)}$ Fig. 15. $G^{(a)}$

Example 6.5. The Kähler graph G given in Fig. 16 has $n_G = 5, d^{(p)} = d^{(a)} = 2$ and has commutative adjacency operators. Here, edges on the circumference are principal and interior edges are auxiliary. Its $G^{(p)}$ and $G^{(a)}$ are isomorphic circuits. The eigenvalues of $\mathcal{A}^{(p)}, \mathcal{A}^{(a)}$ and $\mathcal{A}_{p,q}$ with some (p, q) are as follows:

$$\mathrm{Ev}(\mathcal{A}^{(p)}) = \left\{ 2, \ \frac{\sqrt{5}-1}{2}, \ \frac{\sqrt{5}-1}{2}, -\frac{\sqrt{5}+1}{2}, -\frac{\sqrt{5}+1}{2} \right\},$$

$$\mathrm{Ev}(\mathcal{A}^{(a)}) = \left\{ 2, -\frac{\sqrt{5}+1}{2}, -\frac{\sqrt{5}+1}{2}, \ \frac{\sqrt{5}-1}{2}, \ \frac{\sqrt{5}-1}{2} \right\} = \mathrm{Ev}(\mathcal{A}_{5,1}),$$

$$\mathrm{Ev}(\mathcal{A}_{1,1}) = \mathrm{Ev}(\mathcal{A}_{4,1}) = \left\{ 2, \ -\frac{1}{2}, \ -\frac{1}{2}, \ -\frac{1}{2}, \ -\frac{1}{2} \right\},$$

$$\mathrm{Ev}(\mathcal{A}_{2,1}) = \mathrm{Ev}(\mathcal{A}_{3,1}) = \left\{ 2, \frac{3-\sqrt{5}}{4}, \frac{3-\sqrt{5}}{4}, \frac{3+\sqrt{5}}{4}, \frac{3+\sqrt{5}}{4} \right\}.$$

Fig. 16.

Example 6.6. The regular Kähler graph G given in Figs. 17, 18 has $n_G = 12, d^{(p)} = d^{(a)} = 3$ and has commutative adjacency operators. Its $G^{(p)}$ and its $G^{(a)}$ are connected bipartite graphs having the same vertex-decompositions. They are isomorphic. The eigenvalues of $\mathcal{A}^{(p)}, \mathcal{A}^{(a)}$ and $\mathcal{A}_{p,q}$ with some (p, q) are given as follows:

$$\mathrm{Ev}(\mathcal{A}^{(p)}) = \left\{ 3, \ \frac{\sqrt{17}+1}{2}, \ \frac{\sqrt{17}-1}{2}, 0,0,0,0,0,0, \ \frac{1-\sqrt{17}}{2}, -\frac{\sqrt{17}+1}{2}, -3 \right\},$$

$$\mathrm{Ev}(\mathcal{A}^{(a)}) = \left\{ 3, \ -\frac{\sqrt{17}+1}{2}, \ \frac{1-\sqrt{17}}{2}, 0,0,0,0,0,0, \ \frac{\sqrt{17}-1}{2}, \ \frac{\sqrt{17}+1}{2}, -3 \right\},$$

$$\mathrm{Ev}(\mathcal{A}_{1,1}) = \left\{ 3, \ -\frac{\sqrt{17}+9}{6}, \ \frac{9-\sqrt{17}}{6}, 0,0,0,0,0,0, \ \frac{9-\sqrt{17}}{6}, -\frac{\sqrt{17}+9}{6}, \ 3 \right\},$$

$$\mathrm{Ev}(\mathcal{A}_{2,1}) = \left\{ 6, \ -\frac{\sqrt{17}+5}{3}, \ \frac{5-\sqrt{17}}{3}, 0,0,0,0,0,0, \ \frac{\sqrt{17}-5}{3}, \ \frac{\sqrt{17}+5}{3}, -6 \right\},$$

$$\mathrm{Ev}(\mathcal{A}_{3,1}) = \left\{ 12, -\frac{2\sqrt{17}+2}{3}, \ \frac{2\sqrt{17}-2}{3}, 0,0,0,0,0,0, \ \frac{2\sqrt{17}-2}{3}, -\frac{2\sqrt{17}+2}{3}, \ 12 \right\},$$

$$\mathrm{Ev}(\mathcal{A}_{3,2}) = \left\{ 12, \ \frac{\sqrt{17}+3}{3}, \ \frac{\sqrt{17}-3}{3}, 0,0,0,0,0,0, \ \frac{3-\sqrt{17}}{3}, -\frac{\sqrt{17}+3}{3}, -12 \right\}.$$

Fig. 17. $G^{(p)}$ Fig. 18. $G^{(a)}$ Fig. 19.   Fig. 20.

Example 6.7. The Kähler graph G given in Fig. 19 has $n_G = 12, d^{(p)} = 4, d^{(a)} = 2$, and has commutative adjacency operators. Here edges on the circumference are auxiliary and interior edges are principal. Its $G^{(a)}$ is a circuit and its $G^{(p)}$ is bipartite connected graphs. They have the same vertex-decompositions. Eigenvalues of $\mathcal{A}^{(p)}, \mathcal{A}^{(a)}$ and $\mathcal{A}_{p,q}$ with some (p,q) are as follows:

$$\mathrm{Ev}(\mathcal{A}^{(p)}) = \left\{\, 4, \quad \sqrt{3}, \quad \sqrt{3}, \quad 1, \quad 1, 0, 0, -1, -1, -\sqrt{3}, \ -\sqrt{3}, \ -4\right\},$$

$$\mathrm{Ev}(\mathcal{A}^{(a)}) = \left\{\, 2, \quad -\sqrt{3}, -\sqrt{3}, -1, \ -1, 0, 0, \quad 1, \quad 1, \quad \sqrt{3}, \quad \sqrt{3}, \ -2\right\},$$

$$\mathrm{Ev}(\mathcal{A}_{1,1}) = \left\{\, 4, \quad -\frac{3}{2}, \quad -\frac{3}{2}, \ -1, \ -1, 0, 0, -1, -1, \quad -\frac{3}{2}, \quad -\frac{3}{2}, \quad 4\right\},$$

$$\mathrm{Ev}(\mathcal{A}_{1,2}) = \left\{\, 4, \quad \frac{\sqrt{3}}{2}, \quad \frac{\sqrt{3}}{2}, \ -\frac{1}{2}, \ -\frac{1}{2}, 0, 0, \quad 1, \quad 1, -\frac{\sqrt{3}}{2}, \ -\frac{\sqrt{3}}{2}, \ -4\right\},$$

$$\mathrm{Ev}(\mathcal{A}_{3,2}) = \{36, -2\sqrt{3}, -2\sqrt{3}, \quad 3, \quad 3, 0, 0, -3, -3, \ 2\sqrt{3}, \ 2\sqrt{3}, -36\}.$$

Example 6.8. The regular Kähler graph G given in Fig. 20 has $n_G = 12$, $d^{(p)} = d^{(a)} = 2$ and has commutative adjacency operators, where edges on the circumference are principal and interior edges are auxiliary. Its $G^{(p)}$ and its $G^{(a)}$ are bipartite circuits having the same vertex-decompositions. The eigenvalues of $\mathcal{A}^{(p)}, \mathcal{A}^{(a)}$ and $\mathcal{A}_{p,q}$ with some (p,q) are given as follows:

$$\mathrm{Ev}(\mathcal{A}^{(p)}) = \left\{2, \quad \sqrt{3}, \quad \sqrt{3}, 1, \quad 1, 0, 0, -1, -1, -\sqrt{3}, -\sqrt{3}, -2\right\},$$

$$\mathrm{Ev}(\mathcal{A}^{(a)}) = \left\{3, -\sqrt{3}, \ -\sqrt{3}, 1, \quad 1, 0, 0, -1, -1, \ \sqrt{3}, \ \sqrt{3}, -2\right\},$$

$$\mathrm{Ev}(\mathcal{A}_{1,1}) = \left\{2, \ -\frac{3}{2}, \quad -\frac{3}{2}, \frac{1}{2}, \ \frac{1}{2}, 0, 0, \quad \frac{1}{2}, \ \frac{1}{2}, -\frac{3}{2}, \ -\frac{3}{2}, \quad 2\right\},$$

$$\mathrm{Ev}(\mathcal{A}_{2,1}) = \left\{2, -\frac{\sqrt{3}}{2}, \ -\frac{\sqrt{3}}{2}, \frac{1}{2}, \ \frac{1}{2}, 0, 0, -\frac{1}{2}, -\frac{1}{2}, \frac{\sqrt{3}}{2}, \ \frac{\sqrt{3}}{2}, -2\right\},$$

$$\mathrm{Ev}(\mathcal{A}_{3,1}) = \left\{2, \quad 0, \quad 0, \ -\frac{1}{2}, -\frac{1}{2}, 0, 0, -\frac{1}{2}, -\frac{1}{2}, \quad 0, \quad 0, \quad 2\right\}.$$

Example 6.9. The Kähler graph G given in Fig. 21 has $n_G = 12$, $d^{(p)} = 3$, $d^{(a)} = 2$, and has commutative adjacency operators. Here edges on the circumference are auxiliary and interior edges are principal. Its $G^{(a)}$ is a circuit and its $G^{(p)}$ is bipartite graphs having 2 connected components. Their vertex-decompositions are different. The eigenvalues of $\mathcal{A}^{(p)}, \mathcal{A}^{(a)}$ and $\mathcal{A}_{p,q}$ with some (p,q) are given as follows:

$$\mathrm{Ev}(\mathcal{A}^{(p)}) = \{3, \quad 3, \quad 0, \quad 0, 0, 0, \quad 0, \quad 0, \quad 0, \quad 0, -3, \ -3\},$$

$$\mathrm{Ev}(\mathcal{A}^{(a)}) = \{2, -2, \sqrt{3}, \sqrt{3}, 1, 1, -1, -1, -\sqrt{3}, -\sqrt{3}, \ 0, \quad 0\},$$

$$\mathrm{Ev}(\mathcal{A}_{1,2}) = \{3, \quad 3, \quad 0, \quad 0, 0, 0, \quad 0, \quad 0, \quad 0, \quad 0, \ 3, \quad 3\},$$

$$\mathrm{Ev}(\mathcal{A}_{1,4}) = \{3, \quad 3, \quad 0, \quad 0, 0, 0, \quad 0, \quad 0, \quad 0, \quad 0, -3, \ -3\},$$

$$\mathrm{Ev}(\mathcal{A}_{3,2}) = \{12, 12, \quad 0, \quad 0, 0, 0, \quad 0, \quad 0, \quad 0, \quad 0, \ 12, \ 12\},$$

$$\mathrm{Ev}(\mathcal{A}_{3,4}) = \{12, 12, \quad 0, \quad 0, 0, 0, \quad 0, \quad 0, \quad 0, \quad 0, -12, -12\}.$$

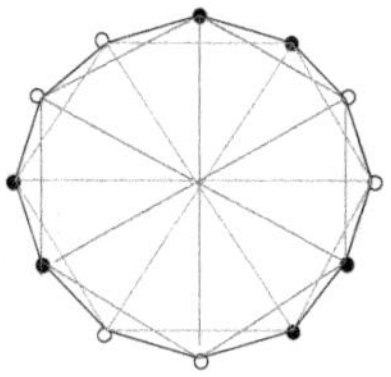

Fig. 21.

References

[1] T. Adachi, Kähler magnetic flows on a manifold of constant holomorphic sectional curvature, *Tokyo J. Math.* **18** (1995), 473–483.

[2] _____, Magnetic mean operators on a Kähler manifold, *Topics in Almost Hermitian Geometry and Related Fields*, Y. Matsushita *et al.* eds., World Scientific 30–40, 2005.

[3] _____, A discrete model for Kähler magnetic fields on a complex hyperbolic space, *Trends in Differential Geometry, Complex Analysis and Mathematical Physics*, K. Sekigawa, V.S. Gerdijikov and S. Dimiev eds., World Scientific, 1–9, 2009.

[4] _____, Laplacians for finite regular Kähler graphs and for their dual graphs, *Current Deveropments in Differential Geometry and its Related Fields*, T. Adachi, H. Hashimoto and M. Haristov eds., World Scientific, 23–44, 2015.

[5] T. Adachi and G. Chen, *Regular and vertex-transitive Kähler graphs having commutative principal and auxiliary adjacency operators*, preprint.

[6] T. Bao and T. Adachi, Characterizations of some homogeneous Hopf real hypersurfaces in a nonflat complex space form by extrinsic shapes of trajectories, *Diff. Geom. Appl.*, 48 (2016), 104–118.

[7] C.-L. Bejan and S.-L. Druta-Romanuic, *F*-geodesics on manifolds, *Filomat* 29 (2015), 2367–2379.

[8] D. Cvetković, P. Rowlinson and S. Simić, *An introduction to the theory of graph spectra*, London Math. Soc. Student Text 75, Cambridge Univ. Press 2010.

[9] S. Maeda and T. Adachi, Sasakian curves on hypersurfaces of type (A) in a complex space form, *Resuts Math.*, 56 (2009), 489–499.

[10] Yaermaimaiti T. and T. Adachi, Laplacians for derived graphs of regular Kähler graphs, *Math. Rep. Acad. Sci. Royal Soc. Canada*, 37 (2015), 141–156.

[11] T. Sunada, *Magnetic flows on a Riemann surface*, Proc. KAIST Math. Workshop 8 (1993), 93–108.

Received November 27, 2018
Revised February 18, 2019

SEMI-RIEMANNIAN D-GENERAL WARPING MANIFOLDS

60 years since the birth of Novica Blažić (1959–2005)

Cornelia-Livia BEJAN

Department of Mathematics, "Gh. Asachi" Technical University,
Bd. Carol I, no. 11, corp A, 700506 Iasi, Romania

Postal address: Seminarul Matematic, Universitatea "Alexandru Ioan Cuza",
Bd. Carol I, no. 11, 700506 Iasi, Romania
E-mail: bejanliv@yahoo.com
URL: http://math.etc.tuiasi.ro/bejan/

Sinem GÜLER

Department of Industrial Engineering, Istanbul Sabahattin Zaim University,
Halkali, Istanbul, Turkey
E-mail: sinem.guler@izu.edu.tr

We introduce here the notion of semi-Riemannian D-general warping ($M = M_1 \times M_2, g$) by extending two geometric notions, one defined by Blair and the other by Tanno. On these kinds of manifolds, we study the behavior of geodesics, magnetic curves and F-geodesics. Then on a semi-Riemannian D-general warping, we investigate some special vector fields, such as Killing, concircular, concurrent and recurrent.

Keywords: Geodesics; Killing vector fields; D-warping manifolds; concircular vector fields.

1. Introduction

The present paper deals with a relatively new idea, namely, that of warping a semi-Riemannian product metric in a way that uses additional structure on the second factor. We continue here the study we began in [4] on this topic. This concept derives from a special class of Riemannian structures, called D-homothetic warping metric, which provides several interesting examples, results and also relations with many other geometrical structures on manifolds ([5]).

For a Riemannian manifold (N, h, ξ) endowed with a unit vector field ξ, the behavior of the Riemann extension (which is a semi-Riemannian metric defined by Patterson and Walker on the cotangent bundle T^*N) was described in [4], under a D-homothetic deformation $h \to h^*$ of the metric [16]. Inspired by D-homothetic deformations, Blair introduced in [5]

D-warping $M_1 \times M_2$ of a Riemannian manifold M_1 with an almost contact metric manifold M_2. Then, by relaxing the condition of the almost contact metric structure on M_2, we extended in [4] the notion of the D-warping to the D-general warping $(M = M_1 \times M_2, g)$, to which we recall here some important features. By computing its Levi-Civita connection which induces on the cotangent bundle the Riemann extension, it shows to be related to the Riemann extensions of both components M_1 and M_2. By obtaining a Laplacian formula given on D-general warping, it happens that the harmonicity of smooth functions and forms with respect to g was related to the one with respect to each component (M_1, g_1), respectively (M_2, g_2). The topics of Einstein manifolds and their generalizations, such as quasi-Einstein and η-Einstein, have been intensively studied in literature. By computing the Riemann and Ricci curvatures, it follows some characterization results for Einsteinian conditions. An interesting topic is the study of gradient Ricci solitons and gradient η-Ricci solitons as well as the scalar and the sectional curvature of the D-general warping (M, g). Some necessary and sufficient conditions for flatness of D-general warping are useful. All these results described above on D-general warping were done in [4]. It is natural to continue here to study some more geometric objects on such D-general warping. Moreover, in what follows, we extend our study from Riemannian to the semi-Riemannian D-general warping and in particular Lorentzian.

In the next section, we obtain some characterizations of geodesics on semi-Riemannian D-general warping manifold. First, we give some necessary and sufficient conditions for an arbitrary smooth curve to be a geodesic on a semi-Riemannian D-general warping manifold. Then, we consider the notion of F-geodesics to extend this study on a semi-Riemannian D-general warping manifold endowed with an F-structure.

Some special vector fields on semi-Riemannian manifolds are frequently used in many papers (e.g. Killing, concircular [3, 8, 9, 17, 18], recurrent [14], concurrent [7]). We deal with these classes of vector fields in the last section and relate these type of vector fields on the semi-Riemannian D-general warping with the ones on each component.

We assume here all manifolds and structures to be smooth.

2. Geodesics on Semi-Riemannian D-General Warping Manifolds

One says that a manifold N carries an almost contact metric structure (φ, ξ, η, g) if it admits a structure tensor field φ of type $(1,1)$, a Reeb vector field ξ, and its dual 1-form η (with respect to the metric g) satisfying the following relations:

$$\varphi^2(X) = -X + \eta(X)\xi, \quad \eta(\xi) = 1,$$
$$g(\varphi X, \varphi Y) = g(X, Y) - \eta(X)\eta(Y) \tag{1}$$

for all vector fields $X, Y \in \mathfrak{X}(N)$. In [16], Tanno defined the D-homothetic deformation h^* of the metric h on an almost contact manifold $(N, \varphi, \xi, \eta, h)$ by

$$h \mapsto h^* = ah + a(a-1)\eta \otimes \eta, \tag{2}$$

for any $a > 0$, where D denotes the contact distribution of N. This notion was related to Betti numbers and harmonic forms in [15].

As was pointed out in Introduction, by extending the concept of D-homothetic deformations, Blair introduced in [5] the notion of D-warping metrics on the product $M_1 \times M_2$, whose components M_1 and M_2 are respectively Riemannian manifold and almost contact metric manifold. This notion was slightly generalized in our recent paper [4] to the notion of D-general warping manifolds. The next step is to go further, from the Riemannian to the semi-Riemannian case.

Definition 2.1. Let (M_j, g_j) $(j = 1, 2)$ be two semi-Riemannian manifolds. We suppose that M_2 carries a unitary vector field ξ_2, and denote by η_2 its dual form with respect to g_2. Then, we call $(M = M_1 \times M_2, g)$ a semi-Riemannian D-general warping manifold, where $D = \mathrm{Ker}\,\eta_2$ and the metric g is given by

$$g = g_1 + f g_2 + f(f-1)\eta_2 \otimes \eta_2, \tag{3}$$

with a positive function f on M_1.

To work in the semi-Riemannian context has the advantage of considering all three cases of Riemannian, Lorentzian and neutral [6]. In a forthcoming paper, we shall give more details on semi-Riemannian metric given in (3) in the context of last two cases.

Notation 2.1. From now on, (M_j, g_j) is a semi-Riemannian manifold of dimension m_j $(j = 1, 2)$ such that M_2 carries a unitary vector field ξ_2

(spacelike or timelike) and f is a positive function on M_1. We denote by η_2 the dual 1-form of ξ_2 and $\nabla^{(1)} f$ the gradient of f. Then $(M = M_1 \times M_2, g)$ denotes the semi-Riemannian D-general warping, where g is given by (3) and $D = \mathrm{Ker}\,\eta_2$. If not otherwise stated, any geometric objects with (resp. without) index j will be assumed to lie on M_j ($j = 1, 2$), (resp. on the product $M = M_1 \times M_2$).

By a straightforward computation, we obtain the following:

Lemma 2.1. *Let* $\nabla^{(j)}$ *be the corresponding Levi-Civita connection of* (M_j, g_j), $j = 1, 2$. *Then, the Levi-Civita connection of the metric* g *of the form* (3) *is given by*

$$\nabla_{X_1} Y_1 = \nabla^{(1)}_{X_1} Y_1, \quad \nabla_{X_1} Y_2 = \nabla_{Y_2} X_1 = \frac{X_1(f)}{2f}[Y_2 + \eta_2(Y_2)\xi_2],$$

$$\nabla_{X_2} Y_2 = \nabla^{(2)}_{X_2} Y_2 + \frac{(f-1)}{2f}[g_2(\nabla^{(2)}_{X_2}\xi_2, Y_2) + g_2(\nabla^{(2)}_{Y_2}\xi_2, X_2)]\xi_2 \quad (4)$$

$$+ f(f-1)[\eta_2(Y_2)(i_{X_2}d\eta_2)^\sharp + \eta_2(X_2)(i_{Y_2}d\eta_2)^\sharp]$$

$$- [g_2(X_2, Y_2) + (2f-1)\eta_2(X_2)\eta_2(Y_2)]\nabla^{(1)} f,$$

for all $X_j, Y_j \in \mathfrak{X}(M_j)$ ($j = 1, 2$), *where* i *denotes the interior product and* $\sharp$ *denotes the musical isomorphism with respect to* g_2.

The following statement will be used later on:

Lemma 2.2. *The distribution* $D = \mathrm{Ker}\,\eta_2$ *on* M_2 *is covariantly invariant with respect to* ξ_2 (*i.e.* $\nabla^{(2)}_{\xi_2} V \in D$ *for any* $V \in D$) *if and only if*

$$i_{\xi_2} d\eta_2 = 0. \quad (5)$$

In the sequel, to handle the very long calculation, we suppose the following condition.

Hypothesis. We assume that ξ_2 is parallel with respect to g_2, that is

$$\nabla^{(2)}_{X_2}\xi_2 = 0 \quad \text{for all } X_2 \in \mathfrak{X}(M_2). \quad (6)$$

Remark 2.1. Under our hypothesis, it follows that η_2 is closed (i.e. $d\eta_2 = 0$) and hence that the relation (5) holds obviously.

The Hypothesis simplifies the long computations, but even with this assumption, our study generalize the classical warped product structure. From Lemma 2.1, we obtain the following:

Lemma 2.3. *Under the Hypothesis, the Levi-Civita connection of the metric g of the form (3) reduces to the following:*

$$\nabla_{X_1}Y_1 = \nabla^{(1)}_{X_1}Y_1,$$

$$\nabla_{X_1}Y_2 = \nabla_{Y_2}X_1 = \frac{X_1(f)}{2f}[Y_2 + \eta_2(Y_2)\xi_2], \tag{7}$$

$$\nabla_{X_2}Y_2 = \nabla^{(2)}_{X_2}Y_2 - [g_2(X_2,Y_2) + (2f-1)\eta_2(X_2)\eta_2(Y_2)]\nabla^{(1)}f$$

for all $X_j, Y_j \in \mathfrak{X}(M_j)$ $(j = 1, 2)$.

The following notion was introduced in [2] in a more general context, namely on affine manifolds.

Definition 2.2. Let N be a semi-Riemannian manifold endowed with a tensor field F of type $(1,1)$. We say that a smooth curve $\gamma : I \to N$, defined on a real interval I, is an F-geodesic on N if $\gamma(u)$ satisfies:

$$\nabla_{\dot\gamma(u)}\dot\gamma(u) = F\big(\dot\gamma(u)\big), \tag{8}$$

where ∇ is the Levi-Civita connection on N.

Remark 2.2. In Definition 2.2, if F is identically zero, then γ becomes a geodesic curve. On the other side, there are several papers which study magnetic curves ([1]). The notion of magnetic curves can be obtained from Definition 2.2 by taking F to be the Lorentz force, which means that F is a divergence free $(1,1)$-tensor field. Hence, F-geodesic curves generalize both geodesic and magnetic curves. The notion of F-geodesic is close to the one of F-planar mapping (see [10, 12, 13]).

Theorem 2.1. *Let $(M = M_1 \times M_2, g)$ be a semi-Riemannian D-general warping and let $\gamma : I \to M$ be a smooth curve. Under the Hypothesis, if γ_1 and γ_2 denote the projections of γ on (M_1, g_1) and (M_2, g_2), respectively, then any two of the following conditions imply the remaining one:*

(i) γ *is a geodesic on (M, g);*
(ii) *both γ_1 and γ_2 of γ are geodesics on (M_1, g_1) and (M_2, g_2), respectively;*
(iii) *either the gradient vector field of f vanishes along γ_1, or γ_2 satisfies the relations*

$$g_2(\dot\gamma_2, \dot\gamma_2) = (1 - 2f)[\eta_2(\dot\gamma_2)]^2. \tag{9}$$

$$\dot\gamma_2 + \eta_2(\dot\gamma_2)\xi_2 = 0. \tag{10}$$

Proof. From Lemma 2.3, we obtain

$$\nabla_{\dot\gamma_1(u)+\dot\gamma_2(u)}\big(\dot\gamma_1(u)+\dot\gamma_2(u)\big)$$

$$= \nabla^{(1)}_{\dot\gamma_1(u)}\dot\gamma_1(u) - [g_2(\dot\gamma_2(u),\dot\gamma_2(u)) + (2f-1)\eta_2(\dot\gamma_2(u))^2]\nabla^{(1)}f \qquad (11)$$

$$+ \nabla^{(2)}_{\dot\gamma_2(u)}\dot\gamma_2(u) + \frac{\dot\gamma_1(u)f}{f}[\dot\gamma_2(u) + \eta_2(\dot\gamma_2(u))\xi_2].$$

From the relation (11), by separating the vectors on M_1 from those on M_2, we complete the proof. $\qquad\qquad\qquad\qquad\qquad\qquad\qquad\qquad\square$

Remark 2.3. If the gradient vector field of f vanishes along γ_1, then f is constant along γ_1.

Theorem 2.1 yields the following two consequences:

Corollary 2.1. *Let $(M = M_1 \times M_2, g)$ be a semi-Riemannian D-general warping and let $\gamma : I \to M$ be a smooth curve which lies in M_1. Under the Hypothesis, γ is a geodesic on (M, g) if and only if it is a geodesic on (M_1, g_1).*

Corollary 2.2. *Let $(M = M_1 \times M_2, g)$ be a semi-Riemannian D-general warping and let $\gamma : I \to M$ be a smooth curve which lies in M_2. We denote as $\gamma(t) = (p_1, \gamma_2(t))$ with $p_1 \in M_1$ and $\gamma_2 : I \to M_2$. Under the Hypothesis, γ is a geodesic on (M, g) if and only if γ_2 is a geodesic on (M_2, g_2) and either*

$$g_2(\dot\gamma_2, \dot\gamma_2) = \big(1 - 2f(p_1)\big)[\eta_2(\dot\gamma_2)]^2 \qquad (12)$$

holds or $(\nabla^{(1)}f)(p_1)$ vanishes.

Remark 2.4. Concerning the condition (12), we suppose $g_2(\dot\gamma_2, \dot\gamma_2) = C[\eta_2(\dot\gamma_2)]^2$ with a constant C. If $\dot\gamma_2$ is orthogonal to ξ_2, then γ_2 should be lightlike. Otherwise (that is, if $\dot\gamma_2$ is not orthogonal to ξ_2), $\dot\gamma_2$ is space-like, timelike or lightlike according as $C > 0$, $C < 0$ or $C = 0$.

Given two $(1,1)$-tensor fields F_1 and F_2 on M_1 and M_2, respectively, we define a $(1,1)$-tensor field F on a semi-Riemannian D-general warping $(M = M_1 \times M_2, g)$ by

$$F(X) = F_1(X_1) + F_2(X_2), \qquad (13)$$

for any vector field $X = X_1 + X_2 \in \mathfrak{X}(M)$, where $X_1 \in \mathfrak{X}(M_1)$ and $X_2 \in \mathfrak{X}(M_2)$ are decomposed components. By using the above notations, we can extend Theorem 2.1 as follows:

Theorem 2.2. *Let* $(M = M_1 \times M_2, F, g)$ *be a semi-Riemannian D-general warping endowed with an F-structure given by* (13) *and let* $\gamma : I \to M$ *be a smooth curve. We denote by* γ_1 *and* γ_2 *the projections of* γ *on* (M_1, F_1, g_1) *and* (M_2, F_2, g_2), *respectively. Under the Hypothesis, any two of the following conditions imply the remaining one:*

(i) γ *is an F-geodesic on* (M, g);
(ii) *both* γ_1 *and* γ_2 *of* γ *are F-geodesics on* (M_1, F_1, g_1) *and* (M_2, F_2, g_2), *respectively;*
(iii) *either the gradient vector field of* f *vanishes along* γ_1, *or* γ_2 *satisfies the relations* (9) *and* (10).

From Theorem 2.2, we obtain the following two consequences:

Corollary 2.3. *Let* $(M = M_1 \times M_2, F, g)$ *be a semi-Riemannian D-general warping endowed with an F-structure given by* (13) *and let* $\gamma : I \to M$ *be a smooth curve which lies in* M_1. *Under the Hypothesis,* γ *is an F-geodesic on* (M, F, g) *if and only if it is an F-geodesic on* (M_1, F_1, g_1).

Corollary 2.4. *Let* $(M = M_1 \times M_2, F, g)$ *be a semi-Riemannian D-general warping endowed with an F-structure given by* (13) *and let* $\gamma : I \to M$ *be a smooth curve which lies in* M_2. *We denote as* $\gamma(t) = (p_1, \gamma_2(t))$ *with* $p_1 \in M_1$ *and* $\gamma_2 : I \to M_2$. *Under the Hypothesis,* γ *is an F-geodesic on* (M, F, g) *if and only if* γ_2 *is an F-geodesic on* (M_2, F_2, g_2) *and either* (12) *holds or* $(\nabla^{(1)} f)(p_1)$ *vanishes.*

Remark 2.5. The statements of Theorem 2.2, Corollary 2.3 and Corollary 2.4 can be particularized to magnetic curves by taking the $(1, 1)$-tensor field F to be the Lorentz force.

3. Special Vector Fields on Semi-Riemannian D-General Warping Manifolds

Some special vector fields on manifolds are well-known for their infinitesimal flow. We recall here some of them.

Definition 3.1. Let ν be a vector field on a semi-Riemannian manifold (N, g) and let ∇ denote the Levi-Civita connection.

(a) The vector field ν is called concircular if it satisfies

$$\nabla_X \nu = \mu X \qquad \text{for all } X \in \mathfrak{X}(N), \tag{14}$$

where μ is a non-trivial function on N (see [9, 11]).

(b) In particular, when $\mu = 1$ in (14), then the vector field ν is called concurrent (see [7, 11]).

(c) The vector field ν is called recurrent if

$$\nabla \nu = \omega \otimes \nu, \tag{15}$$

with some closed 1-form ω (see [14]).

(d) In particular, when the 1-form ω in (15) is identically zero, then the vector field ν is called parallel.

(e) The vector field ν is called a Killing vector field if the Lie derivative of the metric tensor in the direction of ν vanishes identically.

We study such vector fields on semi-Riemannian D-general warpings.

Theorem 3.1. *Let* $(M = M_1 \times M_2, g)$ *be a semi-Riemannian D-general warping and let* $\nu = \nu_1 + \nu_2 \in \mathfrak{X}(M)$ *with the components* $\nu_j \in \mathfrak{X}(M_j)$ $(j = 1, 2)$. *Under the Hypothesis, the followings hold.*

1) *If* ν *is concircular, then both* ν_1 *and* ν_2 *are concircular and f is constant.*

2) *The vector fields* ν_1 *and* ν_2 *are concurrent and f is constant if and only if* ν *is concurrent.*

3) *If* ν *is recurrent, then both* ν_1 *and* ν_2 *are recurrent and f is constant.*

4) *The vector field* ν_1 *and* ν_2 *are parallel and f is constant if and only if* ν *is parallel.*

5) *If a vector field tangent to M_1 is Killing on (M, g), then it is Killing on (M_1, g_1). The converse holds when f is constant along each integral curve of the vector field.*

6) *The unit vector field* ξ_2 *from Notation 2.1 is Killing on (M, g) if and only if it is Killing on (M_2, g_2).*

Proof. We take an arbitrary vector field $X = X_1 + X_2 \in \mathfrak{X}(M)$ with components $X_j \in \mathfrak{X}(M_j)$ $(j = 1, 2)$. Then we have

$$\nabla_X \nu = \nabla_{X_1} \nu_1 + \nabla_{X_1} \nu_2 + \nabla_{X_2} \nu_1 + \nabla_{X_2} \nu_2$$

$$= \nabla_{X_1}^{(1)} \nu_1 + \frac{X_1(f)}{2f}[\nu_2 + \eta_2(\nu_2)\xi_2] + \frac{\nu_1(f)}{2f}[X_2 + \eta_2(X_2)\xi_2] \tag{16}$$

$$+ \nabla_{X_2}^{(2)} \nu_2 - [g_2(\nu_2, X_2) + (2f - 1)\eta_2(\nu_2)\eta_2(X_2)]\nabla^{(1)} f.$$

1) If we assume that ν is concircular, then $\nabla_X \nu = \mu(X_1 + X_2)$ for any $X = X_1 + X_2 \in \mathfrak{X}(M)$, where μ is a non-trivial function on M. Hence we

obtain the following relations for any $X_j \in \mathfrak{X}(M_j)$ $(j = 1, 2)$:

$$\mu X_1 = \nabla^{(1)}_{X_1} \nu_1 - \left[g_2(\nu_2, X_2) + (2f - 1)\eta_2(\nu_2)\eta_2(X_2) \right] \nabla^{(1)} f, \qquad (17)$$

$$\mu X_2 = \nabla^{(2)}_{X_2} \nu_2 + \frac{X_1(f)}{2f} \left[\nu_2 + \eta_2(\nu_2)\xi_2 \right] + \frac{\nu_1(f)}{2f} \left[X_2 + \eta_2(X_2)\xi_2 \right]. \qquad (18)$$

If we take $X_2 = 0$ in (17), then we obtain that ν_1 is concircular. Hence we get

$$\left[g_2(\nu_2, X_2) + (2f - 1)\eta_2(\nu_2)\eta_2(X_2) \right] \nabla^{(1)} f = 0, \qquad (19)$$

for all $X_2 \in \mathfrak{X}(M_2)$. This yields that f is constant. From (18), it follows that ν_2 is concircular.

2) The direct part of the statement follows as a particular case of the statement 1) by taking $\mu = 1$. The converse part is straightforward from the relation (16).

3) The statement follows from an argument similar to that we used in the proof of the statement 1).

4) If the left hand side of (16) is zero for all $X \in \mathfrak{X}(M)$, then we obtain for any $X_j \in \mathfrak{X}(M_j)$ $(j = 1, 2)$ that the following relations hold:

$$\nabla^{(1)}_{X_1} \nu_1 - \left[g_2(\nu_2, X_2) + (2f - 1)\eta_2(\nu_2)\eta_2(X_2) \right] \nabla^{(1)} f = 0, \qquad (20)$$

$$\nabla^{(2)}_{X_2} \nu_2 + \frac{X_1(f)}{2f} \left[\nu_2 + \eta_2(\nu_2)\xi_2 \right] + \frac{\nu_1(f)}{2f} \left[X_2 + \eta_2(X_2)\xi_2 \right] = 0. \qquad (21)$$

If we take $X_2 = 0$ in (20), it follows that ν_1 is parallel. Hence we obtain the relation (19), which yields that f is constant. From (21), it follows that ν_2 is parallel. The converse part of the statement 4) follows straightforward from the relation (16).

5) From (3), the Lie derivative satisfies

$$\begin{aligned} \mathcal{L}_\nu g = \mathcal{L}_\nu g_1 + (\nu f)g_2 + f\mathcal{L}_\nu g_2 + (\nu f(f - 1))\eta_2 \otimes \eta_2 \\ + f(f - 1)\left[(\mathcal{L}_\nu \eta_2) \otimes \eta_2 + \eta_2 \otimes (\mathcal{L}_\nu \eta_2) \right]. \end{aligned} \qquad (22)$$

If we take $\nu = \nu_1$ in (22), then we obtain the relation

$$\mathcal{L}_{\nu_1} g = \mathcal{L}_{\nu_1} g_1 + (\nu_1 f)g_2 + (\nu_1 f(f - 1))\eta_2 \otimes \eta_2. \qquad (23)$$

If $\nu = \nu_1$ is Killing on (M, g), as the left hand side of (23) vanishes identically, we find that ν_1 is Killing on (M_1, g_1). The converse part is clear because both $\nu_1 f$ and $\nu_1 f(f - 1)$ vanish identically.

6) By our Hypothesis we have the relation (5) which yields

$$\mathcal{L}_{\xi_2} \eta_2 = 0. \qquad (24)$$

From (3) and (24), we obtain $\mathcal{L}_{\xi_2} g = f \mathcal{L}_{\xi_2} \eta_2$. This completes the proof.   □

Corollary 3.1. *Under the Hypothesis, if the semi-Riemannian D-general warping* $(M = M_1 \times M_2, g)$ *admits a concircular vector field, then the second component of the metric g from* (3) *reduces to a D-homothetic deformation on* M_2.

Acknowledgments

We thank Professor T. Adachi for organizing the conference and the referee for useful suggestions.

References

[1] T. Adachi, Kähler magnetic fields on a Kähler manifold of negative curvature, *Differential Geom. Appl.* **29** (2011), S2–S8.

[2] C.-L. Bejan and S.-L. Druţă-Romaniuc, *F*-geodesics on manifolds, *Filomat* **29**, 10 (2015), 2367–2379.

[3] C.-L. Bejan and S. Güler, Kähler manifolds of quasi-constant holomorphic sectional curvature and generalized Sasakian space forms, *RACSAM Rev. R. Acad. Cienc. Exactas Fis. Nat. Ser. A Mat.* doi:10.1007/s13398-018-0533-9 (2018).

[4] C.-L. Bejan and S. Güler, Laplace, Einstein and related equations on D-general warping, *Mediterr. J. Math.* **16**, 1 (2019).

[5] D. E. Blair, *D*-homothetic warping, *Publ. Inst. Math. (Beograd)* **94**, 108 (2013), 47–54.

[6] N. Blažić, N. Bokan and Z. Rakić, A Note on the Osserman conjecture and isotropic covariant derivative of curvature, *Proc. Amer. Math. Soc.* **128**, 1 (1999), 245–253.

[7] F. Brickell and K. Yano, Concurrent vector fields and Minkowski structures, *Kodai Math. Sem. Rep.* **26** (1974), 22–28.

[8] B.-Y. Chen, Some results on concircular vector fields and their applications to Ricci solitons, *Bull. Korean Math. Soc.* **52**, 5 (2015), 1535–1547.

[9] A. Fialkow, Conformals geodesics, *Trans. Amer. Math. Soc.* **45**, 3 (1939), 443–473.

[10] I. Hinterleitner and J. Mikes, On *F*-planar mappings of spaces with equiaffine connections, *Note Mat.* **27**, 1 (2007), 111–118.

[11] I. Mihai, R. Rosca and L. Verstraelen, Some aspects of the differential geometry of vector Fields, On skew symmetric Killing and

conformal vector fields, and their relations to various geometrical structures, *Centre for Pure Appl. Diff. Geom. (PADGE)* 2, K.U. Leuven, K.U. Brussel, 1996.

[12] J. Mikes and N. S. Sinyukov, On quasiplanar mappings of spaces of affine connection. (English translation) *Sov. Math.* **27**, 1 (1983), 63–70.

[13] J. Mikes and A. Vanzurova and I. Hinterleitner, *Geodesic mappings and some generalizations*, Olomouc: Palacky University Press, 2009.

[14] J. A. Schouten, *Ricci calculus, an introduction to tensor analysis and its geometrical applications*, Springer-Verlag, Berlin-Gottingen-Heidelberg, 1954.

[15] S. Tanno, Harmonic forms and Betti numbers of certain contact Riemannian manifolds, *J. Math. Soc. Japan* **19**, 3 (1967), 308–316.

[16] S. Tanno, The topology of contact Riemannian manifolds, *Illinois J. Math.* **12**, 4 (1968), 700–717.

[17] Y. Tashiro, Complete Riemannian manifolds and some vector fields, *Trans. Amer. Math. Soc.* **117** (1965), 251–275

[18] K. Yano, Concircular geometry. I, II, III, IV, V, *Proc. Imp. Acad. Tokyo* **16** (1940), **18** (1942), 195–511.

Received February 10, 2019
Revised   February 22, 2019

SYMMETRIC TRIAD WITH MULTIPLICITIES AND GENERALIZED DUALITY WITH APPLICATIONS TO LEUNG'S CLASSIFICATION THEOREMS

Osamu IKAWA*

*Department of Mathematics and Physical Sciences,
Faculty of Arts and Sciences, Kyoto Institute of Technology,
Matsugasaki, Sakyo-ku, Kyoto 606-8585, Japan
E-mail: ikawa@kit.ac.jp*

We introduce the notion of a duality between compact symmetric triads and semisimple pseudo-Riemannian symmetric pairs, which is a generalization of that of the duality between compact/non-compact Riemannian symmetric pairs. Moreover we state a relation between compact symmetric triads and symmetric triads with multiplicities. As its application we state an outline of an alternative proof of Berger's classification of pseudo-Riemannian symmetric pairs. Moreover, we also give an alternative proof of Leung's classification of reflective submanifolds in compact Riemannian symmetric spaces and that of real forms in compact Hermitian symmetric spaces.

Keywords: Symmetric triad; generalized duality; reflective submanifold, real form.

1. Introduction

This article is a joint work with Kurando Baba (Tokyo University of Science) and Atsumu Sasaki (Tokai University). We devote Sections 2–5 to make an announcement of our unpublished papers [1, 3, 4], but §6 is original.

D. S. P. Leung [12] introduced the notion of a reflective submanifold of a Riemannian manifold, and classified reflective submanifolds of simply connected Riemannian symmetric spaces of compact type ([13, 14]). He also introduced the notion of a real form of a Kähler manifold [15], and he and M. Takeuchi [18] independently classified real forms in irreducible Hermitian symmetric spaces of compact type. The purpose of this paper is to give alternative proofs of the Leung's classification theorems using generalized duality and symmetric triad with multiplicities. Here, an imbedded submanifold B of a complete Riemannian manifold M is said to

---

*The author is partially supported by Grant-in-Aid for Scientific Research (C)
(No. 16K05128), Japan Society for the Promotion of Science.

be *reflective* if B is complete with respect to the induced metric and it is a connected component of the fixed point set of an involutive isometry of M ([12]). It is known that a reflective submanifold is totally geodesic [11, p. 61]. An imbedded submanifold of a Kähler manifold M is called a *real form* if it is a connected component of the fixed point set of an involutive anti-holomorphic isometry of M. Thus a real form is a kind of reflective submanifold.

The contents of this paper is as follows. In Section 2 we give a one-to-one correspondence between the set of compact symmetric triads and that of semisimple pseudo-Riemannian symmetric pairs. The correspondence is a generalization of the duality between the set of compact/non-compact Riemannian symmetric pairs. Thus we call the correspondence the generalized duality. In [7], the author defined the notion of a symmetric triad with multiplicities on a finite dimensional vector space with an inner product. The notion of a symmetric triad with multiplicities is a generalization of that of a root system with multiplicities. In Section 3 we review these notions, and introduce an equivalence relation on the set of symmetric triads with multiplicities. A Riemannian symmetric space of compact type assigns a (restricted) root system with multiplicities. Conversely a (restricted) root system with multiplicities in a class, which is called an *admissible* system, assigns a local isomorphic class of a Riemannian symmetric space of compact type. See Definition 3.3 for the definition of admissible system. In Section 4 we state a similar correspondence between the set of compact symmetric triads and that of symmetric triads with multiplicities. In Section 5 we give an outline of an alternative proof of Berger's classification of semisimple pseudo-Riemannian symmetric pairs using the results from Section 2 until Section 4. In Section 6 we give an alternative proof of Leung's classification of reflective submanifolds in compact Riemannian symmetric spaces, and that of real forms in Hermitian symmetric spaces of compact type, which is a main result of this paper.

2. Generalized duality

Let $\mathfrak{g}_u$ be a compact semisimple Lie algebra, and θ_1 and θ_2 be involutive automorphisms on $\mathfrak{g}_u$. In what follows we say involution for involutive automorphism for short. Throughout this paper we assume that θ_1 and θ_2 commute with each other. Such a triplet $(\mathfrak{g}_u, \theta_1, \theta_2)$ is called a *compact symmetric triad*. A compact symmetric triad $(\mathfrak{g}_u, \theta_1, \theta_2)$ is *irreducible*, if $\mathfrak{g}_u$ has no non-trivial ideals which are invariant under both θ_1 and θ_2. If $\mathfrak{g}_u$ is simple, then $(\mathfrak{g}_u, \theta_1, \theta_2)$ is irreducible, but the converse is not true. If $(\mathfrak{g}_u, \theta_1, \theta_2)$ is not irreducible, then it can be decomposed into irreducible factors. Denote by $\mathfrak{A}$ the set of all compact symmetric triads. We introduce two equivalence relations $\equiv$ and $\sim$ on $\mathfrak{A}$ as follows. For $(\mathfrak{g}_u, \theta_1, \theta_2)$, $(\mathfrak{g}'_u, \theta'_1, \theta'_2) \in \mathfrak{A}$, if there exists a Lie algebra isomorphism φ from $\mathfrak{g}_u$ onto $\mathfrak{g}'_u$ such that $\theta'_j = \varphi \theta_j \varphi^{-1}$ $(j = 1, 2)$, then we write $(\mathfrak{g}_u, \theta_1, \theta_2) \equiv (\mathfrak{g}'_u, \theta'_1, \theta'_2)$. We call $\equiv$ the *delicate equivalent relation*. If there exists a Lie algebra isomorphism φ from $\mathfrak{g}_u$ onto $\mathfrak{g}'_u$ and an inner automorphism τ on $\mathfrak{g}'_u$ such that $\theta'_1 = \varphi \theta_1 \varphi^{-1}$ and that $\theta'_2 = \tau \varphi \theta_2 \varphi^{-1} \tau^{-1}$, then we write $(\mathfrak{g}_u, \theta_1, \theta_2) \sim (\mathfrak{g}'_u, \theta'_1, \theta'_2)$. The equivalence relation $\sim$ was introduced by T. Matsuki [17], which we call the *rough equivalence relation*. The condition $(\mathfrak{g}_u, \theta_1, \theta_2) \equiv (\mathfrak{g}'_u, \theta'_1, \theta'_2)$ implies the condition $(\mathfrak{g}_u, \theta_1, \theta_2) \sim (\mathfrak{g}'_u, \theta'_1, \theta'_2)$. But the converse is not true in general. The notation $\theta_1 \sim \theta_2$ is an abbreviation for $(\mathfrak{g}_u, \theta_1, \theta_1) \sim (\mathfrak{g}_u, \theta_1, \theta_2)$. In other words, $\theta_1 \sim \theta_2$ means that there exists an inner automorphism τ on $\mathfrak{g}_u$ such that $\theta_2 = \tau \theta_1 \tau^{-1}$.

Let $\mathfrak{g}$ be a non-compact semisimple Lie algebra over $\mathbb{R}$, and σ be an involution on $\mathfrak{g}$. Such a pair $(\mathfrak{g}, \sigma)$ is called a *semisimple pseodo-Riemannian symmetric pair*. A semisimple pseudo-Riemannian symmetric pair $(\mathfrak{g}, \sigma)$ is *irreducible*, if $\mathfrak{g}$ has no non-trivial ideals which are invariant under σ. If $\mathfrak{g}$ is simple, then $(\mathfrak{g}, \sigma)$ is irreducible, but the converse is not true. If $(\mathfrak{g}, \sigma)$ is not irreducible, then it can be decomposed into irreducible factors. Denote by $\mathfrak{B}$ the set of all semisimple pseodo-Riemannian symmetric pairs. We introduce an equivalence relation $\equiv$ on $\mathfrak{B}$ as follows: For $(\mathfrak{g}, \sigma), (\mathfrak{g}', \sigma') \in \mathfrak{B}$, if there exists a Lie algebra isomorphism ϕ from $\mathfrak{g}$ onto $\mathfrak{g}'$ such that $\sigma' = \phi \sigma \phi^{-1}$, then we write $(\mathfrak{g}, \sigma) \equiv (\mathfrak{g}', \sigma')$. For $(\mathfrak{g}, \sigma) \in \mathfrak{B}$, it is known that there exists a Cartan involution on $\mathfrak{g}$ such that θ and σ commute with each other. Such a $(\mathfrak{g}, \sigma; \theta)$ is called a *semisimple pseudo-Riemannian symmetric pair with a Cartan involution*. A semisimple pseudo-Riemannian symmetric pair $(\mathfrak{g}, \sigma; \theta)$ is *irreducible*, if $\mathfrak{g}$ has no non-trivial ideals which are invariant under both θ and σ. If $\mathfrak{g}$ is simple, then $(\mathfrak{g}, \sigma; \theta)$ is irreducible, but the converse is not true. If $(\mathfrak{g}, \sigma; \theta)$ is not irreducible, then it can be decomposed into irreducible factors. Denote by $\mathfrak{B}_c$ the set of all semisimple pseudo-Riemannian symmetric pairs with Cartan involution. We introduce an equivalence relation $\equiv$ on $\mathfrak{B}_c$ as follows: For $(\mathfrak{g}, \sigma; \theta), (\mathfrak{g}', \sigma'; \theta') \in \mathfrak{B}_c$, if there exists a Lie algebra isomorphism ϕ from $\mathfrak{g}$ onto $\mathfrak{g}'$ such that $\theta' = \phi \theta \phi^{-1}$ and that $\sigma' = \phi \sigma \phi^{-1}$, then we write $(\mathfrak{g}, \sigma; \theta) \equiv (\mathfrak{g}', \sigma'; \theta')$. Denote by $p : \mathfrak{B}_c \to \mathfrak{B}; (\mathfrak{g}, \sigma; \theta) \mapsto (\mathfrak{g}, \sigma)$ the natural projection. Since $(\mathfrak{g}, \sigma; \theta) \equiv (\mathfrak{g}', \sigma'; \theta')$ implies $(\mathfrak{g}, \sigma) \equiv (\mathfrak{g}', \sigma')$ for $(\mathfrak{g}, \sigma; \theta), (\mathfrak{g}', \sigma'; \theta') \in \mathfrak{B}_c$, the natural projection $p : \mathfrak{B}_c \to \mathfrak{B}$ induces a map $\tilde{p} : \mathfrak{B}_c / \equiv \, \to \mathfrak{B} / \equiv; \; [(\mathfrak{g}, \sigma; \theta)] \mapsto [(\mathfrak{g}, \sigma)]$, which is surjective.

Proposition 2.1. *The following relations hold.*

(i) *For* $(\mathfrak{g}, \sigma; \theta) \in \mathfrak{B}_c$, $(\mathfrak{g}, \sigma; \theta)$ *is irreducible if and only if* $(\mathfrak{g}, \sigma)$ *is irreducible.*

(ii) *[16, p. 153] For* $(\mathfrak{g}, \sigma) \in \mathfrak{B}$, *if both* θ *and* θ' *are Cartan involutions of* $\mathfrak{g}$ *which commute with* σ, *then there exists* $X \in \mathfrak{g}$ *such that* $\sigma(X) = X$ *and that* $\theta' = e^{\operatorname{ad}X} \theta e^{-\operatorname{ad}X}$.

By (ii) of Proposition 2.1, the map $\tilde{p} : \mathfrak{B}_c/\equiv \to \mathfrak{B}/\equiv$ is a bijection.

We shall give a natural bijection from $\mathfrak{A}$ onto $\mathfrak{B}_c$. Let $(\mathfrak{g}_u, \theta_1, \theta_2)$ be in $\mathfrak{A}$. Then we have two canonical decompositions of $\mathfrak{g}_u$:

$$\mathfrak{g}_u = \mathfrak{k}_1 \oplus \mathfrak{m}_1 = \mathfrak{k}_2 \oplus \mathfrak{m}_2, \tag{1}$$

where we set $\mathfrak{k}_j = \{X \in \mathfrak{g}_u \mid \theta_j(X) = X\}$ and $\mathfrak{m}_j = \{X \in \mathfrak{g}_u \mid \theta_j(X) = -X\}$. We extend θ_1 and θ_2 to complex linear transformations on the complexification $\mathfrak{g}_u^{\mathbb{C}}$ of $\mathfrak{g}_u$. We denote them by the same symbols θ_1 and θ_2. Then θ_1 and θ_2 are involutions on $\mathfrak{g}_u^{\mathbb{C}}$. If we set $\mathfrak{g} = \mathfrak{k}_1 \oplus \sqrt{-1}\mathfrak{m}_1$, then it is a non-compact real form of $\mathfrak{g}_u^{\mathbb{C}}$. Since θ_1 and θ_2 commute with each other, θ_2 is an involution on $\mathfrak{g}$ and θ_1 is a Cartan involution of $\mathfrak{g}$ that commutes with θ_2. Set $\Phi(\mathfrak{g}_u, \theta_1, \theta_2) = (\mathfrak{g}, \theta_2; \theta_1) \in \mathfrak{B}_c$.

Proposition 2.2. *The mapping*

$$\Phi : \mathfrak{A} \to \mathfrak{B}_c; \ (\mathfrak{g}_u, \theta_1, \theta_2) \mapsto (\mathfrak{g}, \theta_2; \theta_1)$$

is a bijection.

Proof. For $(\mathfrak{g}, \sigma; \theta) \in \mathfrak{B}_c$, denote by $\mathfrak{g} = \mathfrak{k} \oplus \mathfrak{p}$ the Cartan decomposition of $\mathfrak{g}$ with respect to θ. If we set $\mathfrak{g}_u = \mathfrak{k} \oplus \sqrt{-1}\mathfrak{p}$, then $\mathfrak{g}_u$ is a compact semisimple Lie algebra. Denote by the same symbols σ and θ the complex linear extension of σ and θ to $\mathfrak{g}^{\mathbb{C}} = \mathfrak{g}_u^{\mathbb{C}}$. If we set $\theta_1 = \theta$ and $\sigma = \theta_2$, then $(\mathfrak{g}_u, \theta_1, \theta_2)$ is in $\mathfrak{A}$. Set $\Psi(\mathfrak{g}, \sigma; \theta) = (\mathfrak{g}_u, \theta_1, \theta_2)$. Then we have $\Psi\Phi = \mathbf{1}$ and $\Phi\Psi = \mathbf{1}$. $\qquad \square$

In what follows if there is no confusion, we set $\Psi(\mathfrak{g}, \sigma; \theta) = (\mathfrak{g}, \sigma; \theta)^*$ and $\Phi(\mathfrak{g}_u, \theta_1, \theta_2) = (\mathfrak{g}_u, \theta_1, \theta_2)^*$. Then we have $** = \mathbf{1}$.

Proposition 2.3. *The following relations hold.*

(i) $(\mathfrak{g}_u, \theta_1, \theta_2)$ *is irreducible if and only if* $(\mathfrak{g}_u, \theta_1, \theta_2)^*$ *is irreducible.*

(ii) $(\mathfrak{g}_u, \theta_1, \theta_2) \equiv (\mathfrak{g}_u', \theta_1', \theta_2')$ *if and only if* $(\mathfrak{g}_u, \theta_1, \theta_2)^* \equiv (\mathfrak{g}_u', \theta_1', \theta_2')^*$.

The bijection $\Phi : \mathfrak{A} \to \mathfrak{B}_c$ defined by Proposition 2.2 induces a bijection $\tilde{\Phi}$ from $\mathfrak{A}/\equiv$ onto $\mathfrak{B}_c/\equiv$ due to (ii) of Proposition 2.3. Hence we get the following, which is a main result of this section.

Theorem 2.1 (Generalized duality [3]). *The composite mapping* $\tilde{p} \circ \tilde{\Phi} :$ $\mathfrak{A}/\equiv \to \mathfrak{B}/\equiv$ *is a bijection.*

We call the bijection $\tilde{p} \circ \tilde{\Phi}$ the *generalized duality*. If there is no confusion, we write $*$ for $\tilde{p} \circ \tilde{\Phi}$ and its inverse. In particular, the image of $[(\mathfrak{g}_u, \theta, \theta)] \in \mathfrak{A}/\equiv$ by $*$ is the usual duality $[(\mathfrak{g}_u, \theta, \theta)]^* = [(\mathfrak{g}, \theta; \theta)]$ by means of E. Cartan. This is the reason why we call $*$ the generalized duality. See [9] for an application of the generalized duality to calibration's inequalities.

3. Symmetric triad with multiplicities

In this section we review the definition of symmetric triad with multiplicities and introduce an equivalence relation on the set of symmetric triad with multiplicities. These results will be used later. Since the notion of a symmetric triad is a generalization of that of root system, we begin with recalling the definition of a root system.

Let $\mathfrak{a}$ be a finite dimensional vector space over $\mathbb{R}$ with an inner product $\langle \, , \, \rangle$.

Definition 3.1. A finite subset $\Sigma \subset \mathfrak{a} - \{0\}$ is a *root system* of $\mathfrak{a}$, if it satisfies the following two conditions:

(i) $\mathfrak{a} = \mathrm{span}(\Sigma)$.

(ii) If α and β are in Σ, then $s_\alpha \beta := \beta - 2\dfrac{\langle \alpha, \beta \rangle}{\|\alpha\|^2}\alpha \in \Sigma$ and $2\dfrac{\langle \alpha, \beta \rangle}{\|\alpha\|^2} \in \mathbb{Z}$.

The *Weyl group* $W(\Sigma)$ of a root system Σ is defined by a subgroup of $O(\mathfrak{a})$ generated by $\{s_\alpha \mid \alpha \in \Sigma\}$. Since $W(\Sigma)$ induces a permutation on Σ, it is a finite group.

A root system Σ of $\mathfrak{a}$ is called *irreducible* if it cannot be decomposed into two disjoint nonempty orthogonal subsets.

Definition 3.2. Let Σ be a root system of $\mathfrak{a}$. Put $\mathbb{R}_{>0} = \{x \in \mathbb{R} \mid x > 0\}$. Consider a mapping $m : \Sigma \to \mathbb{R}_{>0}; \alpha \mapsto m_\alpha$ which satisfies

$$m_{s_\alpha \beta} = m_\beta \quad \text{for} \quad \alpha, \beta \in \Sigma. \tag{2}$$

We call m_α the *multiplicity* of α. If the multiplicities are given, we call $(\Sigma; m)$ the *root system with multiplicities*.

If Σ is irreducible then it is known that $W(\Sigma)$ acts transitively on each subset $\{\beta \in \Sigma \mid \|\beta\| = \|\alpha\|\}$ for $\alpha \in \Sigma$. Thus the condition (2) means

$$m_\alpha = m_\beta \quad \text{if} \quad \|\alpha\| = \|\beta\|. \tag{3}$$

A Riemannian symmetric pair of compact type assigns a root system with multiplicities in a standard way. But the converse is not true. Based on the fact, we define the notion of admissible root system as follows:

Definition 3.3. A root system with multiplicities is *admissible* if it is a restricted root system with multiplicities of a compact Riemannian symmetric pair.

Example 3.1. When Σ is of type A_r, the Weyl group $W(\Sigma)$ acts on Σ transitively. Thus for any $\alpha \in \Sigma$ and any given $c \in \mathbb{R}$, if we set $m_\alpha = c$, then (Σ, m) is a root system with multiplicities. In what follows we assume that r is greater than or equal to 3 for simplicity. Then the restricted root system with multiplicities of a compact symmetric pair (G, K) is (Σ, m) if and only if

$$(G, K) = \begin{cases} (SU(r+1), SO(r+1)) & \text{(in this case } c = 1), \\ (SU(r+1) \times SU(r+1), SU(r+1)) & \text{(in this case } c = 2), \\ (SU(2(r+1)), Sp(r+1)) & \text{(in this case } c = 4). \end{cases}$$

Hence (Σ, m) is admissible if and only if $c = 1, 2$ or 4.

We define an equivalence relation $\equiv$ on the set of root systems with multiplicities as follows:

Definition 3.4. Let $(\Sigma; m)$ and $(\Sigma'; m')$ be root systems with multiplicities of $\mathfrak{a}$ and $\mathfrak{a}'$, respectively. Then $(\Sigma; m)$ and $(\Sigma'; m')$ are *isomorphic* if there exists a linear isometric (more precisely, homothetic) isomorphism $f : \mathfrak{a} \to \mathfrak{a}'$ such that $f(\Sigma) = \Sigma'$ and $m_\alpha = m'_{f(\alpha)}$ ($\alpha \in \Sigma$). Here a linear transformation f is homothetic means that there exists a positive constant c such that $\langle f(\alpha), f'(\beta) \rangle = c\langle \alpha, \beta \rangle$. If $(\Sigma; m)$ and $(\Sigma'; m')$ are isomorphic, then we write $(\Sigma; m) \equiv (\Sigma'; m')$.

Next, we review the definition of a symmetric triad.

Definition 3.5 ([7, Definition 2.2]). A triple $(\tilde{\Sigma}, \Sigma, W)$ is a *symmetric triad* of $\mathfrak{a}$, if it satisfies the following six conditions:

(i) $\tilde{\Sigma}$ is an irreducible root system of $\mathfrak{a}$.
(ii) Σ is a root system of span(Σ).

(iii) W is a nonempty subset of $\mathfrak{a}$, which is invariant under the multiplication by -1, and $\tilde{\Sigma} = \Sigma \cup W$.

(iv) $\Sigma \cap W$ is a nonempty subset. If we put $l = \max\{\|\alpha\| \mid \alpha \in \Sigma \cap W\}$, then $\Sigma \cap W = \{\alpha \in \tilde{\Sigma} \mid \|\alpha\| \leq l\}$.

(v) For $\alpha \in W, \lambda \in \Sigma - W$, the integer $2\frac{\langle\alpha,\lambda\rangle}{\|\alpha\|^2}$ is odd if and only if $s_\alpha\lambda \in W - \Sigma$.

(vi) For $\alpha \in W, \lambda \in W - \Sigma$, the integer $2\frac{\langle\alpha,\lambda\rangle}{\|\alpha\|^2}$ is odd if and only if $s_\alpha\lambda \in \Sigma - W$.

When $(\tilde{\Sigma}, \Sigma, W)$ is a symmetric triad of $\mathfrak{a}$, then $\text{span}(\Sigma) = \mathfrak{a}$. See [10, Remark 1.13]. It is known that W is invariant under the action of the Weyl group $W(\Sigma)$ of Σ [7, Proposition 2.7]. This fact will be used in (ii) of Definition 3.6. For a symmetric triad $(\tilde{\Sigma}, \Sigma, W)$ of $\mathfrak{a}$ we define a lattice Γ of $\mathfrak{a}$ by

$$\Gamma = \left\{ X \in \mathfrak{a} \mid \langle\lambda, X\rangle \in \frac{\pi}{2}\mathbb{Z} \quad (\lambda \in \tilde{\Sigma}) \right\}. \tag{4}$$

Definition 3.6 ([7, Definition 2.13]). Let $(\tilde{\Sigma}, \Sigma, W)$ be a symmetric triad of $\mathfrak{a}$. Put $\mathbb{R}_{\geq 0} = \{x \in \mathbb{R} \mid x \geq 0\}$. Consider two mappings $m, n : \tilde{\Sigma} \to \mathbb{R}_{\geq 0}$ which satisfy the following four conditions:

(i) $m(\lambda) = m(-\lambda)$, $n(\alpha) = n(-\alpha)$ for $\lambda, \alpha \in \tilde{\Sigma}$ and

$$m(\lambda) > 0 \Leftrightarrow \lambda \in \Sigma, \quad n(\alpha) > 0 \Leftrightarrow \alpha \in W.$$

(ii) When $\lambda \in \Sigma, \alpha \in W, s \in W(\Sigma)$ then $m(\lambda) = m(s\lambda), n(\alpha) = n(s\alpha)$.

(iii) When $\sigma \in W(\tilde{\Sigma})$, the Weyl group of $\tilde{\Sigma}$, and $\lambda \in \tilde{\Sigma}$ then $n(\lambda) + m(\lambda) = n(\sigma\lambda) + m(\sigma\lambda)$.

(iv) Let $\lambda \in \Sigma \cap W$ and $\alpha \in W$.
If $\frac{2\langle\alpha,\lambda\rangle}{\|\alpha\|^2}$ is even then $m(\lambda) = m(s_\alpha\lambda)$.
If $\frac{2\langle\alpha,\lambda\rangle}{\|\alpha\|^2}$ is odd then $m(\lambda) = n(s_\alpha\lambda)$.

We call $m(\lambda)$ and $n(\alpha)$ the *multiplicities* of λ and α, respectively. If multiplicities are given, we call $(\tilde{\Sigma}, \Sigma, W; m, n)$ the *symmetric triad with multiplicities*.

The author defined an equivalence relation on the set of symmetric triads in [7]. We extend it to an equivalence relation on the set of symmetric triads with multiplicities as follows:

Definition 3.7. Let $(\tilde{\Sigma}, \Sigma, W; m, n)$ and $(\tilde{\Sigma}', \Sigma', W'; m', n')$ be symmetric triads with multiplicities of $\mathfrak{a}$ and $\mathfrak{a}'$, respectively. Then two symmetric

triads $(\tilde{\Sigma}, \Sigma, W)$ and $(\tilde{\Sigma}', \Sigma', W')$ are *equivalent*, if there exist a linear isometric (more precisely, homothetic) isomorphism $f : \mathfrak{a} \to \mathfrak{a}'$ and $Y \in \Gamma$
such that $f(\tilde{\Sigma}) = \tilde{\Sigma}'$ and

$$\begin{cases} \Sigma' - W' = \{f(\alpha) \mid \alpha \in \Sigma - W, \langle \alpha, 2Y \rangle \in 2\pi\mathbb{Z}\} \\ \qquad\qquad \cup \{f(\alpha) \mid \alpha \in W - \Sigma, \langle \alpha, 2Y \rangle \in \pi + 2\pi\mathbb{Z}\}, \\ W' - \Sigma' = \{f(\alpha) \mid \alpha \in W - \Sigma, \langle \alpha, 2Y \rangle \in 2\pi\mathbb{Z}\} \\ \qquad\qquad \cup \{f(\alpha) \mid \alpha \in \Sigma - W, \langle \alpha, 2Y \rangle \in \pi + 2\pi\mathbb{Z}\}. \end{cases} \tag{5}$$

In addition if f and Y satisfy the following condition, we say that
$(\tilde{\Sigma}, \Sigma, W; m, n)$ and $(\tilde{\Sigma}', \Sigma', W'; m', n')$ are *equivalent*. For $\alpha \in \tilde{\Sigma}$,

$$m(\alpha) = m'(f(\alpha)), \ n(\alpha) = n'(f(\alpha)) \text{ if } \langle \alpha, 2Y \rangle \in 2\pi\mathbb{Z},$$
$$m(\alpha) = n'(f(\alpha)), \ n(\alpha) = m'(f(\alpha)) \text{ if } \langle \alpha, 2Y \rangle \in \pi + 2\pi\mathbb{Z}.$$

If $(\tilde{\Sigma}, \Sigma, W; m, n)$ and $(\tilde{\Sigma}', \Sigma', W'; m', n')$ are equivalent, we write

$$(\tilde{\Sigma}, \Sigma, W; m, n) \sim (\tilde{\Sigma}', \Sigma', W'; m', n').$$

The relation $\sim$ is an equivalence relation.

We know the set of all symmetric triads with multiplicities explicitly.
And we also know which of two symmetric triads with multiplicities are
equivalent.

4. Compact symmetric triad and symmetric triad with multiplicities

In this section, we construct symmetric triads with multiplicities from compact symmetric triads. Further, we give a correspondence between compact
symmetric triads and symmetric triads with multiplicities.

Let $(\mathfrak{g}_u, \theta_1, \theta_2)$ be a compact symmetric triad, and $\langle \, , \, \rangle$ be an invariant
inner product on $\mathfrak{g}_u$. We retain the notations as in the previous sections.
Since we have assumed that θ_1, θ_2 commute with each other, we have a
direct sum decomposition

$$\mathfrak{g}_u = (\mathfrak{k}_1 \cap \mathfrak{k}_2) \oplus (\mathfrak{m}_1 \cap \mathfrak{m}_2) \oplus (\mathfrak{k}_1 \cap \mathfrak{m}_2) \oplus (\mathfrak{m}_1 \cap \mathfrak{k}_2),$$

where $\mathfrak{k}_j = \{X \in \mathfrak{g}_u \mid \theta_j(X) = X\}$ and $\mathfrak{m}_j = \{X \in \mathfrak{g}_u \mid \theta_j(X) = -X\}$.
Take a maximal abelian subspace $\mathfrak{a}$ of $\mathfrak{m}_1 \cap \mathfrak{m}_2$. For $\alpha \in \mathfrak{a}$ we define a
subspace $\mathfrak{g}_u(\mathfrak{a}, \alpha)$ of the complexification $\mathfrak{g}_u^{\mathbb{C}}$ of $\mathfrak{g}_u$ by

$$\mathfrak{g}_u(\mathfrak{a}, \alpha) = \{X \in \mathfrak{g}_u^{\mathbb{C}} \mid [H, X] = \sqrt{-1}\langle \alpha, H \rangle X \ (H \in \mathfrak{a})\}.$$

If we set $\tilde{\Sigma} = \{\alpha \in \mathfrak{a} - \{0\} \mid \mathfrak{g}_u(\mathfrak{a}, \alpha) \neq \{0\}\}$ then

$$\mathfrak{g}_u^{\mathbb{C}} = \mathfrak{g}_u(\mathfrak{a}, 0) \oplus \sum_{\alpha \in \tilde{\Sigma}} \mathfrak{g}_u(\mathfrak{a}, \alpha).$$

For $\epsilon = \pm 1$ we define a subspace $\mathfrak{g}_u(\mathfrak{a}, \alpha, \epsilon)$ of $\mathfrak{g}_u(\mathfrak{a}, \alpha)$ by

$$\mathfrak{g}_u(\mathfrak{a}, \alpha, \epsilon) = \{X \in \mathfrak{g}_u(\mathfrak{a}, \alpha) \mid \theta_1 \theta_2 X = \epsilon X\}.$$

Since $\mathfrak{g}_u(\mathfrak{a}, \alpha)$ is invariant under the action of $\theta_1 \theta_2$, we have $\mathfrak{g}_u(\mathfrak{a}, \alpha) = \mathfrak{g}_u(\mathfrak{a}, \alpha, 1) \oplus \mathfrak{g}_u(\mathfrak{a}, \alpha, -1)$. Thus if we set

$$\Sigma = \{\lambda \in \tilde{\Sigma} \mid \mathfrak{g}_u(\mathfrak{a}, \lambda, 1) \neq \{0\}\}, \quad W = \{\lambda \in \tilde{\Sigma} \mid \mathfrak{g}_u(\mathfrak{a}, \lambda, -1) \neq \{0\}\},$$

then $\tilde{\Sigma} = \Sigma \cup W$ holds. Consider two mappings $m, n : \tilde{\Sigma} \to \mathbb{R}_{\geq 0}$ defined by

$$m(\alpha) = \dim_{\mathbb{C}} \mathfrak{g}_u(\mathfrak{a}, \alpha, 1), \quad n(\alpha) = \dim_{\mathbb{C}} \mathfrak{g}_u(\mathfrak{a}, \alpha, -1).$$

From now on we assume that $\mathfrak{g}_u$ is simple and that $\theta_1 \not\sim \theta_2$ until the end of this section. With the notation above, the author [7] proved that, $(\tilde{\Sigma}, \Sigma, W; m, n)$ satisfies the axiom a symmetric triad with multiplicities of $\mathfrak{a}$. Clearly $(\mathfrak{g}_u, \theta_1, \theta_2)$ and $(\mathfrak{g}_u, \theta_2, \theta_1)$ define the same symmetric triad with multiplicities. The following theorem is needed later.

Theorem 4.1 (Hermann [6]). *Let $(\mathfrak{g}_u, \theta_1, \theta_2)$ be a compact symmetric triad. Let G_u be a connected Lie group whose Lie algebra is $\mathfrak{g}_u$. Denote by K_1 and K_2 the analytic subgroups of G_u whose Lie algebras are $\mathfrak{k}_1$ and $\mathfrak{k}_2$, respectively. Then we have the following:*

$$G_u = K_1(\exp \mathfrak{a})K_2 = K_2(\exp \mathfrak{a})K_1. \tag{6}$$

The following theorem is a main result of this section, which is a refinement of [7, Theorem 4.33, (2)].

Theorem 4.2. *Let $(\mathfrak{g}_u, \theta_1, \theta_2)$ and $(\mathfrak{g}_u, \theta_1', \theta_2')$ be two compact symmetric triads, where $\mathfrak{g}_u$ is a compact simple Lie algebra and $\theta_1 \theta_2 = \theta_2 \theta_1$ and $\theta_1' \theta_2' = \theta_2' \theta_1'$. We assume that $\theta_1 \not\sim \theta_2$ and $\theta_1' \not\sim \theta_2'$. Denote by $(\tilde{\Sigma}, \Sigma, W; m, n)$ and $(\tilde{\Sigma}', \Sigma', W'; m', n')$ the corresponding symmetric triads with multiplicities, respectively. If $(\mathfrak{g}_u, \theta_1, \theta_2) \sim (\mathfrak{g}_u, \theta_1', \theta_2')$ then $(\tilde{\Sigma}, \Sigma, W; m, n) \sim (\tilde{\Sigma}', \Sigma', W'; m', n')$. Conversely if $(\tilde{\Sigma}, \Sigma, W; m, n) \sim (\tilde{\Sigma}', \Sigma', W'; m', n')$ then $(\mathfrak{g}_u, \theta_1, \theta_2) \sim (\mathfrak{g}_u, \theta_1', \theta_2')$ or $(\mathfrak{g}_u, \theta_1, \theta_2) \sim (\mathfrak{g}_u, \theta_2', \theta_1')$.*

Proof. Assume that $(\mathfrak{g}_u, \theta_1, \theta_2) \sim (\mathfrak{g}_u, \theta_1', \theta_2')$. There exist $\varphi \in \mathrm{Aut}(\mathfrak{g}_u)$ and $\tau \in \mathrm{Int}(\mathfrak{g}_u)$ such that $\theta_1' = \varphi \theta_1 \varphi^{-1}, \theta_2' = \tau \varphi \theta_2 \varphi^{-1} \tau^{-1}$. Without loss of generality we may assume that $\theta_1' = \theta_1, \theta_2' = \tau \theta_2 \tau^{-1}$ since $(\mathfrak{g}_u, \theta_1, \theta_2) \equiv (\mathfrak{g}_u, \varphi \theta_1 \varphi^{-1}, \varphi \theta_2 \varphi^{-1})$. Let G_u be a simply connected Lie group with Lie algebra $\mathfrak{g}_u$. Then G_u is compact. Since G_u is simply connected, θ_i and θ_i' can be extended to involutions on G_u. By the same reason, τ can be extended to an inner automorphism on G_u. We denote these extensions by the same symbols. Then we have

$$\theta_1' = \theta_1, \quad \theta_2' = \tau\theta_2\tau^{-1}, \quad \theta_1\theta_2 = \theta_2\theta_1, \quad \theta_1'\theta_2' = \theta_2'\theta_1'$$

on G_u. Since τ is inner, there exists g in G_u such that $\tau = \tau_g$, where $\tau_g(x) = gxg^{-1}$. We have two canonical decompositions of $\mathfrak{g}_u$ using θ_1 and θ_2 as in (1). Take a maximal abelian subspace $\mathfrak{a}$ of $\mathfrak{m}_1 \cap \mathfrak{m}_2$. There exist $Y \in \mathfrak{a}$ and $k_i \in K_i$ $(i = 1, 2)$ such that $g = k_1(\exp Y)k_2$ by (6). Since $\tau_{k_2}\theta_2\tau_{k_2}^{-1} = \theta_2$ we have

$$\theta_2' = \tau_g\theta_2\tau_g^{-1} = \tau_{k_1}\tau_{\exp Y}\theta_2\tau_{\exp Y}^{-1}\tau_{k_1}^{-1}.$$

Since $\tau_{k_1}\theta_1\tau_{k_1}^{-1} = \theta_1$ we have

$$(\mathfrak{g}_u, \theta_1', \theta_2') = (\mathfrak{g}_u, \theta_1, \tau_{k_1}\tau_{\exp Y}\theta_2\tau_{\exp Y}^{-1}\tau_{k_1}^{-1}) \equiv (\mathfrak{g}_u, \theta_1, \tau_{\exp Y}\theta_2\tau_{\exp Y}^{-1}).$$

Hence we may assume that $\theta_1' = \theta_1, \theta_2' = \tau_{\exp Y}\theta_2\tau_{\exp Y}^{-1}$. Since $\theta_1\theta_2 = \theta_2\theta_1, \theta_1'\theta_2' = \theta_2'\theta_1'$ and $\tau_{\exp Y}\theta_i = \theta_i\tau_{\exp(-Y)}$ $(i = 1, 2)$, we have $\tau_{\exp 4Y} = 1$, that is, $\exp 4Y$ is in the center of G_u. This means $e^{\mathrm{ad}4Y} = 1$, which is equivalent to $Y \in \Gamma$. Here Γ is a lattice defined in (4). In order to study the relation between the symmetric triads $(\tilde{\Sigma}, \Sigma, W; m, n)$ with multiplicities defined by $(\mathfrak{g}_u, \theta_1, \theta_2)$ and $(\tilde{\Sigma}', \Sigma', W'; m', n')$ defined by $(\mathfrak{g}_u, \theta_1, \tau_{\exp Y}\theta_2\tau_{\exp(-Y)})$ we decompose $\mathfrak{g}_u$ by $\tau_{\exp Y}\theta_2\tau_{\exp(-Y)}$ to

$$\mathfrak{g}_u = \mathfrak{k}_2' \oplus \mathfrak{m}_2'.$$

Then

$$\mathfrak{m}_2' = \{X \in \mathfrak{g}_u \mid \tau_{\exp(-Y)}X \in \mathfrak{m}_2\} = e^{\mathrm{ad}Y}\mathfrak{m}_2.$$

Hence $\mathfrak{a}$ is also a maximal abelian subspace of $\mathfrak{m}_1 \cap \mathfrak{m}_2'$. We denote by $\mathfrak{a}'$ the subspace $\mathfrak{a}$ considered as the maximal abelian subspace of $\mathfrak{m}_1 \cap \mathfrak{m}_2'$. For $X \in \mathfrak{g}(\mathfrak{a}, \alpha, \epsilon)$ we have

$$\theta_1\tau_{\exp Y}\theta_2\tau_{\exp(-Y)}X = \theta_1\tau_{\exp 2Y}\theta_2 X = \tau_{\exp(-2Y)}\theta_1\theta_2 X$$

$$= \tau_{\exp(-2Y)}\epsilon X = \epsilon e^{-2\sqrt{-1}\langle\alpha, Y\rangle}X,$$

which implies that $\tilde{\Sigma}' = \tilde{\Sigma}$ and

$$\mathfrak{g}_u(\mathfrak{a}, \alpha, \epsilon) = \mathfrak{g}_u(\mathfrak{a}', \alpha, \epsilon e^{-2\sqrt{-1}\langle\alpha, Y\rangle}).$$

Thus we have

$$\Sigma' = \{\alpha \in \tilde{\Sigma} \mid \epsilon e^{-2\sqrt{-1}\langle \alpha, Y\rangle} = 1\}$$
$$= \{\alpha \in \Sigma \mid e^{2\sqrt{-1}\langle \alpha, Y\rangle} = 1\} \cup \{\alpha \in W \mid e^{2\sqrt{-1}\langle \alpha, Y\rangle} = -1\}$$
$$= \{\alpha \in \Sigma \mid 2\langle \alpha, Y\rangle \in 2\pi\mathbb{Z}\} \cup \{\alpha \in W \mid 2\langle \alpha, Y\rangle \in \pi + 2\pi\mathbb{Z}\},$$

and

$$W' = \{\alpha \in \tilde{\Sigma} \mid \epsilon e^{-2\sqrt{-1}\langle \alpha, Y\rangle} = -1\}$$
$$= \{\alpha \in W \mid e^{2\sqrt{-1}\langle \alpha, Y\rangle} = 1\} \cup \{\alpha \in \Sigma \mid e^{2\sqrt{-1}\langle \alpha, Y\rangle} = -1\}$$
$$= \{\alpha \in W \mid 2\langle \alpha, Y\rangle \in 2\pi\mathbb{Z}\} \cup \{\alpha \in \Sigma \mid 2\langle \alpha, Y\rangle \in \pi + 2\pi\mathbb{Z}\}.$$

Hence we have $\Sigma' \cap W' = \Sigma \cap W$ and

$$\Sigma' - W' = \{\alpha \in \Sigma - W \mid 2\langle \alpha, Y\rangle \in 2\pi\mathbb{Z}\}$$
$$\cup \{\alpha \in W - \Sigma \mid 2\langle \alpha, Y\rangle \in \pi + 2\pi\mathbb{Z}\},$$
$$W' - \Sigma' = \{\alpha \in W - \Sigma \mid 2\langle \alpha, Y\rangle \in 2\pi\mathbb{Z}\}$$
$$\cup \{\alpha \in \Sigma - W \mid 2\langle \alpha, Y\rangle \in \pi + 2\pi\mathbb{Z}\}.$$

We study the multiplicities. When $2\langle \alpha, Y\rangle \in 2\pi\mathbb{Z}$, since $e^{2\sqrt{-1}\langle \alpha, Y\rangle} = 1$ we have $\mathfrak{g}_u(\mathfrak{a}', \alpha, \epsilon) = \mathfrak{g}_u(\mathfrak{a}, \alpha, \epsilon)$. Thus

$$m'(\alpha) = \dim \mathfrak{g}_u(\mathfrak{a}', \alpha, 1) = \dim \mathfrak{g}_u(\mathfrak{a}, \alpha, 1) = m(\alpha),$$
$$n'(\alpha) = \dim \mathfrak{g}_u(\mathfrak{a}', \alpha, -1) = \dim \mathfrak{g}_u(\mathfrak{a}, \alpha, -1) = n(\alpha).$$

When $2\langle \alpha, Y\rangle \in \pi + 2\pi\mathbb{Z}$, since $e^{2\sqrt{-1}\langle \alpha, Y\rangle} = -1$ we have $\mathfrak{g}_u(\mathfrak{a}', \alpha, \epsilon) = \mathfrak{g}_u(\mathfrak{a}, \alpha, -\epsilon)$. Thus

$$m'(\alpha) = \dim \mathfrak{g}_u(\mathfrak{a}', \alpha, 1) = \dim \mathfrak{g}_u(\mathfrak{a}, \alpha, -1) = n(\alpha),$$
$$n'(\alpha) = \dim \mathfrak{g}_u(\mathfrak{a}', \alpha, -1) = \dim \mathfrak{g}_u(\mathfrak{a}, \alpha, 1) = m(\alpha).$$

Hence we get $(\tilde{\Sigma}, \Sigma, W; m, n) \sim (\tilde{\Sigma}', \Sigma', W'; m', n')$ if $(\mathfrak{g}_u, \theta_1, \theta_2) \sim (\mathfrak{g}_u, \theta_1', \theta_2')$.

The converse follows from the explicit description of the space of rough equivalence classes $\mathfrak{A}/\sim$ by T. Matsuki. See [4] for the details. $\square$

It is known that for any symmetric triad $(\tilde{\Sigma}, \Sigma, W)$ there exists a compact symmetric triad $(\mathfrak{g}_u, \theta_1, \theta_2)$ such that the corresponding symmetric triad coincides with $(\tilde{\Sigma}, \Sigma, W)$. But it is not true that for any symmetric triad $(\tilde{\Sigma}, \Sigma, W; m, n)$ with multiplicities there exists a compact symmetric triad $(\mathfrak{g}_u, \theta_1, \theta_2)$ such that the corresponding symmetric triad with multiplicities coincides with $(\tilde{\Sigma}, \Sigma, W; m, n)$. A symmetric triad with multiplicities is *admissible*, if it is obtained from a compact symmetric pair. We know the set of all admissible symmetric triads with multiplicities explicitly ([1]).

In the next two sections, we apply the generalized duality and the symmetric triad with multiplicities mentioned in the previous sections to giving alternative proofs of classification theorems: the classification of semisimple pseudo-Riemannian symmetric pair by Berger, the classification of reflective submanifolds in compact symmetric spaces by Leung, and the classification of real forms in compact Hermitian symmetric spaces by Leung and M. Takeuchi.

5. Berger's classification theorem

The contents of this section are a part of [4]. See also [2] for the reference.

Berger [5] explicitly described the space of equivalence classes $\mathfrak{B}/\equiv$ of semisimple pseudo-Riemannian symmetric pairs. This is called Berger's classification theorem. We can give an alternative proof of Berger's classification theorem using the generalized duality and the symmetric triad with multiplicities. We shall give an outline of the proof. Since each element in $\mathfrak{B}/\equiv$ can be decomposed into irreducible factors, we can focus our attention to the irreducible ones in $\mathfrak{B}/\equiv$. If we know the space of delicate equivalence classes $\mathfrak{A}/\equiv$ explicitly, then we get $\mathfrak{B}/\equiv$ since we have an algorithm to determine the image by the generalized duality $*$ of each element in $\mathfrak{A}/\equiv$. T. Matsuki [17] explicitly described the space of rough equivalence classes $\mathfrak{A}/\sim$. Thus, if we get the space of delicate equivalence classes $\mathfrak{A}/\equiv$ from the space of rough equivalence classes $\mathfrak{A}/\sim$, then we could give an alternative proof of Berger's classification theorem. We assume that $\mathfrak{g}_u$ is simple and $\theta_1 \not\sim \theta_2$ for a while. For an explicitly given $[(\mathfrak{g}_u, \theta_1, \theta_2)]_\sim \in \mathfrak{A}/\sim$, we have a method of knowing the corresponding equivalence class $[(\tilde{\Sigma}, \Sigma, W; m, n)]$ of multiplicities. This method is based on the notion of a *double Satake diagram*, which consists of two kinds of Satake diagrams obtained from the compact symmetric pairs $(\mathfrak{g}_u, \theta_1)$ and $(\mathfrak{g}_u, \theta_2)$. The list of the correspondence between the set of $[(\mathfrak{g}_u, \theta_1, \theta_2)]$ and $[(\tilde{\Sigma}, \Sigma, W; m, n)]$ is given in [1] and [8].[a] We have already known the symmetric triads $(\tilde{\Sigma}', \Sigma', W'; m', n')$

---

[a] *Errata: A note on symmetric triad and Hermann action, by O. Ikawa, Proceedings of the workshop on differential geometry of submanifolds and its related topics, Saga, August 4–6 (2012), 220–229.* The following list should be added to the table on p. 228:
$$(\text{II-}BC_r) \mid (SO(4r+2), U(2r+1), S(O(2r+1) \times O(2r+1)))$$

with multiplicities which satisfy $(\tilde{\Sigma}', \Sigma', W'; m', n') \sim (\tilde{\Sigma}, \Sigma, W; m, n)$. By Theorem 4.2 and its proof, we explicitly know the compact symmetric triad $(\mathfrak{g}'_u, \theta'_1, \theta'_2)$ whose symmetric triad with multiplicities is $(\tilde{\Sigma}', \Sigma', W'; m', n')$. When $\mathfrak{g}_u$ is not simple or $\theta_1 \sim \theta_2$, similar results as Theorem 4.2 hold. In this case we can discuss a similar argument. When $[(\mathfrak{g}_u, \theta_1, \theta_2)]_\sim$ runs through $\mathfrak{A}/\sim$, then we get Berger's classification theorem.

6. Leung's classification theorems

6.1. *Reflective submanifolds of symmetric spaces*

Let G_u be a compact simply connected semisimple Lie group, and θ_1 be an involution on G_u. Let K_1 be a closed subgroup of G_u such that it lies between $F(\theta_1, G_u)(:= \{g \in G_u \mid \theta_1(g) = g\})$ and the identity component $F(\theta_1, G_u)_0$ of $F(\theta_1, G_u)$. Then the coset manifold G_u/K_1 has a structure of a Riemannian symmetric space of compact type. Under the assumption the following theorems are known:

Theorem 6.1 ([12, Theorem 3]). *Let θ_2 be an involution of G_u which commutes with θ_1, and set $K_2 = F(\theta_2, G_u)$. Then the identity component $(K_2/(K_1 \cap K_2))_0$ of $K_2/(K_1 \cap K_2)$ has a structure of a reflective submanifold of G_u/K_1 through the origin o. Conversely every reflective submanifold in G_u/K_1 through o is obtained in this way.*

It follows from Theorem 6.1 that the classification for reflective submanifolds of G_u/K_1 is reduced to that of involutions θ_2 of G_u commuting with θ_1.

Next we consider an infinitesimal model of a reflective submanifold, which was introduced by Leung [12]. For an involution ν of G_u, we use the same letter ν to denote the involution of $\mathfrak{g}_u$. We write $\mathfrak{g}_u = \mathfrak{k}_1 \oplus \mathfrak{m}_1$ for the canonical decomposition corresponding to θ_1. Take an invariant inner product $\langle \, , \, \rangle$ on $\mathfrak{g}_u$. A linear subspace $\mathfrak{b}$ of $\mathfrak{m}_1$ is said to be *reflective* if both $\mathfrak{b}$ and its orthogonal complement $\mathfrak{b}^\perp$ in $\mathfrak{m}_1$ are Lie triple systems which satisfy $[[\mathfrak{b}, \mathfrak{b}^\perp], \mathfrak{b}] \subset \mathfrak{b}^\perp$, $[[\mathfrak{b}^\perp, \mathfrak{b}], \mathfrak{b}^\perp] \subset \mathfrak{b}$. It follows easily from the definition that $\mathfrak{b}^\perp$ is also reflective. Then $\{\mathfrak{b}, \mathfrak{b}^\perp\}$ is called a *complementary pair of reflective subspaces*.

Theorem 6.2 ([12, Theorem 3]). *Let B be a totally geodesic submanifold of G_u/K_1 through the origin o and set $\mathfrak{b} = T_o(B) \subset T_o(G_u/K_1) = \mathfrak{m}_1$.*

(III-BC_r) | $(SU(2(2r+1)), S(U(2r+1) \times U(2r+1)), Sp(2r+1))$.

Then B is a reflective submanifold if and only if $\mathfrak{b}$ is a reflective subspace.

Let B and $B^{\perp}$ be two reflective submanifolds of G_u/K_1 through the origin o. Then $\{B, B^{\perp}\}$ is a *complementary pair of reflective submanifolds* if $\{\mathfrak{b}, \mathfrak{b}^{\perp}\}$ is a complementary pair of reflective subspaces, where we set $\mathfrak{b} = T_o(B) \subset \mathfrak{m}_1$ and $\mathfrak{b}^{\perp} = T_o(B^{\perp}) \subset \mathfrak{m}_1$ (cf. [13, p. 328]). Further B is called *self-complementary space* if $B^{\perp}$ is congruent to B. The above two theorems immediately imply the following corollary.

Corollary 6.1 ([13, Cor. 1.2]). *Let $\{B, B^{\perp}\}$ be a complementary pair of reflective submanifolds of G_u/K_1. We denote by θ_1 the involution on G_u which defines K_1, and by θ_2 the involution on G_u which is induced from B. Then $B^{\perp} = (G_{12}/(K_1 \cap K_2))_o$, where we set $K_i = F(\theta_i, G_u)$ and $G_{12} = F(\theta_1 \theta_2, G_u)$.*

From Corollary 6.1 we can determine all complementary pairs of reflective submanifolds of G_u/K_1. Further we can determine which reflective submanifolds are self-complementary spaces.

Example 6.1. When $(G_u, K_1) = (Sp(n), U(n))$: If $\theta_1 \sim \theta_2$ then $(G_{12}, K_1 \cap K_2) = (Sp(n-i) \times Sp(i), U(n-i) \times U(i))$ or $(G_{12}, K_1 \cap K_2) = (U(n), O(n))$. Thus the obtained reflective submanifolds $(K_2/(K_1 \cap K_2))_o$ are

$$\frac{U(n)}{U(n-i) \times U(i)} = \frac{SU(n)}{S(U(n-i) \times U(i))},$$

$$\frac{U(n)}{O(n)} \quad \text{(self-complementary space)}.$$

If $\theta_1 \not\sim \theta_2$ then $K_2 = Sp(n-i) \times Sp(i), (G_{12}, K_1 \cap K_2) = (U(n), U(i) \times U(n-i))$ or, $K_2 = Sp(p) \times Sp(p), (G_{12}, K_1 \cap K_2) = (Sp(p) \times Sp(p), Sp(p))$ when $n = 2p$. Thus the obtained reflective submanifolds $(K_2/(K_1 \cap K_2))_o$ are

$$\frac{Sp(n-i)}{U(n-i)} \times \frac{Sp(i)}{U(i)}, \quad Sp(p) \quad \text{(self-complementary space)},$$

and $\left\{ \frac{SU(n)}{S(U(n-i) \times U(i))}, \frac{Sp(n-i)}{U(n-i)} \times \frac{Sp(i)}{U(i)} \right\}$ is a complementary pair. The above reflective submanifolds exhaust of all reflective submanifolds of $Sp(n)/U(n)$ through the origin.

Here, we show the self-complementarity of the reflective submanifolds $\frac{U(n)}{O(n)}$ and $Sp(n)$. In the case when $B = (K_2/(K_1 \cap K_2))_o = U(n)/O(n)$, we have $B^{\perp} = (G_{12}/(K_1 \cap K_2))_o = U(n)/O(n)$ by Corollary 6.1. Since any reflective submanifolds of G_u/K_1 is congruent to one of the above reflective submanifolds, $B^\perp$ is congruent to B, i.e., B is self-complementary. Similarly, in the case when $B = (K_2/(K_1 \cap K_2))_o = Sp(n)$, we can verify that B is self-complementary.

In a similar manner in the example above, we can give an alternative proof of the Leung's classification theorem [13, Theorem 4.2] of reflective submanifolds of simply connected Riemannian symmetric spaces of compact type. Here we observe a difference between our proof and Leung's one for the classification of reflective submanifolds. Due to the above argument our proof is based on the classification of commuting involutions on compact semisimple Lie algebras. On the other hand, Leung introduced the notion of *reflective subalgebra* [13], which is another infinitesimal model for a reflective submanifold, and classified reflective submanifolds in terms of these subalgebras.

6.2. *Real forms of Hermitian symmetric spaces*

In this subsection, we will give an alternative proof of the classification for real forms of irreducible Hermitian symmetric spaces of compact type. For this purpose, we prepare two lemmas (cf. [15, Lemmas 3.1 and 3.2]).

Lemma 6.1. *Let $(M_1, J) = G_u/K_1$ be an irreducible Hermitian symmetric space of compact type, and $B = (K_2/(K_1 \cap K_2))_o$ a reflective submanifold of M_1 through the origin o. Let Z be the element of the center $\mathfrak{z}(\mathfrak{k}_1)$ of $\mathfrak{k}_1$ with $J = \mathrm{ad}\, Z$. Then Z is in $\mathfrak{k}_1 \cap \mathfrak{m}_2$ or in $\mathfrak{k}_1 \cap \mathfrak{k}_2$. Further*

(i) $Z \in \mathfrak{k}_1 \cap \mathfrak{m}_2$ *if and only if B is a real form of M_1.*

(ii) $Z \in \mathfrak{k}_1 \cap \mathfrak{k}_2$ *if and only if B is a complex submanifold. In this case B has a structure of a compact Hermitian symmetric space.*

Proof. Since $\mathfrak{k}_1$ is θ_2-invariant, $\mathfrak{z}(\mathfrak{k}_1)$ is also θ_2-invariant. Since $\dim \mathfrak{z}(\mathfrak{k}_1) = 1$, we have $Z \in \mathfrak{k}_1 \cap \mathfrak{m}_2$ or $Z \in \mathfrak{k}_1 \cap \mathfrak{k}_2$. Set $\mathfrak{b} = T_o(B) \subset \mathfrak{m}_1$ and $\mathfrak{b}^\perp = T_o^\perp(B) \subset \mathfrak{m}_1$. Then we have $\mathfrak{b} = \mathfrak{m}_1 \cap \mathfrak{k}_2$, $\mathfrak{b}^\perp = \mathfrak{m}_1 \cap \mathfrak{m}_2$ and an orthogonal direct sum decomposition $\mathfrak{m}_1 = \mathfrak{b} \oplus \mathfrak{b}^\perp$. Denote by σ the reflection which defines B. Since $\sigma_* = \mathrm{id}$ on $\mathfrak{b}$ and $\sigma_* = -\mathrm{id}$ on $\mathfrak{b}^\perp$, we have the following:

(i) B is a real form $\Leftrightarrow \sigma_* J = -J\sigma_* \Leftrightarrow J\mathfrak{b} = \mathfrak{b}^\perp, J\mathfrak{b}^\perp = \mathfrak{b} \Leftrightarrow Z \in \mathfrak{k}_1 \cap \mathfrak{m}_2$;

(ii) B is a complex submanifold $\Leftrightarrow \sigma_* J = J\sigma_* \Leftrightarrow Z \in \mathfrak{k}_1 \cap \mathfrak{k}_2$.

In (ii) B has a structure of a compact Hermitian symmetric space since so is M_1. $\qquad\square$

Lemma 6.2. *A real form B in a Hermitian symmetric space of compact type is a self-complementary reflective submanifold.*

Proof. It is clear that B is a reflective submanifold. We retain the notation as in the proof of Lemma 6.1. Then $J\mathfrak{b} = \mathfrak{b}^{\perp}$. Denote by $\langle \, , \, \rangle$ and R the Kähler metric and the Riemannian curvature tensor of M_1 respectively. Then $\langle JX, JY \rangle = \langle X, Y \rangle$ and $R(JX, JY)JW = J(R(X, Y)W)$. Since M_1 is simply connected and complete, we can extend J to an isometry on M_1. Then $B^{\perp} = JB$. Thus B is a self-complementary space. □

Due to Lemma 6.2, in order to classify real forms of irreducible Hermitian symmetric spaces M of compact type, it is sufficient to pick up all real forms among the set of self-complementary spaces. It is easy to pick up all self-complementary spaces B. And it is easy to determine whether an explicitly given self-complementary space B is a Hermitian symmetric space or not. If B is not a Hermitian symmetric space, then B is a real form. When B is a Hermitian symmetric space, we shall give a criterion to determine whether B is a real form or not. Since both $[\mathfrak{k}_1, \mathfrak{k}_1]$ and $\mathfrak{z}(\mathfrak{k}_1)$ are θ_2-invariant, we have $\mathfrak{k}_1 \cap \mathfrak{k}_2 = ([\mathfrak{k}_1, \mathfrak{k}_1] \cap \mathfrak{k}_2) \oplus (\mathfrak{z}(\mathfrak{k}_1) \cap \mathfrak{k}_2)$. Since $\mathfrak{z}(\mathfrak{k}_1) \cap \mathfrak{k}_2$ is abelian, we have $\mathfrak{z}(\mathfrak{k}_1 \cap \mathfrak{k}_2) = \mathfrak{z}([\mathfrak{k}_1, \mathfrak{k}_1] \cap \mathfrak{k}_2) \oplus (\mathfrak{z}(\mathfrak{k}_1) \cap \mathfrak{k}_2)$. Counting these dimensions, we get

$$\dim(\mathfrak{z}(\mathfrak{k}_1) \cap \mathfrak{k}_2) = \dim \mathfrak{z}(\mathfrak{k}_1 \cap \mathfrak{k}_2) - \dim \mathfrak{z}([\mathfrak{k}_1, \mathfrak{k}_1] \cap \mathfrak{k}_2).$$

By Lemma 6.1, B is a real form if and only if

$$\dim \mathfrak{z}(\mathfrak{k}_1 \cap \mathfrak{k}_2) - \dim \mathfrak{z}([\mathfrak{k}_1, \mathfrak{k}_1] \cap \mathfrak{k}_2) = 0.$$

Here we can count $\dim \mathfrak{z}(\mathfrak{k}_1 \cap \mathfrak{k}_2)$ since we have determined $B = (K_2/(K_1 \cap K_2))_0$ explicitly. We can also count $\dim \mathfrak{z}([\mathfrak{k}_1, \mathfrak{k}_1] \cap \mathfrak{k}_2)$ since we have determined the algebraic structure of $[\mathfrak{k}_1, \mathfrak{k}_1] \cap \mathfrak{k}_2$ using the symmetric triad $(\tilde{\Sigma}, \Sigma, W; m, n)$ with multiplicities induced from $(\mathfrak{g}_u, \theta_1, \theta_2)$. Hence we can determine whether a given self-complementary space is a real form or not. See [4] for the detail.

Example 6.2. When $M_1 = SU(p + q)/S(U(p) \times U(q))$, the self-complementary spaces through o are

(i) $B_1 = U(p)$ when $p = q$,
(ii) $B_2 = Sp((p + q)/2)/(Sp(p/2) \times Sp(q/2))$ when both p and q are even,
(iii) $B_3 = SO(p + q)/S(O(p) \times O(q))$ where $3 \leq p \leq q$,
(iv) $B_4 = SO(2 + q)/S(O(2) \times O(q))$ where $2 = p \leq q$,

(v) $B_5 = \dfrac{S(U(a) \times U(2k + q - a))}{S(U(k) \times U(a - k) \times U(q - a + k) \times U(k))}$

$\cong \dfrac{SU(a)}{S(U(k) \times U(a - k))} \times \dfrac{SU(2k + q - a)}{S(U(q - a + k) \times U(k))}$,

when $2k = p \leq q$ and $k < a < q + k$.

Then

- the spaces B_1, B_2 and B_3 are real forms in M_1 since these are not Hermitian symmetric spaces.
- In the case of (iv), since $\mathfrak{k}_1 \cap \mathfrak{k}_2 = [\mathfrak{k}_1, \mathfrak{k}_1] \cap \mathfrak{k}_2 = \mathfrak{so}(2) \oplus \mathfrak{so}(q)$, we have $\dim \mathfrak{z}(\mathfrak{k}_1 \cap \mathfrak{k}_2) - \dim \mathfrak{z}([\mathfrak{k}_1, \mathfrak{k}_1] \cap \mathfrak{k}_2) = 0$. Hence B_4 is a real form.
- In the case of (v), since

$$\mathfrak{k}_1 \cap \mathfrak{k}_2 = \mathfrak{s}(\mathfrak{u}(k) \oplus \mathfrak{u}(a - k) \oplus \mathfrak{u}(q - a + k) \oplus \mathfrak{u}(k)),$$

$$[\mathfrak{k}_1, \mathfrak{k}_1] \cap \mathfrak{k}_2 = \mathfrak{s}(\mathfrak{u}(k) \oplus \mathfrak{u}(k)) \oplus \mathfrak{s}(\mathfrak{u}(a - k) \oplus \mathfrak{u}(q - a + k)),$$

we have $\dim \mathfrak{z}(\mathfrak{k}_1 \cap \mathfrak{k}_2) - \dim \mathfrak{z}([\mathfrak{k}_1, \mathfrak{k}_1] \cap \mathfrak{k}_2) = 3 - 2 = 1$. Hence B_5 is not a real form.

The following table is the lists of all self-complementary spaces $B = (K_2/(K_1 \cap K_2))_0$ in irreducible Hermitian symmetric spaces $M_1 = G_u/K_1$ of compact type. When G_u is of exceptional type, we indicate Lie algebra structures $(\mathfrak{k}_2, \mathfrak{k}_1 \cap \mathfrak{k}_2)$ and $(\mathfrak{g}_u, \mathfrak{k}_1)$ instead of global structures $(K_2/(K_1 \cap K_2))_0$ and G_u/K_1. And we indicate which complementary spaces are real forms. We also describe the algebraic structure of $[\mathfrak{k}_1, \mathfrak{k}_1] \cap \mathfrak{k}_2$ when B is a Hermitian symmetric space.

Here we observe a difference between our proof and Takeuchi's one for the classification of real forms of Hermitian symmetric spaces of compact type. Our proof is based on the classification of self-complementary spaces. On the other hand, Takeuchi reduced the classification of real forms to that of positive definite symmetric graded Lie algebras. See [18] for the details.

Acknowledgment

The author is very indebted to both the anonymous referee and the editor for careful reading of the manuscript and useful suggestions.

Table

$M_1 = G_u/K_1$	$B = (K_2/(K_1 \cap K_2))_0$	$[\mathfrak{k}_1, \mathfrak{k}_1] \cap \mathfrak{k}_2$	Real form
$SO(4m)/U(2m)$	$U(2m)/Sp(m)$	/	Yes
	$SO(2m)$	/	Yes
$SO(4m+2)/U(2m+1)$	$SO(2m+1)$	/	Yes
$Sp(n)/U(n)$	$U(n)/O(n)$	/	Yes
	$Sp(n/2)$ $(n:\text{even})$	/	Yes
$SO(2+q)/(SO(2) \times SO(q))$	$S^a \times S^{q-a}$	/	Yes
$(q \geq 3)$	$U(r+1)/(U(1) \times U(r))$ $(q=2r)$	$\mathfrak{u}(r)$	No
	$\dfrac{SO(2+r)}{SO(2) \times SO(r)}\left(\times \dfrac{SO(r)}{SO(r)}\right)$ $(q=2r)$	$\mathfrak{so}(r) \oplus \mathfrak{so}(r)$	No
$SU(p+q)/S(U(p) \times U(q))$	See Example 6.2		
$(\mathfrak{e}_6, \mathfrak{so}(10) \oplus \mathfrak{so}(2))$	$(\mathfrak{f}_4, \mathfrak{so}(9))$	/	Yes
	$(\mathfrak{sp}(4), \mathfrak{sp}(2) \oplus \mathfrak{sp}(2))$	/	Yes
	$(\mathfrak{su}(6), \mathfrak{s}(\mathfrak{u}(2) \oplus \mathfrak{u}(4))(\oplus(\mathfrak{su}(2), \mathfrak{su}(2))))$	$\mathfrak{so}(4) \oplus \mathfrak{so}(6)$	No
	$(\mathfrak{so}(10), \mathfrak{so}(8) \oplus \mathfrak{so}(2))(\oplus(\mathfrak{so}(2), \mathfrak{so}(2)))$	$\mathfrak{so}(8) \oplus \mathfrak{so}(2)$	No
$(\mathfrak{e}_7, \mathfrak{e}_6 \oplus \mathfrak{so}(2))$	$(\mathfrak{e}_6, \mathfrak{f}_4) \oplus (\mathfrak{so}(2), \{0\})$	/	Yes
	$(\mathfrak{su}(8), \mathfrak{sp}(4))$	/	Yes

References

[1] K. Baba and O. Ikawa, Symmetric triads and double Satake diagrams for compact symmetric triads, in preparation.

[2] K. Baba, O. Ikawa and A. Sasaki, A duality between compact symmetric triads and semisimple pseudo-Riemannian symmetric pairs with applications to geometry of Hermann type actions, *Hermitian-Grassmannian Submanifolds*, 211–221, Springer Proc. Math. Stat., 203, Springer, Singapore, 2017.

[3] K. Baba, O. Ikawa and A. Sasaki, A duality between symmetric pairs and compact symmetric triads, in preparation.

[4] K. Baba, O. Ikawa and A. Sasaki, An alternative proof for Berger's classification of semisimple pseudo-Riemannian symmetric pairs from the view point of compact symmetric triads, in preparation.

[5] M. Berger, Les espaces symetriques noncompacts, *Ann. Sci. Ecole Norm. Sup.* **74** (1957), 85–177.

[6] R. Hermann, Totally geodesic orbits of groups of isometries, *Nerderl. Akad. Wetensch. Proc. Ser. A* (1962), 291–298.

[7] O. Ikawa, The geometry of symmetric triad and orbit spaces of Hermann actions, *J. Math. Soc. Japan* **63** (2011) 70–136, DOI: 10.2969/jmsj/06310079.

[8] O. Ikawa, A note on symmetric triad and Hermann action, *Proceedings of the workshop on differential geometry of submanifolds and its related topics*, Saga, August 4–6 (2012), 220–229.

[9] O. Ikawa, The geometry of orbits of Hermann type actions, *Contemporary Perspectives in Differential Geometry and its Related Fields*, T. Adachi, H. Hashimoto and M. Hristov eds. (2018) 67–78, World Scientific.

[10] O. Ikawa, σ-actions and symmetric triads, *Tohoku Math. J.* **70** (2018) 547–565.

[11] S. Kobayashi and K. Nomizu, *Foundations of differential geometry Vol. II*, John Wiley & Sons, Inc., New York (1996).

[12] D. S. P. Leung, The reflection principle for minimal submanifolds of Riemannian symmetric spaces, *J. Differential Geom.* **8** (1973) 153–160.

[13] D. S. P. Leung, On the classification of reflective submanifolds of Riemannian symmetric spaces, *Indiana Univ. Math. J.* **24** (1974) 327–339.

[14] D. S. P. Leung, Errata: On the classification of reflective submanifolds of Riemannian symmetric spaces, *Indiana Univ. Math. J.* **24** (1975) 1199.

[15] D. S. P. Leung, Reflective submanifolds IV. Classification of real forms of Hermitian symmetric spaces, *J. Differential Geom.* **14** (1979) 179–185.

[16] O. Loos, Symmetric spaces I: General theory, Benjamin, Inc., New York-Amsterdam, 1969.

[17] T. Matsuki, Classification of two involutions on compact semisimple Lie groups and root systems, *J. Lie Theory* **12** (2002), 41–68.

[18] M. Takeuchi, Stability of certain minimal submanifolds of compact Hermitian symmetric spaces, *Tohoku Math. J.* **36** (1984), 293–314.

Received November 30, 2018
Revised February 25, 2019

AN INDECOMPOSABLE REPRESENTATION
AND THE COMPLEX VECTOR SPACE
OF HOLOMORPHIC VECTOR FIELDS
ON A PSEUDO-HERMITIAN SYMMETRIC SPACE

Nobutaka BOUMUKI*

*Division of Mathematical Sciences,
Faculty of Science and Technology, Oita University,
700 Dannoharu, Oita-shi, Oita 870-1192, Japan
E-mail: boumuki@oita-u.ac.jp*

In this paper we clarify a property of the complex vector space of holomorphic
vector fields on a simple irreducible pseudo-Hermitian symmetric space by
means of the representation theory.

Keywords: Simple irreducible pseudo-Hermitian symmetric space; indecomposable representation; canonical central element.

1. Introduction

For a complex manifold $M = (M, J)$, we suppose that a Lie group G acts
on M holomorphically, $G \times M \ni (g, p) \mapsto \tau_g(p) \in M$. Then, one can set
the complex vector space $\mathcal{O}(T^{1,0}M)$ of holomorphic vector fields on M, and
can naturally define a representation ϱ of G on $\mathcal{O}(T^{1,0}M)$ as follows:

$$\big(\varrho(g)Z\big)_p := d\tau_g(Z_{\tau_{g^{-1}}(p)}) \text{ for } (g, Z) \in G \times \mathcal{O}(T^{1,0}M) \text{ and } p \in M.$$

In this paper we deal with the case that M is an effective, simple irreducible
pseudo-Hermitian symmetric space and clarify a property of the complex
vector space $\mathcal{O}(T^{1,0}M)$ in view of the representation ϱ.

Let us give an outline of the main result in this paper. For an effective,
simple irreducible pseudo-Hermitian symmetric space G/L, one can regard
it as a domain in a compact Hermitian symmetric space $G_{\mathbb{C}}/Q_-$ via the
generalized Borel embedding $\iota : G/L \to G_{\mathbb{C}}/Q_-$, $gL \mapsto gQ_-$. This allows
us to identify

$$\mathcal{V} := \left\{ \psi : GQ_- \to \mathfrak{u}_+ \,\middle|\, \begin{array}{l} \psi \text{ is holomorphic,} \\ \psi(xq) = \rho(q)^{-1}\big(\psi(x)\big) \text{ for all } (x, q) \in GQ_- \times Q_- \end{array} \right\}$$

---
*This work was supported by JSPS KAKENHI Grant Number JP 17K05229.

with the complex vector space $\mathcal{O}(T^{1,0}(G/L))$ of holomorphic vector fields on G/L. Then, we get a group homomorphism $\varrho : G \to GL(\mathcal{V})$, $g \mapsto \varrho(g)$, by setting

$$\big(\varrho(g)\psi\big)(x) := \psi(g^{-1}x) \quad \text{for } (g, \psi) \in G \times \mathcal{V} \text{ and } x \in GQ_-,$$

and conclude that the mapping $G \times \mathcal{V} \ni (g, \psi) \mapsto \varrho(g)\psi \in \mathcal{V}$ is continuous, where the topology for $\mathcal{V}$ is the topology of uniform convergence on compact sets. In this paper, we establish the following statement: *Under the above conditions, there is a* $\dim_{\mathbb{C}} G/L$-*dimensional closed* $\varrho(L)$-*invariant complex vector subspace* $\mathcal{W}$ *of* $\mathcal{V}$ *having the following two properties*:

(1) *The representation module* $\mathcal{W}$ *of* $\varrho|_L$ *is isomorphic to the representation module* $\mathfrak{u}_+$ *of* $\mathrm{Ad}\,|_L$;
(2) *The inclusion* $\mathcal{W} \subset \mathcal{V}_1$ *always holds for an arbitrary closed* $\varrho(G)$-*invariant complex vector subspace* $\mathcal{V}_1$ *of* $\mathcal{V}$ *with* $\mathcal{V}_1 \neq \{0\}$

(see Theorem 4.1 for detail). Remark here, (2) implies that the continuous representation ϱ of G on $\mathcal{V}$ is indecomposable.

This paper consists of four sections. In §2 we fix the notations utilized in this paper, and recall the definition of indecomposable representation. In §3 we review fundamental facts about pseudo-Hermitian symmetric spaces. In §4 we first review fundamental facts about holomorphic tangent bundles and afterwards establish Theorem 4.1.

2. Notations and indecomposable representations

We fix the notations utilized in this paper, and recall the definition of indecomposable representation.

Throughout this paper, for a Lie group G, we denote its Lie algebra by the corresponding Fraktur small letter $\mathfrak{g}$, and utilize the following notations:

(n1) $i := \sqrt{-1}$,
(n2) Ad, ad : the adjoint representation of G, $\mathfrak{g}$,
(n3) $C_G(T) := \{g \in G \,|\, \mathrm{Ad}\,g(T) = T\}$ for an element $T \in \mathfrak{g}$,
(n4) $N_G(\mathfrak{m}) := \{g \in G \,|\, \mathrm{Ad}\,g(\mathfrak{m}) \subset \mathfrak{m}\}$ for a vector subspace $\mathfrak{m} \subset \mathfrak{g}$,
(n5) $\mathfrak{m} \oplus \mathfrak{n}$: the direct sum of vector spaces $\mathfrak{m}$ and $\mathfrak{n}$,
(n6) $M^f := \{p \in M \,|\, f(p) = p\}$ for a set M and a mapping $f : M \to M$,
(n7) $f|_S$: the restriction of a mapping f to a set S.

Now, let us recall the definition of indecomposable representation. Let G be a topological group, let $\mathcal{V}$ be a complex topological vector space which satisfies the first separation axiom, and let $\varrho : G \to GL(\mathcal{V})$, $g \mapsto \varrho(g)$, be a group homomorphism, where $GL(\mathcal{V})$ denotes the general linear group of $\mathcal{V}$ and it does not matter whether ϱ is continuous here. Then, the definition of indecomposable representation is as follows:

Definition 2.1 (cf. Kirillov [3, p. 112]).
(i) ϱ is called a *continuous representation* of G on $\mathcal{V}$, if the mapping $G \times \mathcal{V} \ni (g, \psi) \mapsto \varrho(g)\psi \in \mathcal{V}$ is continuous.
(ii) We say that a continuous representation ϱ of G on $\mathcal{V}$ is (*topologically*) *indecomposable*, if $\mathcal{V}$ cannot be decomposed into the direct sum of two closed $\varrho(G)$-invariant complex vector subspaces different from $\mathcal{V}$ and $\{0\}$.

Remark 2.1. If G is a connected semisimple Lie group and $\dim_{\mathbb{C}} \mathcal{V} < \infty$, then any indecomposable representation ϱ of G on $\mathcal{V}$ is irreducible by Weyl's theorem on semisimplicity of representations.

3. Pseudo-Hermitian symmetric spaces

We review fundamental facts about pseudo-Hermitian symmetric spaces.

Let us recall the definition of symmetric space, and then state the definition of pseudo-Hermitian symmetric space.

Definition 3.1 (cf. Nomizu [4, p. 52, p. 56]).
(i) Let G be a connected Lie group, and let L be a closed subgroup of G. The homogeneous space G/L is said to be an *affine symmetric space* or a *symmetric space*, if there exists an involutive automorphism σ of G satisfying

$$(G^\sigma)_0 \subset L \subset G^\sigma,$$

where $(G^\sigma)_0$ denotes the identity component of G^σ.
(ii) A symmetric space G/L is called *effective* (resp. *almost effective*), if G is effective (resp. almost effective) on G/L as transformation group.
(iii) A symmetric space G/L is said to be *semisimple* (resp. *simple*), if $\mathfrak{g}$ is semisimple (resp. simple).
(iv) An almost effective, semisimple symmetric space $(G/L, \sigma)$ is called *irreducible*, if the action of $\mathrm{ad}\,\mathfrak{l}$ on $\mathfrak{u}$ is irreducible. Here, $\mathfrak{u} := \mathfrak{g}^{-\sigma_*}$ and we denote by σ_* the differential homomorphism of $\sigma : G \to G$.

Definition 3.2 (cf. Berger [1, p. 94]). A symmetric space G/L is said to be *pseudo-Hermitian*, if it admits a G-invariant complex structure J and a G-invariant pseudo-Hermitian metric $\mathbf{g}$ (with respect to J).

Remark 3.1. A simple pseudo-Hermitian symmetric space G/L is irreducible if and only if $\mathfrak{g}$ is a real form of a complex simple Lie algebra (cf. Shapiro [5, p. 532]).

Let us review known results about pseudo-Hermitian symmetric spaces.

Proposition 3.1 (cf. Shapiro [5, pp. 533–534]). *Let* $(G/L, \sigma, J, \mathrm{g})$ *be an almost effective, semisimple pseudo-Hermitian symmetric space, and let* $\mathfrak{u} := \mathfrak{g}^{-\sigma*}$. *Then, there exists a unique* $T \in \mathfrak{l}$ *such that*

$$\text{(i) } L = C_G(T) = (G^\sigma)_0, \quad \text{(ii) } J_o = \operatorname{ad} T|_{\mathfrak{u}}.$$

Here we identify $\mathfrak{u}$ *with the tangent space of* G/L *at the origin* o.

Remark 3.2. The element T in Proposition 3.1 is called the *canonical central element* of $\mathfrak{l}$ (cf. Shapiro [5, p. 533]).

In addition to Remark 3.2, we pay attention to

Remark 3.3. Proposition 3.1-(i) implies that a simple irreducible pseudo-Hermitian symmetric space G/L is effective if and only if the center $Z(G)$ of G is trivial. Consequently, for any effective simple irreducible pseudo-Hermitian symmetric space G/L, there exists a connected complex simple Lie group $G_\mathbb{C}$ such that

(1) $Z(G_\mathbb{C})$ is trivial,
(2) G is a connected closed subgroup of $G_\mathbb{C}$,
(3) $\mathfrak{g}$ is a real form of $\mathfrak{g}_\mathbb{C}$.

Indeed, take the complexification $\mathfrak{g}_\mathbb{C}$ of $\mathfrak{g}$, and let $G_\mathbb{C}$ be the adjoint group of $\mathfrak{g}_\mathbb{C}$. Then one can realize G as the connected Lie subgroup of $G_\mathbb{C}$ corresponding to the subalgebra $\operatorname{ad}\mathfrak{g} \subset \operatorname{ad}\mathfrak{g}_\mathbb{C}$, because the Lie group G is isomorphic to the adjoint group of $\mathfrak{g}$.

Now, let G/L be an effective, simple irreducible pseudo-Hermitian symmetric space, let T denote the canonical central element of $\mathfrak{l}$, and let $G_\mathbb{C}$ be a connected complex simple Lie group such that

(1) the center $Z(G_\mathbb{C})$ of $G_\mathbb{C}$ is trivial,
(2) G is a connected closed subgroup of $G_\mathbb{C}$,
(3) $\mathfrak{g}$ is a real form of $\mathfrak{g}_\mathbb{C}$.

Setting

$$\begin{aligned}
&L_\mathbb{C} := C_{G_\mathbb{C}}(T), \quad \mathfrak{u}_\pm := \{A \in \mathfrak{g}_\mathbb{C} \mid \operatorname{ad} T(A) = \pm iA\}, \\
&U_\pm := \exp \mathfrak{u}_\pm, \quad Q_- := N_{G_\mathbb{C}}(\mathfrak{l}_\mathbb{C} \oplus \mathfrak{u}_-),
\end{aligned} \tag{3.1}$$

one has $\mathfrak{g}_\mathbb{C} = \mathfrak{u}_+ \oplus \mathfrak{q}_-$, $Q_- = L_\mathbb{C} \ltimes U_-$ (semi-direct), $L = G \cap Q_-$, and obtains a G-equivariant holomorphic embedding $\iota : G/L \to G_\mathbb{C}/Q_-$, $gL \mapsto gQ_-$, whose image is a simply connected domain in $G_\mathbb{C}/Q_-$. This mapping $\iota : G/L \to G_\mathbb{C}/Q_-$ is called the *generalized Borel embedding*, cf. Shapiro [5, Theorem 3.1, p. 535]. Here, $G_\mathbb{C}/Q_-$ is biholomorphic to a Hermitian symmetric space of the compact type.

4. Holomorphic tangent bundles

We first review fundamental facts about holomorphic tangent bundles and afterwards establish Theorem 4.1.

We are going to mention a holomorphic vector field on the symmetric space G/L (here, we obey the setting and notations in the latter part of the previous section). For a $q \in Q_- = L_\mathbb{C} \ltimes U_-$, there exists a unique $(l, u) \in L_\mathbb{C} \times U_-$ such that $q = lu$. This, together with $\operatorname{Ad} L_\mathbb{C}(\mathfrak{u}_+) \subset \mathfrak{u}_+$, enables us to define a holomorphic homomorphism $\rho : Q_- \to GL(\mathfrak{u}_+)$, $q \mapsto \rho(q)$, by

$$\rho(q)X := \operatorname{Ad} l(X) \quad \text{for } q = lu \in Q_- = L_\mathbb{C} \ltimes U_- \text{ and } X \in \mathfrak{u}_+. \qquad (4.1)$$

The representation module $\mathfrak{u}_+$ of ρ and the representation module $\mathfrak{g}_\mathbb{C}/\mathfrak{q}_-$ of the linear isotropy representation $\operatorname{Ad}_{\mathfrak{g}_\mathbb{C}/\mathfrak{q}_-} : Q_- \to GL(\mathfrak{g}_\mathbb{C}/\mathfrak{q}_-)$ are isomorphic. Accordingly, the holomorphic tangent bundle $T^{1,0}(G_\mathbb{C}/Q_-)$ over the compact Hermitian symmetric space $G_\mathbb{C}/Q_-$ is the homogeneous holomorphic vector bundle over $G_\mathbb{C}/Q_-$ associated with $\rho : Q_- \to GL(\mathfrak{u}_+)$. Thus, for the complex vector space $\mathcal{O}(T^{1,0}(G_\mathbb{C}/Q_-))$ of holomorphic vector fields on $G_\mathbb{C}/Q_-$, we may assume that

$$\mathcal{O}(T^{1,0}(G_\mathbb{C}/Q_-))$$
$$= \left\{ h : G_\mathbb{C} \to \mathfrak{u}_+ \;\middle|\; \begin{array}{l} h \text{ is holomorphic,} \\ h(xq) = \rho(q)^{-1}\big(h(x)\big) \text{ for all } (x, q) \in G_\mathbb{C} \times Q_- \end{array} \right\}.$$

That allows us to identify

$$\mathcal{V} := \left\{ \psi : GQ_- \to \mathfrak{u}_+ \;\middle|\; \begin{array}{l} \psi \text{ is holomorphic,} \\ \psi(xq) = \rho(q)^{-1}\big(\psi(x)\big) \text{ for all } (x, q) \in GQ_- \times Q_- \end{array} \right\}$$
$$\tag{4.2}$$

with the complex vector space $\mathcal{O}(T^{1,0}(G/L))$ of holomorphic vector fields on G/L, where we regard G/L as a domain in $G_\mathbb{C}/Q_-$ via the generalized Borel embedding $\iota : G/L \to G_\mathbb{C}/Q_-$, $gL \mapsto gQ_-$.

Remark 4.1. Taking an $X \in \mathfrak{g}_\mathbb{C}$, one can define a smooth vector field X^* on $\iota(G/L)$ by

$$X_p^* f := \frac{d}{dt}\Big|_{t=0} f\big(\tau_{\exp(-tX)}(p)\big)$$

for a point $p \in \iota(G/L) \subset G_\mathbb{C}/Q_-$ and a smooth function f around p, where $\tau_x(aQ_-) := xaQ_-$ for $x \in G_\mathbb{C}$ and $aQ_- \in G_\mathbb{C}/Q_-$. This correspondence $X \mapsto X^*$ is one-to-one (because $G_\mathbb{C}$ is effective on $G_\mathbb{C}/Q_-$) and X^* is an infinitesimal automorphism of the complex structure on $\iota(G/L)$. For this reason, one may assume that $\mathfrak{g}_\mathbb{C} \subset \mathcal{V}$.

We want to define a topology for the vector space $\mathcal{V}$ and a group homomorphism $\varrho : G \to GL(\mathcal{V})$. First, let us consider a topology for $\mathcal{V}$. For a non-empty compact subset $E \subset GQ_-$ and $\psi_1, \psi_2 \in \mathcal{V}$, we put

$$d_E(\psi_1, \psi_2) := \sup\big\{\|\psi_1(a) - \psi_2(a)\| : a \in E\big\},$$

where $\|X\|$ denotes an arbitrary norm of $X \in \mathfrak{u}_+$. Since $G_\mathbb{C}$ is connected and GQ_- is open in $G_\mathbb{C}$, there exist non-empty open subsets $W_n \subset GQ_-$ such that

(1) $GQ_- = \bigcup_{n=1}^\infty W_n$ (countable union),
(2) the closure $\overline{W_n}$ in GQ_- is compact for each $n \in \mathbb{N}$.

Then, we define $E_n := \overline{W_n}$ for $n \in \mathbb{N}$ and moreover define

$$d(\psi_1, \psi_2) := \sum_{n=1}^\infty \frac{1}{2^n} \frac{d_{E_n}(\psi_1, \psi_2)}{1 + d_{E_n}(\psi_1, \psi_2)} \tag{4.3}$$

for $\psi_1, \psi_2 \in \mathcal{V}$. This d is called the *Fréchet metric*. Next, let us set a group homomorphism $\varrho : G \to GL(\mathcal{V})$, $g \mapsto \varrho(g)$, as follows:

$$\big(\varrho(g)\psi\big)(x) := \psi(g^{-1}x) \quad \text{for } \psi \in \mathcal{V},\ x \in GQ_-. \tag{4.4}$$

In the setting above, one can show

Proposition 4.1. *The following four items hold*:

(1) *$(\mathcal{V}, d)$ is a complete metric space.*
(2) *The metric topology for $(\mathcal{V}, d)$ coincides with the topology of uniform convergence on compact sets; and besides it also coincides with the locally convex topology determined by a countable number of seminorms $\{p_n\}_{n \in \mathbb{N}}$, where $p_n(\psi) := d_{E_n}(\psi, 0)$ for $n \in \mathbb{N}$, $\psi \in \mathcal{V}$.*
(3) *Both mappings $\mathcal{V} \times \mathcal{V} \ni (\psi_1, \psi_2) \mapsto \psi_1 + \psi_2 \in \mathcal{V}$ and $\mathbb{C} \times \mathcal{V} \ni (\alpha, \psi) \mapsto \alpha\psi \in \mathcal{V}$ are continuous.*

(4) $G \times \mathcal{V} \ni (g, \psi) \mapsto \varrho(g)\psi \in \mathcal{V}$ *is a continuous mapping.*

Proof. Refer to the arguments in [2, Paragraph 2.4.4]. □

We end the review about bundles with the following remark:

Remark 4.2. In accordance with the correspondence between $\mathcal{V}$ and $\mathcal{O}(T^{1,0}(G/L))$, the representation ϱ of G on $\mathcal{V}$ in (4.4) corresponds to the representation ϱ of G on $\mathcal{O}(T^{1,0}(G/L))$ at page 139.

Now, the main result in this paper is as follows:

Theorem 4.1. *Let G/L be an effective, simple irreducible pseudo-Hermitian symmetric space, and let ϱ be the continuous representation of G on the complex vector space $\mathcal{V}$ of holomorphic vector fields on G/L defined by (4.4). Then, there exists a $\dim_{\mathbb{C}} G/L$-dimensional closed $\varrho(L)$-invariant complex vector subspace $\mathcal{W}$ of $\mathcal{V}$ having the following two properties:*

(1) *The representation module $\mathcal{W}$ of $\varrho|_L$ is isomorphic to the representation module $\mathfrak{u}_+$ of $\mathrm{Ad}|_L$;*
(2) *The inclusion $\mathcal{W} \subset \mathcal{V}_1$ always holds for an arbitrary closed $\varrho(G)$-invariant complex vector subspace $\mathcal{V}_1$ of $\mathcal{V}$ with $\mathcal{V}_1 \neq \{0\}$.*

Here, the topology for $\mathcal{V}$ is the topology of uniform convergence on compact sets. cf. (3.1), (4.1), (4.2).

Theorem 4.1-(2) implies that the continuous representation ϱ of G on $\mathcal{V}$ is indecomposable. So, one has

Corollary 4.1. *In the setting of Theorem 4.1; suppose that G is compact. Then, we have the following:*

(i) *The representation $\varrho : G \to GL(\mathcal{V})$ is irreducible;*
(ii) *$\mathcal{V}$ is isomorphic to $\mathfrak{g}_{\mathbb{C}}$ as complex vector space;*
(iii) *The identity component of the group $\mathrm{Hol}(G/L)$ of holomorphic automorphisms of G/L is isomorphic to $G_{\mathbb{C}}$ as complex Lie group.*

Here $G_{\mathbb{C}}$ is a connected complex simple Lie group such that (1) its center $Z(G_{\mathbb{C}})$ is trivial, (2) G is a connected closed subgroup of $G_{\mathbb{C}}$ and (3) $\mathfrak{g}$ is a real form of $\mathfrak{g}_{\mathbb{C}}$.

From now on, our aim is to complete the proof of Theorem 4.1. For the aim, we are going to give Lemma 4.1 and Proposition 4.2 first. One can deduce the lemma and the proposition by arguments similar to those in the proof of two Lemmas 3.5-(iv) and 3.11, and two Lemmas 3.4-(iii) and 3.7 in [2], respectively.

Lemma 4.1.

(i) *If $\psi_1, \psi_2 \in \mathcal{V}$ and $\psi_1 = \psi_2$ on $(U_+ \cap GQ_-)_e$, then $\psi_1 = \psi_2$ on the whole GQ_-.*

(ii) *For $\varphi \in \mathcal{V}$, suppose that φ is constant on $(U_+ \cap GQ_-)_e$. Then it follows that $\varphi(\ell x \ell^{-1}) = \varphi(x)$ for all $(\ell, x) \in L \times GQ_-$.*

Here $(U_+ \cap GQ_-)_e$ denotes the connected component of $U_+ \cap GQ_-$ containing the unit element $e \in G_{\mathbb{C}}$.

By use of the canonical central element $T \in \mathfrak{l}$, we define

$$\hat{\psi}_\lambda(x) := \psi\big((\exp \lambda T)x \exp(-\lambda T)\big) \tag{4.5}$$

for $(\psi, \lambda) \in \mathcal{V} \times [0, 2\pi]$ and $x \in GQ_-$. cf. Remark 3.2. Note that $\hat{\psi}_\lambda = \varrho(\exp(-\lambda T))(e^{i\lambda}\psi)$ in terms of (4.1), (4.2) and (4.4).

Proposition 4.2. *The following two items hold for each $\psi \in \mathcal{V}$:*

(1) $\displaystyle\int_0^{2\pi} \hat{\psi}_\lambda d\lambda$ *belongs to $\mathcal{V}$.*

(2) $\displaystyle\int_0^{2\pi} \hat{\psi}_\lambda d\lambda$ *is the constant mapping with the value $2\pi\psi(e)$ on $(U_+ \cap GQ_-)_e$.*

Now, we define a complex vector subspace $\mathcal{W} \subset \mathcal{V}$ by

$$\mathcal{W} := \{\varphi \in \mathcal{V} \mid \varphi \text{ is constant on } (U_+ \cap GQ_-)_e\}. \tag{4.6}$$

Let us confirm Lemma 4.2, which will play a role later.

Lemma 4.2. *For the $\mathcal{W}$ in (4.6), the following four items hold:*

(1) *$\mathcal{W}$ is a $\varrho(L)$-invariant complex vector subspace of $\mathcal{V}$.*

(2) *$\mathcal{W} \neq \{0\}$.*

(3) *Let $F : \mathcal{W} \to \mathfrak{u}_+$ be a linear mapping defined by*

$$F(\varphi) := \varphi(e) \text{ for } \varphi \in \mathcal{W}.$$

Then, the representation module $\mathcal{W}$ of $\varrho|_L$ is isomorphic to the representation module $\mathfrak{u}_+$ of $\mathrm{Ad}\,|_L$ via F. In particular, $\mathcal{W}$ is an irreducible module of $\varrho|_L$ and $\dim_{\mathbb{C}} \mathcal{W} = \dim_{\mathbb{C}} G/L$.

(4) *$\mathcal{W}$ is a closed subset in $\mathcal{V}$.*

Proof. (1) Fix an element $(\ell, \varphi) \in L \times W$. For any $u \in (U_+ \cap GQ_-)_e$ we see that

$$\big(\varrho(\ell)\varphi\big)(u) = \varphi(\ell^{-1}u) = \varphi(u\ell^{-1}) = \operatorname{Ad}\ell\big(\varphi(u)\big) = \operatorname{Ad}\ell\big(\varphi(e)\big)$$

in view of (4.4), $\varphi \in W$, Lemma 4.1-(ii), (4.1) and (4.2). This tells us that $\big(\varrho(\ell)\varphi\big)(u)$ does not depend on $u \in (U_+ \cap GQ_-)_e$, so that $\varrho(\ell)\varphi \in W$ due to (4.6). Hence, W is $\varrho(L)$-invariant.

(2) Since $\mathfrak{g}_{\mathbb{C}} \hookrightarrow V$, there exists a $\psi_a \in V - \{0\}$. So, it is immediate from (4.2) that there exists a $g \in G$ satisfying $\psi_a(g) \neq 0$. Setting $\psi_b := \varrho(g^{-1})\psi_a$ we obtain

$$\psi_b \in V, \ \psi_b(e) \neq 0.$$

Hence, $0 \neq \displaystyle\int_0^{2\pi} (\hat{\psi}_b)_\lambda d\lambda \in W$ follows by Proposition 4.2 and (4.6).

(3) Lemma 4.1-(i) assures that F is injective. By arguments in (1) we see that $F \circ \varrho(\ell) = \operatorname{Ad}\ell \circ F$ for all $\ell \in L$. Consequently $F(W)$ is an $\operatorname{Ad}L$-invariant complex vector subspace of $\mathfrak{u}_+$ and $F(W) \neq \{0\}$ due to (1), (2). Thus $F(W) = \mathfrak{u}_+$, namely F is surjective, because $\mathfrak{u}_+$ is an irreducible module of $\operatorname{Ad}|_L$. Therefore $(W, \varrho|_L)$ and $(\mathfrak{u}_+, \operatorname{Ad}|_L)$ are isomorphic via F. The last statement comes from $\dim_{\mathbb{C}} \mathfrak{u}_+ = \dim_{\mathbb{C}} G/L$.

(4) W is a finite-dimensional vector subspace of V and V satisfies the first separation axiom. Hence we conclude (4). □

We are in a position to prove Theorem 4.1.

Proof of Theorem 4.1. Let V_1 be an arbitrary closed $\varrho(G)$-invariant complex vector subspace of V with $V_1 \neq \{0\}$. By Lemma 4.2, it suffices to prove that

$$W \subset V_1,$$

where W is given in (4.6). From $V_1 \neq \{0\}$ we obtain a non-zero $\psi_1 \in V_1$, and there exists a $g \in G$ such that $\psi_1(g) \neq 0$. Therefore $\psi_2 := \varrho(g^{-1})\psi_1$ satisfies $\psi_2(e) \neq 0$, and it belongs to V_1 because V_1 is $\varrho(G)$-invariant. Moreover, $(\hat{\psi}_2)_\lambda = \varrho(\exp(-\lambda T))\big(e^{i\lambda}\psi_2\big) \in V_1$ for all $\lambda \in [0, 2\pi]$. Now, put

$$\psi_3 := \int_0^{2\pi} (\hat{\psi}_2)_\lambda d\lambda.$$

For this ψ_3 it follows that

(1) $0 \neq \psi_3 \in W$ because of Proposition 4.2, $\psi_2(e) \neq 0$ and (4.6);

(2) $\psi_3 \in \mathcal{V}_1$ because $(\mathcal{V}_1, d)$ is a complete, topological vector space (recall (4.3) for d) and $(\hat{\psi}_2)_\lambda \in \mathcal{V}_1$ for all $\lambda \in [0, 2\pi]$.

Let $\mathcal{W}'$ be the complex vector subspace of $\mathcal{V}_1$ generated by $\{\varrho(\ell)\psi_3 \mid \ell \in L\}$. Then, $\mathcal{W}'$ is $\varrho(L)$-invariant and $\{0\} \neq \mathcal{W}' \subset \mathcal{W}$ because $\psi_3 \in \mathcal{W}$, Lemma 4.2-(1). Consequently Lemma 4.2-(3) yields $\mathcal{W} = \mathcal{W}' \subset \mathcal{V}_1$. $\qquad\square$

Acknowledgments

The author is sincerely grateful to the organizers of the 6th International Colloquium on Differential Geometry and its Related Fields, 3–9 September 2018, University of Veliko Tarnovo. Many thanks are also due to the referee for his valuable comments on an earlier version of this paper.

References

[1] M. Berger, Les espaces symétriques noncompacts, *Ann. Sci. École Norm. Sup.* (3), **74**, no. 2 (1957), 85–177.

[2] N. Boumuki & T. Noda, Irreducible unitary representations concerning homogeneous holomorphic line bundles over elliptic orbits, *J. Math. Res.*, **10**, no. 4 (2018), 62–88.

[3] A. A. Kirillov, *Elements of the theory of representations*, Grundlehren der mathematischen Wissenschaften **220**, Springer-Verlag, Berlin, Heidelberg, New York, 1976.

[4] K. Nomizu, Invariant affine connections on homogeneous spaces, *Amer. J. Math.*, **76**, no. 1 (1954), 33–65.

[5] R. A. Shapiro, Pseudo-Hermitian symmetric spaces, *Comment. Math. Helv.*, **46** (1971), 529–548.

Received October 30, 2018
Revised   February 11, 2019

NON-FLAT TOTALLY GEODESIC SURFACES
OF $SU(4)/SO(4)$
AND FIBRE BUNDLE STRUCTURES RELATED TO $SU(4)$

Hideya HASHIMOTO*

Department of Mathematics, Meijo University,
Nagoya 468-8502, Japan
E-mail: hhashi@meijo-u.ac.jp

Misa OHASHI

Department of Mathematics, Nagoya Institute of Technology,
Nagoya 466-8555, Japan
E-mail: ohashi.misa@nitech.ac.jp

Kazuhiro SUZUKI

Division of Mathematics and Mathematical Science,
Nagoya Institute of Technology,
Nagoya 466-8555, Japan
E-mail: cjv17505@nitech.ac.jp

The purpose of this paper is to give some fibre bundle structures related to the special unitary group $SU(4)$. The classical Lie group $SU(4)$ is isomorphic to the spinor group $Spin(6)$ which is a double covering group of the special orthogonal group $SO(6)$. This isomorphism gives rise to some fibre bundle structures on some homogeneous spaces related to $SU(4)$. By using this structure, we give a relationship between a certain non-flat totally geodesic surface in $SU(4)/SO(4)$ and in $Sp(2)/U(2) \simeq Spin(5)/U(2)$.

Keywords: Octonions; $Spin(7)$; $SU(4)$; $Spin(6)$.

1. Introduction

It is known that the classical Lie group $SU(4)$ is isomorphic to the spinor group $Spin(6)$. We obtain some isomorphisms between the symmetric spaces and some homogeneous spaces related to $SU(4)$, and obtain fibre bundle structures on these spaces. For example, we obtain a diffeomorphism from the symmetric space $SU(4)/SO(4)$ to the Grassmannian manifold $Gr_+^3(\mathbb{R}^6)$ of all oriented 3-planes of a 6-dimensional Euclidean space

---
*The first author is partially supported by Grant-in-Aid for Scientific Research (C) No. 15K04860 and 19K03482, Japan Society for the Promotion of Science.

$\mathbb{R}^6$ (see [9]), that is,

$$Gr_+^3(\mathbb{R}^6) = SO(6)/SO(3) \times SO(3) \simeq SU(4)/SO(4).$$

Also we have an isomorphism of homogeneous spaces

$$Spin(6)/Spin(4) \simeq SU(4)/Spin(4) \simeq SO(6)/SO(4) = V_2(\mathbb{R}^6),$$

where $V_2(\mathbb{R}^6)$ is a Stiefel manifold of all orthonormal 2-frames of $\mathbb{R}^6$. We can show that the Grassmannian manifold $Gr_+^3(\mathbb{R}^6) = SO(6)/SO(3) \times SO(3)$ is not homeomorphic to the Stiefel manifold $V_2(\mathbb{R}^6)$. Moreover we have the following double fibrations

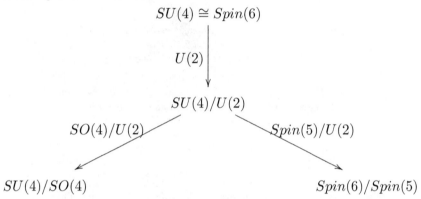

where $U(2) = Spin(5) \cap SO(4)$. From this fibration, we give a relationship between a certain non-flat totally geodesic surface of $SU(4)/SO(4)$ and that of $Sp(2)/U(2) \simeq Spin(5)/U(2) \simeq Gr_+^2(\mathbb{R}^5)$.

The authors are grateful to the referees for giving them many valuable comments.

2. $Spin(7)$

In order to obtain the isomorphism from $SU(4)$ to the spinor group $Spin(6)$, in [3], we give the representation of $Spin(7)$ by using octonions, which is obtained by R.L. Bryant in [1]. We recall this method. The octonions $\mathbb{O}$ is defined as a direct sum of quaternions (as a vector space) $\mathbb{H} \oplus \mathbb{H}$ with the following multiplication

$$(a + b\varepsilon)(c + d\varepsilon) = ac - \bar{d}b + (da + b\bar{c})\varepsilon,$$

where $\varepsilon = (0, 1) \in \mathbb{H} \oplus \mathbb{H} \simeq \mathbb{O}$. Then $\mathbb{O}$ is a non-commutative, non-associative, alternative normed division algebra.

A double covering Lie group $Spin(7)$ of $SO(7)$ is defined by

$$Spin(7) = \{g \in SO(8) \mid g(uv) = g(u)\chi_g(v) \text{ for all } u, v \in \mathbb{O}\} \subset SO(\mathbb{O})$$

where $\chi_g(v) = g(g^{-1}(1)v)$. If $g \in Spin(7)$, then $\chi_g(1) = 1$. Therefore $\chi_g \in SO(7) = SO(\operatorname{Im}\mathbb{O})$. Since $\chi_g = \chi_{(-g)}$ for all $g \in Spin(7)$, we see that $Spin(7)$ is a double covering Lie group of $SO(7)$ with the covering map $\chi : Spin(7) \to SO(7)$. From this we can show that $Spin(7)$ is simply connected. We note that

$$\big(\chi_g(i), \chi_g(j), \chi_g(k), \chi_g(\varepsilon), \chi_g(i\varepsilon), \chi_g(j\varepsilon), \chi_g(k\varepsilon)\big)$$

is a $SO(7)$-frame field on $\operatorname{Im}\mathbb{O}$.

In order to give a precise representation of Maurer-Cartan form of $Spin(7)$, we need the complexification $\mathbb{C} \otimes_{\mathbb{R}} \mathbb{O}$ of $\mathbb{O}$. We set N, $\bar{N} \in \mathbb{C} \otimes_{\mathbb{R}} \mathbb{O}$ as

$$N = \frac{1}{2}(1 - \sqrt{-1}\varepsilon), \qquad \bar{N} = \frac{1}{2}(1 + \sqrt{-1}\varepsilon),$$

and set

$$\mathscr{E} = \big(E_0, E_1, E_2, E_3\big) = (N, iN, jN, -kN),$$
$$\bar{\mathscr{E}} = \big(\bar{E}_0, \bar{E}_1, \bar{E}_2, \bar{E}_3\big) = (\bar{N}, i\bar{N}, j\bar{N}, -k\bar{N}).$$

Then $\mathscr{E}$ is a unitary basis of $(\mathbb{C} \otimes_{\mathbb{R}} \mathbb{O})^{1,0} \cong \mathbb{C}^4$ with respect to the right multiplication R_ε, that is, $R_\varepsilon(E_\ell) = \sqrt{-1}E_\ell$. Also we have $R_\varepsilon(\bar{E}_\ell) = -\sqrt{-1}\bar{E}_\ell$. For $g \in Spin(7)$, we set

$$f_\ell = g(E_\ell), \quad \bar{f}_\ell = g(\bar{E}_\ell)$$

for $\ell \in \{0, 1, 2, 3\}$, then f_ℓ and $\bar{f}_\ell$ are $\mathbb{C} \otimes_{\mathbb{R}} \mathbb{O}$-valued functions on $Spin(7)$. We call

$$(\mathscr{F}, \bar{\mathscr{F}}) = \big(f_0, f_1, f_2, f_3, \bar{f}_0, \bar{f}_1, \bar{f}_2, \bar{f}_3\big)$$

the $Spin(7)$-moving frame.

Proposition 2.1. *Let $(\mathscr{F}, \bar{\mathscr{F}})$ be a $Spin(7)$-moving frame field. Then the Maurer-Cartan form Φ, which is a $\mathfrak{spin}(7)$-valued left invariant 1-form of $Spin(7)$ is given by*

$$d\big(\mathscr{F}, \bar{\mathscr{F}}\big) = \big(\mathscr{F}, \bar{\mathscr{F}}\big)\Phi$$

with

$$\Phi = \begin{pmatrix} \sqrt{-1}\rho & -{}^t\bar{\mathfrak{h}} & 0 & -{}^t\theta \\ \mathfrak{h} & \kappa & \theta & [\bar{\theta}] \\ \hline 0 & -{}^t\bar{\theta} & -\sqrt{-1}\rho & -{}^t\mathfrak{h} \\ \bar{\theta} & [\theta] & \bar{\mathfrak{h}} & \bar{\kappa}, \end{pmatrix}.$$

Here ρ is a $\mathbb{R}$-valued 1-form, $\mathfrak{h}$, θ are $M_{3\times 1}(\mathbb{C})$-valued 1-forms, κ is a $\mathfrak{u}(3)$-valued 1-form satisfying

$$\sqrt{-1}\rho + \operatorname{tr} \kappa = 0, \tag{1}$$

and

$$[\theta] = \begin{pmatrix} 0 & \theta^3 & -\theta^2 \\ -\theta^3 & 0 & \theta^1 \\ \theta^2 & -\theta^1 & 0 \end{pmatrix}.$$

The subgroups G_2, $SU(4)$, $Sp(2)$ of $Spin(7)$ in [3] are given by

$$G_2 = \{g \in Spin(7) \mid g(1) = 1\} \quad = Aut(\mathbb{O}),$$
$$SU(4) = \{g \in Spin(7) \mid \chi_g(\varepsilon) = \varepsilon\} \cong Spin(6),$$
$$Sp(2) = \{g \in SU(4) \mid \chi_g(i) = i\} \quad \cong Spin(5).$$

From Proposition 2.1, we find that the action of $SU(4)$ is given by

$$h(\mathscr{E}\bar{\mathscr{E}}) = (\mathscr{E} \ \bar{\mathscr{E}}) \begin{pmatrix} h & O \\ O & \bar{h} \end{pmatrix} = \mathscr{E}h + \bar{\mathscr{E}}\bar{h} \tag{2}$$

for $h \in SU(4)$.

First we prepare the following multiplication formula. Its proof is a straightforward calculation.

Lemma 2.1. *For*

$$u = \sum_{\ell=0}^{3}(z_\ell E_\ell + \bar{z}_\ell \bar{E}_\ell), \quad v = \sum_{\ell=0}^{3}(w_\ell E_\ell + \bar{w}_\ell \bar{E}_\ell) \in \mathbb{O} \quad (\subset \mathbb{C} \otimes_{\mathbb{R}} \mathbb{O}),$$

the multiplication $uv \in \mathbb{O}$ is represented by

$$uv = (\mathscr{E} \ \bar{\mathscr{E}}) \begin{pmatrix} z_0 & -{}^t\bar{z} & O_{1\times 1} & O_{1\times 3} \\ z & \bar{z}_0 I_3 & O_{3\times 1} & [\bar{z}] \\ \hline O_{1\times 1} & O_{1\times 3} & \bar{z}_0 & -{}^t z \\ O_{3\times 1} & [z] & \bar{z} & z_0 I_3 \end{pmatrix} \begin{pmatrix} w_0 \\ w \\ \bar{w}_0 \\ \bar{w} \end{pmatrix}$$

$$= (\mathscr{E} \ \bar{\mathscr{E}}) \begin{pmatrix} z_0 w_0 - {}^t\bar{z}w \\ zw_0 + \bar{z}_0 w + [\bar{z}]\bar{w} \\ \bar{z}_0 \bar{w}_0 - {}^t z\bar{w} \\ \bar{z}\bar{w}_0 + z_0\bar{w} + [z]w \end{pmatrix},$$

where $z_0, w_0 \in \mathbb{C}$, and $z = \begin{pmatrix} z_1 \\ z_2 \\ z_3 \end{pmatrix}$, $w = \begin{pmatrix} w_1 \\ w_2 \\ w_3 \end{pmatrix} \in \mathbb{C}^3$.

Next we give the double covering map from $SU(4)$ to $SO(6)$, explicitly. From this we can identify $SU(4)$ with $Spin(6)$.

Proposition 2.2. *Let* $h = (h_{ij}) \in SU(4) \subset M_{4\times4}(\mathbb{C})$. *Then the map* $\chi : SU(4) \to SO(6)(\subset SO(8))$ *is given by*

$$\left(\chi_h(1)\ \chi_h(i)\ \chi_h(j)\ \chi_h(k)\ \chi_h(\varepsilon)\ \chi_h(\varepsilon i)\ \chi_h(\varepsilon j)\ \chi_h(\varepsilon k)\right)$$

$$= \begin{pmatrix} \mathscr{E} & \bar{\mathscr{E}} \end{pmatrix} \begin{pmatrix} h & O \\ O & \bar{h} \end{pmatrix} \begin{pmatrix} M & N \\ \bar{M} & \bar{N} \end{pmatrix}$$

$$= \begin{pmatrix} 1\ i\ j\ k\ \varepsilon\ i\varepsilon\ j\varepsilon\ k\varepsilon \end{pmatrix} \left(\frac{1}{2} \begin{pmatrix} I_{3,1} & I_{3,1} \\ -\sqrt{-1}I_{3,1} & \sqrt{-1}I_{3,1} \end{pmatrix} \right) \begin{pmatrix} h & O \\ O & \bar{h} \end{pmatrix} \begin{pmatrix} M & N \\ \bar{M} & \bar{N} \end{pmatrix},$$

where $I_{3,1} = \begin{pmatrix} I_3 & O_{3\times1} \\ O_{1\times3} & -1 \end{pmatrix} \in M_{4\times4}(\mathbb{C})$,

$$M = M(h) = \begin{pmatrix} \bar{h}_{11} & -h_{12} & -h_{13} & h_{14} \\ \bar{h}_{12} & h_{11} & h_{14} & h_{13} \\ \bar{h}_{13} & -h_{14} & h_{11} & -h_{12} \\ \bar{h}_{14} & h_{13} & -h_{12} & -h_{11} \end{pmatrix},$$

and

$$N = N(h) = \sqrt{-1} \begin{pmatrix} \bar{h}_{11} & -h_{12} & -h_{13} & h_{14} \\ \bar{h}_{12} & h_{11} & -h_{14} & -h_{13} \\ \bar{h}_{13} & h_{14} & h_{11} & h_{12} \\ \bar{h}_{14} & -h_{13} & h_{12} & -h_{11} \end{pmatrix}.$$

Proof. Since $h^{-1} = {}^t\bar{h}$, we get

$$h^{-1}(1) = \begin{pmatrix} \mathscr{E} & \bar{\mathscr{E}} \end{pmatrix} \begin{pmatrix} {}^t\bar{h} & O \\ O & {}^t h \end{pmatrix} \begin{pmatrix} 1 \\ 0 \\ 0 \\ 0 \\ 1 \\ 0 \\ 0 \\ 0 \end{pmatrix} = \begin{pmatrix} \mathscr{E} & \bar{\mathscr{E}} \end{pmatrix} \begin{pmatrix} \bar{h}_{11} \\ \bar{h}_{12} \\ \bar{h}_{13} \\ \bar{h}_{14} \\ h_{11} \\ h_{12} \\ h_{13} \\ h_{14} \end{pmatrix}.$$

From this and Lemma 2.1, we have

$(h^{-1}(1))v$

$$
= \left(\mathscr{E} \ \bar{\mathscr{E}}\right)
\left(\begin{array}{cccc|cccc}
\bar{h}_{11} & -h_{12} & -h_{13} & -h_{14} & 0 & 0 & 0 & 0 \\
\bar{h}_{12} & h_{11} & 0 & 0 & 0 & 0 & h_{14} & -h_{13} \\
\bar{h}_{13} & 0 & h_{11} & 0 & 0 & -h_{14} & 0 & h_{12} \\
\bar{h}_{14} & 0 & 0 & h_{11} & 0 & h_{13} & -h_{12} & 0 \\
\hline
0 & 0 & 0 & 0 & h_{11} & -\bar{h}_{12} & -\bar{h}_{13} & -\bar{h}_{14} \\
0 & 0 & \bar{h}_{14} & -\bar{h}_{13} & h_{12} & \bar{h}_{11} & 0 & 0 \\
0 & -\bar{h}_{14} & 0 & \bar{h}_{12} & h_{13} & 0 & \bar{h}_{11} & 0 \\
0 & \bar{h}_{13} & -\bar{h}_{12} & 0 & h_{14} & 0 & 0 & \bar{h}_{11}
\end{array}\right)
\left(\begin{array}{c}
w_0 \\ w_1 \\ w_2 \\ w_3 \\ \hline \bar{w}_0 \\ \bar{w}_1 \\ \bar{w}_2 \\ \bar{w}_3
\end{array}\right). \tag{3}
$$

If we put $v = i \in \mathbb{O}$ in (3), then we have

$$(h^{-1}(1))i = h^{-1}(1)(E_1 + \bar{E}_1)$$

$$
= \left(\mathscr{E} \ \bar{\mathscr{E}}\right)
\left(\begin{array}{cccc|cccc}
\bar{h}_{11} & -h_{12} & -h_{13} & -h_{14} & 0 & 0 & 0 & 0 \\
\bar{h}_{12} & h_{11} & 0 & 0 & 0 & 0 & h_{14} & -h_{13} \\
\bar{h}_{13} & 0 & h_{11} & 0 & 0 & -h_{14} & 0 & h_{12} \\
\bar{h}_{14} & 0 & 0 & h_{11} & 0 & h_{13} & -h_{12} & 0 \\
\hline
0 & 0 & 0 & 0 & h_{11} & -\bar{h}_{12} & -\bar{h}_{13} & -\bar{h}_{14} \\
0 & 0 & \bar{h}_{14} & -\bar{h}_{13} & h_{12} & \bar{h}_{11} & 0 & 0 \\
0 & -\bar{h}_{14} & 0 & \bar{h}_{12} & h_{13} & 0 & \bar{h}_{11} & 0 \\
0 & \bar{h}_{13} & -\bar{h}_{12} & 0 & h_{14} & 0 & 0 & \bar{h}_{11}
\end{array}\right)
\left(\begin{array}{c}
0 \\ 1 \\ 0 \\ 0 \\ \hline 0 \\ 1 \\ 0 \\ 0
\end{array}\right)
$$

$$
= \mathscr{E}
\left(\begin{array}{c}
-h_{12} \\ h_{11} \\ -h_{14} \\ h_{13}
\end{array}\right)
+ \bar{\mathscr{E}}
\left(\begin{array}{c}
-\bar{h}_{12} \\ \bar{h}_{11} \\ -\bar{h}_{14} \\ \bar{h}_{13}
\end{array}\right).
$$

In the same way, since

$$1 = E_0 + \bar{E}_0, \quad j = E_2 + \bar{E}_2, \quad k = -(E_3 + \bar{E}_3), \varepsilon = \sqrt{-1}(E_0 - \bar{E}_0),$$

$$i\varepsilon = \sqrt{-1}(E_1 - \bar{E}_1), \quad j\varepsilon = \sqrt{-1}(E_2 - \bar{E}_2), \quad k\varepsilon = -\sqrt{-1}(E_3 + \bar{E}_3),$$

we obtain

$$(h^{-1}(1))1 = \mathscr{E} \begin{pmatrix} \bar{h}_{11} \\ \bar{h}_{12} \\ \bar{h}_{13} \\ \bar{h}_{14} \end{pmatrix} + \bar{\mathscr{E}} \begin{pmatrix} h_{11} \\ h_{12} \\ h_{13} \\ h_{14} \end{pmatrix},$$

$$(h^{-1}(1))j = \mathscr{E} \begin{pmatrix} -h_{13} \\ h_{14} \\ h_{11} \\ -h_{12} \end{pmatrix} + \bar{\mathscr{E}} \begin{pmatrix} -\bar{h}_{13} \\ \bar{h}_{14} \\ \bar{h}_{11} \\ -\bar{h}_{12} \end{pmatrix},$$

$$(h^{-1}(1))k = \mathscr{E} \begin{pmatrix} h_{14} \\ h_{13} \\ -h_{12} \\ -h_{11} \end{pmatrix} + \bar{\mathscr{E}} \begin{pmatrix} \bar{h}_{14} \\ \bar{h}_{13} \\ -\bar{h}_{12} \\ \bar{h}_{11} \end{pmatrix},$$

$$(h^{-1}(1))\varepsilon = \mathscr{E}\sqrt{-1} \begin{pmatrix} \bar{h}_{11} \\ \bar{h}_{12} \\ \bar{h}_{13} \\ \bar{h}_{14} \end{pmatrix} + \bar{\mathscr{E}}\sqrt{-1} \begin{pmatrix} h_{11} \\ h_{12} \\ h_{13} \\ h_{14} \end{pmatrix},$$

$$(h^{-1}(1))i\varepsilon = \mathscr{E}\sqrt{-1} \begin{pmatrix} -h_{12} \\ h_{11} \\ h_{14} \\ -h_{13} \end{pmatrix} + \bar{\mathscr{E}}\sqrt{-1} \begin{pmatrix} -\bar{h}_{12} \\ \bar{h}_{11} \\ \bar{h}_{14} \\ -\bar{h}_{13} \end{pmatrix},$$

$$(h^{-1}(1))j\varepsilon = \mathscr{E}\sqrt{-1} \begin{pmatrix} -h_{13} \\ -h_{14} \\ h_{11} \\ h_{12} \end{pmatrix} + \bar{\mathscr{E}}\sqrt{-1} \begin{pmatrix} -\bar{h}_{13} \\ -\bar{h}_{14} \\ \bar{h}_{11} \\ \bar{h}_{12} \end{pmatrix},$$

$$(h^{-1}(1))k\varepsilon = \mathscr{E}\sqrt{-1} \begin{pmatrix} h_{14} \\ -h_{13} \\ h_{12} \\ -h_{11} \end{pmatrix} + \bar{\mathscr{E}}\sqrt{-1} \begin{pmatrix} \bar{h}_{14} \\ -\bar{h}_{13} \\ \bar{h}_{12} \\ -\bar{h}_{11} \end{pmatrix}.$$

Therefore we see that

$$\chi_h(i) = h\big((h^{-1}(1))i\big) = \mathscr{E}\, h \begin{pmatrix} -h_{12} \\ h_{11} \\ -h_{14} \\ h_{13} \end{pmatrix} + \bar{\mathscr{E}}\, \bar{h} \begin{pmatrix} -\bar{h}_{12} \\ \bar{h}_{11} \\ -\bar{h}_{14} \\ \bar{h}_{13} \end{pmatrix}.$$

In the same way, we obtain the representation

$$\chi_h(j), \chi_h(k), \chi_h(\varepsilon), \cdots, \chi_h(k\varepsilon).$$

Hence, we get the desired result. $\square$

By Proposition 2.2, we obtain

Corollary 2.1. *The above map* χ_h *satisfies*

$$\chi_h(1) = 1 \quad and \quad \chi_h(\varepsilon) = \varepsilon.$$

for $h \in SU(4)$. *Therefore,* χ_h *acts on* $\mathbb{R}^6 = \operatorname{Im}\mathbb{H} \oplus (\operatorname{Im}\mathbb{H})\varepsilon$. *Thus, we obtain the double covering map* $\chi|_{SU(4)} : SU(4) \to SO(6)$.

3. The symmetric space $SU(4)/SO(4)$

First we recall the definition of the symmetric space $SU(4)/SO(4)$. Let σ_{I} be the Cartan involution of $SU(4)$, which is defined by

$$\sigma_{\mathrm{I}}(g) = \bar{g}$$

for $g \in SU(4)$. Then we can easily see that σ_{I} is not the identity of $SU(4)$ and satisfies $\sigma_{\mathrm{I}}^2 = I_4$. An isotropy subgroup of σ_{I} is

$$SO(4) = \{g \in SU(4) \mid \bar{g} = g\}.$$

Then $(SU(4)/SO(4), \sigma_{\mathrm{I}})$ is a symmetric space of type AI with rank 3.

Next we show that $SU(4)/SO(4) \cong \operatorname{Gr}_3^+(\mathbb{R}^6)$. Since $SU(4) \cong Spin(6)$, we obtain

$$
\begin{aligned}
SU(4)/SO(4) &\cong Spin(6)/(Spin(4)/\mathbb{Z}_2) \\
&\cong (Spin(6)/\mathbb{Z}_2)/(Spin(4)/\mathbb{Z}_2 \times \mathbb{Z}_2) \\
&\cong SO(6)/(SO(3) \times SO(3)) \cong \operatorname{Gr}_3^+(\mathbb{R}^6),
\end{aligned}
$$

where $\operatorname{Gr}_3^+(\mathbb{R}^6)$ is a Grassmannian manifold of all oriented 3-dimensional subspaces of $\mathbb{R}^6$. From the above diffeomorphism, we have the representation $Spin(4) \subset Spin(7)$

$$Spin(4) = \{h \in Spin(7) \mid \chi_h(\varepsilon) = \varepsilon, \ \chi_h(i) \wedge \chi_h(j) \wedge \chi_h(k) = i \wedge j \wedge k\}.$$

We can easily see that

$$Spin(6)/Spin(4) \cong SO(6)/SO(4) \cong V_2(\mathbb{R}^6).$$

From these representations, we can see that $\operatorname{Gr}_3^+(\mathbb{R}^6)$ is not homemorphic to $V_2(\mathbb{R}^6)$. In fact, by using the long exact sequence of homotopy groups, we obtain the 2nd homotopy group $\pi_2(SU(4)/SO(4))$ of $SU(4)/SO(4)$ as

$$\pi_2(SU(4)/SO(4)) \simeq \pi_1(SO(4)) \simeq Z_2,$$

where $\pi_1(SO(4))$ is the fundamental group of $SO(4)$. Similarly,

$$\pi_2(Spin(6)/Spin(4)) \simeq \pi_1(Spin(4)) \simeq \{1\}.$$

Since $\pi_2(SU(4)/SO(4))$ is different from $\pi_2(Spin(6)/Spin(4))$, we proved the above result.

4. On totally geodesic surfaces in $SU(4)/SO(4)$

Let $V(3)$ be the 4-dimensional vector space over $\mathbb{C}$ of homogeneous polynomials of two variables (z, w) of degree 3. We set the Hermitian inner product on $V(3)$ such that the set of the following polynomials (P_0, P_1, P_2, P_3) is an orthonormal basis of $V(3)$, where

$$(P_0, P_1, P_2, P_3) = (P_0(z, w), P_1(z, w), P_2(z, w), P_3(z, w))$$
$$= \left(\frac{w^3}{\sqrt{6}}, \frac{zw^2}{\sqrt{2}}, \frac{z^2 w}{\sqrt{2}}, \frac{z^3}{\sqrt{6}} \right).$$

If we set $g = \begin{pmatrix} a & -\bar{b} \\ b & \bar{a} \end{pmatrix} \in SU(2)$ $(|a|^2 + |b|^2 = 1)$, then the matrix representation $\mu_4(g)$ of the irreducible representation $\rho_3(g) : SU(2) \to SU(4)$ with respect to the basis (P_0, P_1, P_2, P_3) is given by

$$\mu_4(g) = \begin{pmatrix} a^3 & \sqrt{3}a^2\bar{b} & \sqrt{3}a\bar{b}^2 & \bar{b}^3 \\ -\sqrt{3}a^2 b & a(|a|^2 - 2|b|^2) & \bar{b}(2|a|^2 - |b|^2) & \sqrt{3}\bar{a}\bar{b}^2 \\ \sqrt{3}ab^2 & -b(2|a|^2 - |b|^2) & \bar{a}(|a|^2 - 2|b|^2) & \sqrt{3}\bar{a}^2\bar{b} \\ -b^3 & \sqrt{3}\bar{a}b^2 & -\sqrt{3}\bar{a}^2 b & \bar{a}^3 \end{pmatrix},$$

where

$$\rho_3(g)(P_0, P_1, P_2, P_3) = (P_0, P_1, P_2, P_3)\mu_4(g)$$

We set a base of $\mathfrak{su}(2)$ by

$$E_1 = \begin{pmatrix} 0 & -1 \\ 1 & 0 \end{pmatrix}, \qquad E_2 = \begin{pmatrix} \sqrt{-1} & 0 \\ 0 & -\sqrt{-1} \end{pmatrix}, \qquad E_3 = \begin{pmatrix} 0 & \sqrt{-1} \\ \sqrt{-1} & 0 \end{pmatrix}.$$

Since

$$\mu_4(\exp(tE_1)) = \mu_4\left(\begin{pmatrix} \cos t & -\sin t \\ \sin t & \cos t \end{pmatrix} \right)$$

is a real valued function, by the definition of the isotropy group $SO(4)$, we see that $\mu_4(\exp(tE_1)) \in SO(4)$. Moreover the 2-dimensional subspace

$$\mathfrak{m} = \mathrm{span}_{\mathbb{R}} \left\{ \frac{d}{dt}\Big|_{t=0} \mu_4(\exp(tE_2)), \ \frac{d}{dt}\Big|_{t=0} \mu_4(\exp(tE_3)) \right\} \subset \mathfrak{su}(4)$$

is a Lie triple system, that is, it satisfies $[[\mathfrak{m}, \mathfrak{m}], \mathfrak{m}] \subset \mathfrak{m}$. Therefore the representation matrix μ_4 implies that we can construct a totally geodesic surface in $SU(4)/SO(4)$ (see [8, 2, 4–7]).

Lemma 4.1. *For* $g = \begin{pmatrix} a & -\bar{b} \\ b & \bar{a} \end{pmatrix} \in SU(2)$, *the representation* $\chi_{\mu_4(g)}$ *is given by*

$$\left(\chi_{\mu_4(g)}(1)\ \chi_{\mu_4(g)}(i)\ \chi_{\mu_4(g)}(j)\ \chi_{\mu_4(g)}(k)\ \chi_{\mu_4(g)}(\varepsilon) \cdots \chi_{\mu_4(g)}(k\varepsilon)\right)$$

$$= \left(\mathscr{E}\ \bar{\mathscr{E}}\right) \begin{pmatrix} \mu_4(g) & O \\ O & \overline{\mu_4(g)} \end{pmatrix} \left(\frac{M(\mu_4(g))\ N(\mu_4(g))}{\overline{M(\mu_4(g))}\ \overline{N(\mu_4(g))}}\right)$$

$$= \left(1\ i\ j\ k\ \varepsilon\ i\varepsilon\ j\varepsilon\ k\varepsilon\right) \left(\frac{1}{2}\begin{pmatrix} I_{3,1} & I_{3,1} \\ -\sqrt{-1}I_{3,1} & \sqrt{-1}I_{3,1} \end{pmatrix}\right) \left(\frac{K(g)\ L(g)}{\overline{K(g)}\ \overline{L(g)}}\right),$$

where

$$K(g) = \mu_4(g)M(\mu_4(g))$$

$$= \begin{pmatrix} 1 & 0 & 0 & 0 \\ 0 & a^4+\bar{b}^4 & 2a\bar{b}(a^2-\bar{b}^2) & 2\sqrt{3}a^2\bar{b}^2 \\ 0 & -2(a^3b-\bar{a}\bar{b}^3) & a^2(|a|^2-3|b|^2)-\bar{b}^2(3|a|^2-|b|^2) & 2\sqrt{3}a\bar{b}(|a|^2-|b|^2) \\ 0 & \sqrt{3}(a^2b^2+\bar{a}^2\bar{b}^2) & -\sqrt{3}(|a|^2-|b|^2)(ab+\bar{a}\bar{b}) & |a|^4-4|ab|^2+|b|^4 \end{pmatrix},$$

$$L(g) = \mu_4(g)N(\mu_4(g))$$

$$= \begin{pmatrix} \sqrt{-1} & 0 & 0 & 0 \\ 0 & \sqrt{-1}(a^4-\bar{b}^4) & 2\sqrt{-1}a\bar{b}(a^2+\bar{b}^2) & 0 \\ 0 & -2\sqrt{-1}(a^3b+\bar{a}\bar{b}^3) & \sqrt{-1}\{a^2(|a|^2-3|b|^2)+\bar{b}^2(3|a|^2-|b|^2)\} & 0 \\ 0 & \sqrt{3}\sqrt{-1}(a^2b^2-\bar{a}^2\bar{b}^2) & -\sqrt{3}\sqrt{-1}(|a|^2-|b|^2)(ab-\bar{a}\bar{b}) & -\sqrt{-1} \end{pmatrix}.$$

In particular, we have $\chi_{\mu_4(g)}(k\varepsilon) = k\varepsilon$.

Proof. Since we have

$$\mu_4(g)M(\mu_4(g))\begin{pmatrix} 0 \\ 1 \\ 0 \\ 0 \end{pmatrix}$$

$$
= \begin{pmatrix} a^3 & \sqrt{3}a^2\bar{b} & \sqrt{3}a\bar{b}^2 & \bar{b}^3 \\ -\sqrt{3}a^2b & a(|a|^2-2|b|^2) & \bar{b}(2|a|^2-|b|^2) & \sqrt{3}\bar{a}\bar{b}^2 \\ \sqrt{3}ab^2 & -b(2|a|^2-|b|^2) & \bar{a}(|a|^2-2|b|^2) & \sqrt{3}\bar{a}^2\bar{b} \\ -b^3 & \sqrt{3}\bar{a}b^2 & -\sqrt{3}\bar{a}^2b & \bar{a}^3 \end{pmatrix} \begin{pmatrix} -\sqrt{3}a^2\bar{b} \\ a^3 \\ -\bar{b}^3 \\ \sqrt{3}a\bar{b}^2 \end{pmatrix}
$$

$$
= \begin{pmatrix} 0 \\ a^4\{3|b|^2+(|a|^2-2|b|^2)\}+\bar{b}^4\{-(2|a|^2-|b|^2)+3|a|^2\} \\ -a^3b\{3|b|^2+(2|a|^2-|b|^2)\}-\bar{a}\bar{b}^3\{(|a|^2-2|b|^2)-3|a|^2\} \\ \sqrt{3}\{a^2b^2(|b|^2+|a|^2)+\bar{a}^2\bar{b}^2(|b|^2+|a|^2)\} \end{pmatrix}
$$

$$
= \begin{pmatrix} 0 \\ a^4+\bar{b}^4 \\ -2(a^3b-\bar{a}\bar{b}^3) \\ \sqrt{3}(a^2b^2+\bar{a}^2\bar{b}^2) \end{pmatrix},
$$

from Proposition 2.2, we get

$$
\chi_{\mu_4(g)}(i) = \mathscr{E}\begin{pmatrix} 0 \\ a^4+\bar{b}^4 \\ -2(a^3b-\bar{a}\bar{b}^3) \\ \sqrt{3}(a^2b^2+\bar{a}^2\bar{b}^2) \end{pmatrix} + \bar{\mathscr{E}}\begin{pmatrix} 0 \\ \bar{a}^4+b^4 \\ -2(\bar{a}^3\bar{b}-ab^3) \\ \sqrt{3}(\bar{a}^2\bar{b}^2+a^2b^2) \end{pmatrix}.
$$

Similarly, other elements of $\chi_{\mu_4(g)}$ can be computed. In particular, we obtain

$$
\mu_4(g)N(\mu_4(g))\begin{pmatrix} 0 \\ 0 \\ 0 \\ 1 \end{pmatrix}
$$

$$
= \begin{pmatrix} a^3 & \sqrt{3}a^2\bar{b} & \sqrt{3}a\bar{b}^2 & \bar{b}^3 \\ -\sqrt{3}a^2b & a(|a|^2-2|b|^2) & \bar{b}(2|a|^2-|b|^2) & \sqrt{3}\bar{a}\bar{b}^2 \\ \sqrt{3}ab^2 & -b(2|a|^2-|b|^2) & \bar{a}(|a|^2-2|b|^2) & \sqrt{3}\bar{a}^2\bar{b} \\ -b^3 & \sqrt{3}\bar{a}b^2 & -\sqrt{3}\bar{a}^2b & \bar{a}^3 \end{pmatrix}\left(\sqrt{-1}\begin{pmatrix} \bar{b}^3 \\ -\sqrt{3}a\bar{b}^2 \\ \sqrt{3}a^2\bar{b} \\ -a^3 \end{pmatrix}\right)
$$

$$
= \sqrt{-1}\begin{pmatrix} 0 \\ -\sqrt{3}a^2\bar{b}^2\{|b|^2+(|a|^2-2|b|^2)-(2|a|^2-|b|^2)+|a|^2\} \\ \sqrt{3}a\bar{b}\{|b|^4+|b|^2(2|a|^2-|b|^2)+|a|^2(|a|^2-2|b|^2)-|a|^4\} \\ -|b|^6-3|a|^2|b|^4-3|a|^4|b|^2-|a|^6 \end{pmatrix}
$$

$$= \begin{pmatrix} 0 \\ 0 \\ 0 \\ -\sqrt{-1} \end{pmatrix}.$$

Therefore we get the desired result. □

Theorem 4.1. *The non-flat totally geodesic surface $\mu_4(SU(2))/\mu_4(SO(2))$ in $SU(4)/SO(4)$ can be considered as a non-flat totally geodesic surface in $Spin(5)/U(2)$. Also, the surface $\chi_{\mu_4(SU(2))}/\chi_{\mu_4(SO(2))}$ can be considered as a non-flat totally geodesic surface in the Grassmannian manifold $G_2^+(\mathbb{R}^5)$ of all oriented 2-planes of $\mathbb{R}^5$.*

Proof. From Lemma 4.1, we obtain,

$$\chi_{\mu_4(g)} \in SO(5) \ \ (\subset SO(6))$$

for every $g \in SU(2)$, where

$$SO(5) = \{\chi_g \in SO(6) \mid g \in SU(4), \chi_g(k\varepsilon) = k\varepsilon\}.$$

In particular, for $\tau = \begin{pmatrix} \cos\theta & -\sin\theta \\ \sin\theta & \cos\theta \end{pmatrix} \in SO(2)$, we have

$$\chi_{\mu_4(\tau)} \in SO(5) \cap (SO(3) \times SO(3)) = SO(2) \times SO(3).$$

In fact, $\chi_{\mu_4(\tau)}$ satisfies

$$\chi_{\mu_4(\tau)}(k\varepsilon) = k\varepsilon \qquad \text{and} \qquad \chi_{\mu_4(\tau)}(i\varepsilon) \wedge \chi_{\mu_4(\tau)}(j\varepsilon) = i\varepsilon \wedge j\varepsilon.$$

From this $\chi_{\mu_4(g)}$ induces a surface in $SO(5)/(SO(2) \times SO(3)) \cong \mathrm{Gr}_2^+(\mathbb{R}^5)$. Therefore $\mu_4(g)$ induces a surface in $Spin(5)/(SO(2) \times Spin(3))$, where

$$SO(2) \times Spin(3)$$
$$= \{h \in Spin(6) \mid \chi_{\mu_4(\tau)}(k\varepsilon) = k\varepsilon, \chi_{\mu_4(\tau)}(i\varepsilon) \wedge \chi_{\mu_4(\tau)}(j\varepsilon) = i\varepsilon \wedge j\varepsilon\}.$$

□

We note that there exist the following diffeomorphisms

$$G_2^+(\mathbb{R}^5) = SO(5)/(SO(2) \times SO(3))$$
$$\simeq Spin(5)/(SO(2) \times Spin(3)) \simeq Sp(2)/U(2).$$

References

[1] R. L. Bryant. Submanifolds and special structures on the octonions. *J. Differential Geom.*, **17** (1982), 185–232.

[2] T. Fujimaru, A. Kubo and H. Tamaru, On totally geodesic surfaces in symmetric spaces of type AI. *Springer Proc. Math. Stat.*, 106, Springer, Tokyo, 2014, 211–227.

[3] H. Hashimoto and M. Ohashi, Realizations of subgroups of G_2, $Spin(7)$ and their applications, in *Recent progress in Differential Geometry and its Related Fields*, 2010, 159–175.

[4] H. Hashimoto and M. Ohashi, Fundamental relationship between Cartan imbeddings of type A and Hopf fibrations, in *Contemporary Perspectives in Differential Geometry and its Related Fields*, 2017, 79–94.

[5] H. Hashimoto, M. Ohashi and K. Suzuki, Relationships among non-flat totally geodesic surfaces in symmetric spaces of type A and their polynomial representations, to appear in *Kodai Math. J.*

[6] H. Hashimoto and K. Suzuki, Hopf fibration and Cartan imbeddings of type AI, in *Current Developments in Differential Geometry and its Related Fields*, 2015, 155–163.

[7] S. Helgason, *Differential geometry, Lie group, and symmetric spaces*, Amer. Math. Soc., 1978.

[8] K. Mashimo, Non-flat totally geodesic surfaces of symmetric space of classical type. *Osaka J. Math.*, **56** (2019), 1–32.

[9] J. A. Wolf, *Spaces of constant curvature*. sixth edition. Amer. Math. Soc. Chelsa Publishing, Providence, RI, 2011.

Received March 12, 2019
Revised April 10, 2019

AN ALGORITHM FOR COMPUTING
THE WEIGHT DISTRIBUTION OF A LINEAR CODE
OVER A COMPOSITE FINITE FIELD
WITH CHARACTERISTIC 2

Paskal PIPERKOV* and Iliya BOUYUKLIEV**

*Institute of Mathematics and Informatics,
Bulgarian Academy of Sciences, Veliko Tarnovo, Bulgaria
* E-mail: ppiperkov@math.bas.bg   ** E-mail: iliyab@math.bas.bg
www.math.bas.bg

Stefka BOUYUKLIEVA

Faculty of Mathematics and Informatics,
St. Cyril and St. Methodius University of Veliko Tarnovo,
2 Teodosii Tarnovski Str., 5000 Veliko Tarnovo, Bulgaria
E-mail: stefka@ts.uni-vt.bg

We consider the problem of computing the weight distribution of a linear code
of dimension k over a composite finite field $\mathbb{F}_q$ where $q = 2^m$. Due to the trace
function we reduce the arithmetic operations over the composite field to those
over a prime field. This allows us to apply a transform of Walsh-Hadamard
type. The codes are represented by their characteristic vector with respect to
a given generator matrix and a generator matrix of the k-dimensional simplex
code $\mathcal{S}_{q,k}$. The developed algorithm has complexity $O(kmq^{k-1})$.

Keywords: Linear code; weight distribution; Walsh-Hadamard transform.

1. Introduction

Many computational problems in Coding Theory, including the computation of the weight distribution of a linear code, are NP-hard [1–3]. Never-
theless, algorithms for these computations are very useful for theory and
practice. Our aim is to develop a new efficient algorithm for computation the weight distribution of linear codes over composite finite fields with
characteristic 2. A technique that uses trace and characteristic functions
has been proposed by Karpovsky [4, 5]. The key idea is to apply a modified Walsh-Hadamard transform and a butterfly algorithm for its imple-
mentation. The term "butterfly algorithm" is introduced in [6], due to
its similarity to the butterfly stage in the Fast Fourier Transform (FFT).
The algorithm permits a fast matrix-vector multiplication by decomposing
the matrix into blocks. The multiplication of each block by a subvector

is executed using a multilevel scheme. These algorithms can be visualized through a butterfly type diagram and rely on well-known algebraic techniques. A detailed description of a butterfly algorithm for numerical applications to arbitrary vectors of several special function transforms is presented in [7].

Let $\mathbb{F}_q$ be a Galois field with q elements where $q = 2^m$. An $[n, k]_q$ linear code C is a k dimensional linear subspace of the n dimensional vector space $\mathbb{F}_q^n$. The integers n and k are called *length* and *dimension* of C, respectively, and the vectors in C are called *codewords*. The *Hamming weight* wt(w) of a codeword w is the number of its nonzero coordinates. The *weight distribution* of a code is the sequence $(A_0, A_1, \ldots, A_n)$ where A_u is the number of the codewords with Hamming weight u. The *weight enumerator* of the code is the polynomial $W(y) = \sum_{i=1}^{n} A_i y^i$. Usually, a linear code is represented by its *generator matrix*. The rows of a generator matrix form a basis of the code as a linear space. Here we use a representation of the codes by their characteristic vectors. To define such a vector, we use the q-ary simplex code of dimension k. For more details on the parameters and properties of linear codes we refer to [8].

We give preliminary information about the simplex codes, characteristic vectors, and the field elements in Section 2. Section 3 is devoted to Walsh-Hadamard transform. The butterfly algorithm is presented in Section 4.

In the paper, if some number (or element) is written in bold it means that this is a matrix or vector with suitable size whose elements are equal to this number (element).

2. Preliminaries

2.1. *Generator matrices of the simplex codes*

The q-ary simplex code $S_{q,k}$ is a linear code over $\mathbb{F}_q$ generated by a $k \times \theta(q, k)$ matrix having as columns a maximal set of nonproportional vectors from the vector space $\mathbb{F}_q^k$, $\theta(q, k) = (q^k - 1)/(q - 1)$. In other words, the columns of the matrix represent all points in the projective geometry $PG(k - 1, q)$. All nonzero codewords of $S_{q,k}$ have the same Hamming weight q^{k-1}, so the q-ary simplex code is a linear constant weight code with parameters $[\theta(q, k), k, q^{k-1}]$.

Let $\alpha_0 = 0, \alpha_1, \ldots, \alpha_{q-1}$ be the elements of the field $\mathbb{F}_q$. Note that α_1 may not be equal to 1. We define the sequence of matrices G_k as follows:

$$G_1 = (1), \quad G_k = \begin{pmatrix} \mathbf{0} & \alpha_1 & \cdots & \alpha_{q-1} & 1 \\ G_{k-1} & G_{k-1} & \cdots & G_{k-1} & \mathbf{0} \end{pmatrix}, \quad k \in \mathbb{Z}, \, k \geq 2 \quad (1)$$

The size of the matrix G_k is $k \times \theta(q, k)$. All columns in G_k are pairwise linearly independent, so G_k is a generator matrix of $\mathcal{S}_{q,k}$. The columns of the matrix

$$G'_k = (\alpha_1 G_k | \alpha_2 G_k | \ldots | \alpha_{q-1} G_k) \tag{2}$$

are all nonzero vectors from $\mathbb{F}_q^k$. Let $\bar{G}_k = (\mathbf{0}|G'_k)$ where $\mathbf{0}$ is the zero column. Then the codewords of the simplex code $\mathcal{S}_{q,k}$ are the rows of the matrix $\bar{G}_k^{\mathrm{T}} \cdot G_k$, and the matrix $G_k^{\mathrm{T}} \cdot G_k$ represents a maximal set of nonproportional codewords of $\mathcal{S}_{q,k}$. If a $k \times n$ matrix G is a generator matrix of a linear $[n, k]_q$ code C, then the matrix $\bar{G}_k^{\mathrm{T}} \cdot G$ consists of all codewords of C as rows.

2.2. Characteristic vector

Let G be a generator matrix of the $[n, k]_q$ linear code C. Without loss of generality we can suppose that G has no zero columns. The characteristic vector of the code C with respect to its generator matrix G is the vector

$$\chi(C, G) = (\chi_1, \chi_2, \ldots, \chi_{\theta(q,k)}) \in \mathbb{Z}^{\theta(q,k)} \tag{3}$$

where χ_u is the number of the columns of G that are equal or proportional to the u-th column of G_k, $u = 1, \ldots, \theta(q, k)$. The extended characteristic vector of the code C with respect to its generator matrix G is

$$\overline{\chi}(C, G) = (0, \underbrace{\chi|\chi|\cdots|\chi}_{q-1}) \tag{4}$$

When C and G are clear from the context, we will briefly write χ or $\overline{\chi}$.

2.3. Ordering the elements of $\mathbb{F}_q$

The composite finite field $\mathbb{F}_{q^m}$ with characteristic p, p-prime, can be considered as a linear space over the field $\mathbb{F}_q$. We need a self-dual basis of this space, namely a basis $\beta_1, \ldots, \beta_m$ such that $\mathrm{Tr}(\beta_i \beta_j) = \delta_{ij}$ for all $i, j = 1, 2, \ldots, m$, where δ_{ij} is the Kronecker delta, and $\mathrm{Tr} : \mathbb{F}_{q^m} \to \mathbb{F}_q$ is the trace map. The following theorem gives a sufficient and necessary condition for the existence of a self-dual basis.

Theorem 2.1 ([9], [10, Theorem 5.1.18]). *There exists a self-dual basis of $\mathbb{F}_{q^m}$ over $\mathbb{F}_q$ if and only if either q is even or both q and m are odd.*

We consider fields with characteristic 2, so q is even and according to Theorem 2.1 there exists a self-dual basis $\beta_1, \ldots, \beta_m$ of $\mathbb{F}_{2^m}$ over $\mathbb{F}_2$.

Denote by $v^\alpha = (v_1^\alpha, \ldots, v_m^\alpha)$ the vector $v^\alpha \in \mathbb{F}_2^m$ that corresponds to the element $\alpha = v_1^\alpha \beta_1 + \cdots + v_m^\alpha \beta_m \in \mathbb{F}_q$. We order the elements $\alpha_0 = 0, \alpha_1, \ldots, \alpha_{q-1}$ of the considered field $\mathbb{F}_q$, $q = 2^m$, such that the binary vectors $v^0, v^{\alpha_1}, \ldots, v^{\alpha_{q-1}}$ are ordered lexicographically.

Proposition 2.1. *The matrix*

$$\Lambda = \left((-1)^{Tr(\alpha_i \alpha_j)} \right)_{ij}, \quad i = 0, \ldots, q-1, \; j = 0, \ldots, q-1$$

is a Sylvester-Hadamard matrix.

Proof. For the trace of $\alpha_i \alpha_j$ we have

$$\mathrm{Tr}(\alpha_i \alpha_j) = \mathrm{Tr}\left(\sum_{r,s=1}^m v_r^{\alpha_i} v_s^{\alpha_j} \beta_r \beta_s \right) = \sum_{r,s=1}^m \mathrm{Tr}(v_r^{\alpha_i} v_s^{\alpha_j} \beta_r \beta_s).$$

Since $v_r^{\alpha_i} v_s^{\alpha_j} = 0$ or 1, then $\mathrm{Tr}(v_r^{\alpha_i} v_s^{\alpha_j} \beta_r \beta_s) = \mathrm{Tr}(\beta_i \beta_j)$ or $\mathrm{Tr}(0) = 0$. It follows that $\mathrm{Tr}(v_r^{\alpha_i} v_s^{\alpha_j} \beta_r \beta_s) = 0$ for $r \neq s$, and $\mathrm{Tr}(v_s^{\alpha_i} v_s^{\alpha_j} \beta_s \beta_s) = v_s^{\alpha_i} v_s^{\alpha_j}$, $s = 1, \ldots, m$. Hence

$$\mathrm{Tr}(\alpha_i \alpha_j) = \sum_{s=1}^m v_s^{\alpha_i} v_s^{\alpha_j} = v^{\alpha_i} \cdot v^{\alpha_j} = \bar{i} \cdot \bar{j},$$

where $\bar{i}$ is the binary vector of length m that correspond to the binary representation of the integer i. In other words, if $i = i_0.2^{m-1} + \cdots + i_{m-2}.2 + i_{m-1}$ then $\bar{i} = (i_0, \ldots, i_{m-1})$, $i_s \in \mathbb{F}_2$, $s = 0, \ldots, m-1$. It turns out that $\Lambda = ((-1)^{\bar{i} \cdot \bar{j}})_{ij}$. Denote by Λ_n the matrix, defined as Λ but if the considered field has 2^n elements, $n \geq 1$.

Taking in mind that the first coordinate of $\bar{i}$ is 0 if $i < 2^{n-1}$ and 1 otherwise, we have

$$\bar{i} \cdot \bar{j} = \begin{cases} \bar{i} \cdot \overline{(j - 2^{n-1})}, & \text{if } i < 2^{n-1}, \; j \geq 2^{n-1} \\ \overline{(i - 2^{n-1})} \cdot \bar{j}, & \text{if } i \geq 2^{n-1}, \; j < 2^{n-1} \\ 1 \oplus \left(\overline{(i - 2^{n-1})} \cdot \overline{(j - 2^{n-1})} \right), & \text{if } i \geq 2^{n-1}, \; j \geq 2^{n-1} \end{cases}$$

It follows that

$$\Lambda_n = \begin{pmatrix} \Lambda_{n-1} & \Lambda_{n-1} \\ \Lambda_{n-1} & -\Lambda_{n-1} \end{pmatrix},$$

and the matrices Λ_n can be defined recursively as

$$\Lambda_1 = \begin{pmatrix} 1 & 1 \\ 1 & -1 \end{pmatrix}, \quad \Lambda_n = \begin{pmatrix} \Lambda_{n-1} & \Lambda_{n-1} \\ \Lambda_{n-1} & -\Lambda_{n-1} \end{pmatrix} = \Lambda_1 \otimes \Lambda_{n-1} \text{ for } n \geq 2.$$

It turns out that Λ_n are Hadamard matrices of Sylvester type called also Sylvester matrices or Walsh matrices [11]. $\qquad\square$

The above proposition shows that the following equality holds

$$\Lambda = \left((-1)^{\mathrm{Tr}(\alpha_i \alpha_j)} \right)_{ij} = \otimes^m \begin{pmatrix} 1 & 1 \\ 1 & -1 \end{pmatrix}, \tag{5}$$

where $\otimes^m$ means the m-th Kronecker power.

3. Walsh-Hadamard Transform

Let $h(x) = h(x_1, x_2, \ldots, x_k)$ be a Boolean function in k variables. Discrete Walsh-Hadamard transform $\hat{h}$ of h is the integer valued function $\hat{h} : \mathbb{F}_2^k \to \mathbb{Z}$, defined by

$$\hat{h}(\omega) = \sum_{x \in \mathbb{F}_2^k} h(x)(-1)^{x \cdot \omega}, \quad \omega \in \mathbb{F}_2^k \tag{6}$$

where $x \cdot \omega$ is the Euclidean inner product. This transform is equivalent to the multiplication of the True Table of h by the matrix $\otimes^k \begin{pmatrix} 1 & 1 \\ 1 & -1 \end{pmatrix}$. The Kronecker power of a matrix can be represented as a product of sparse matrices [12] that leads to the more effective butterfly algorithm for calculation (fast transform).

There is a method based on the fast Walsh-Hadamard transform for the computation of the weight distribution of a given binary linear code ($q = 2$). The complexity of this computation is $O(k2^k)$ [4].

The definition of Walsh-Hadamard transform can be generalized over the functions $h : \mathbb{F}_2^k \to \mathbb{Z}$. Further, if G is a generator matrix of an $[n, k]$ binary code, we can take h to be the characteristic function in sense that $h(x)$ is the number of the columns in G that are equal to x. In that case the Walsh spectrum $\hat{h}$ is related to the weight of the codeword ωG as follows

$$\mathrm{wt}(\omega G) = \frac{n - \hat{h}(\omega)}{2}. \tag{7}$$

A generalization of Walsh-Hadamard transform for linear codes over composite finite fields was proposed by Karpovsky [5]. Karpovsky uses this transform to compute the weight distribution of translates and the covering radius of a linear code. The essence of his algorithm is as follows.

Let G be a generator matrix of an $[n, k]_q$ code, $q = 2^m$. The characteristic function $h_G(x)$ is the number of columns in G that are proportional to x. We extend the matrix G to the matrix

$$G' = (\alpha_1 G | \alpha_2 G | \ldots | \alpha_{q-1} G) \tag{8}$$

where $\mathbb{F}_q^* = \{\alpha_1, \alpha_2, \ldots, \alpha_{q-1}\}$. Let h' be the characteristic function of G'. The new transform is

$$\hat{h}'(\omega) = \sum_{x \in \mathbb{F}_q^k} h'(x)(-1)^{\mathrm{Tr}(x \cdot \omega)}, \quad \omega \in \mathbb{F}_q^k \tag{9}$$

where $x \cdot \omega$ is the Euclidean inner product of x and ω over $\mathbb{F}_q$ and Tr is the trace function from $\mathbb{F}_q$ to $\mathbb{F}_2$. The connection between this transform and the weight of a codeword is

$$\mathrm{wt}(\omega G) = \frac{(q-1)n - \hat{h}'(\omega)}{q} \tag{10}$$

If the elements of $\mathbb{F}_q$ are ordered as described in Section 2.3, and ω_0, $\omega_1, \ldots, \omega_{q^k-1}$ are all vectors in $\mathbb{F}_q^k$ in lexicographic order, then

$$\left((-1)^{\mathrm{Tr}(\omega_i \cdot \omega_j)}\right)_{ij} = \otimes^k \left((-1)^{\mathrm{Tr}(\alpha_i \alpha_j)}\right)_{ij} = \otimes^{km} \begin{pmatrix} 1 & 1 \\ 1 & -1 \end{pmatrix} \tag{11}$$

and one can use a butterfly algorithm to compute the transform (9). A pseudocode of the butterfly algorithm is presented in Algorithm 3.1. A similar pseudo-code is presented in [13, Algorithm 9.3].

Algorithm 3.1 Butterfly Algorithm for Transform (9) when the elements of the field are ordered as in Section 2.3

Require: Positive integers k, m, $q = 2^m$, and an array h with length q^k – the value table of a characteristic function h'

Ensure: An updated array h – the result of the transform

1:   $s = 1$;
2: **for** $l = 1$ **to** km **do**
3:     **for** $t = 0$ **to** $2^{km-l} - 1$ **do**
4:       **for** $i = 0$ **to** $s - 1$ **do**
5:         $j = 2st + i$;
6:         $A = h[j]$;
7:         $B = h[j + s]$;
8:         $h[j] = A + B$;
9:         $h[j + s] = A - B$;
10:      **end for**
11:    **end for**
12:    $s = 2s$;
13: **end for**

Example 3.1. Let $G = \begin{pmatrix} 1 & 0 & 1 & \omega \\ 0 & 1 & \omega & \overline{\omega} \end{pmatrix}$ be a generator matrix of a $[4,2]_4$ quaternary code C. The elements $\{\omega, \overline{\omega} = \omega + 1 = \omega^2\}$ form a self-dual basis of $\mathbb{F}_4$ and so we take $\alpha_0 = 0$, $\alpha_1 = \omega$, $\alpha_2 = \overline{\omega}$, $\alpha_3 = 1$. According to the considered ordering of the vectors in $\mathbb{F}_4^2$, we take the characteristic function of C to be $h' = (0,1,1,1,1,0,2,0,1,0,0,2,1,2,0,0)$. Applying Algorithm 3.1, we obtain $\hat{h}' = (12,0,0,0,0,-4,-4,4,0,4,-4,-4,0,-4,4,-4)$. Hence the weight enumerator of C is $W(y) = 1 + 3y^2 + 6y^3 + 6y^4$.

Some improvements on the computations can be done when the characteristic function is defined using a generator matrix of the simplex code $\mathcal{S}_{q,k}$. By definition the columns of a generator matrix of $\mathcal{S}_{q,k}$ form a maximal collection of nonproportional vectors belonging to $\mathbb{F}_q^k$. We use the generator matrix G_k of $\mathcal{S}_{q,k}$, inductively defined in (1). The characteristic vector of a linear code with respect to a generator matrix G shows how many columns of G are equal or proportional (with nonzero coefficient) to a given column of G_k. We develop an algorithm for computing the weight distribution of the code when the input is its characteristic vector. The core of our idea is that we use a characteristic vector of length $\theta(q, k)$ instead of q^k. The complexity of this algorithm is $O(mkq^{k-1})$.

4. The Butterfly Algorithm

4.1. *Matrix description of the algorithm*

Consider the matrices $H_k = G_k^{\mathrm{T}} \cdot G_k$ and $\bar{H}_k = \bar{G}_k^{\mathrm{T}} \cdot \bar{G}_k$ in more details. Using (1) we obtain the following recurrence relation

$$H_k = \begin{pmatrix} H_{k-1} & H_{k-1} & \cdots & H_{k-1} & \mathbf{0} \\ H_{k-1} & \alpha_1^2 J + H_{k-1} & \cdots & \alpha_1\alpha_{q-1}J + H_{k-1} & \alpha_1 \\ H_{k-1} & \alpha_2\alpha_1 J + H_{k-1} & \cdots & \alpha_2\alpha_{q-1}J + H_{k-1} & \alpha_2 \\ \vdots & \vdots & & \vdots & \vdots \\ H_{k-1} & \alpha_{q-1}\alpha_1 J + H_{k-1} & \cdots & \alpha_{q-1}^2 J + H_{k-1} & \alpha_{q-1} \\ \mathbf{0} & \alpha_1 & \cdots & \alpha_{q-1} & 1 \end{pmatrix}, \qquad (12)$$

where J is the all 1's matrix of the corresponding size. Note that

$$\bar{H}_k = \begin{pmatrix} 0 & \mathbf{0} & \mathbf{0} & \cdots & \mathbf{0} \\ \mathbf{0} & \alpha_1^2 H_k & \alpha_1\alpha_2 H_k & \cdots \alpha_1\alpha_{q-1}H_k \\ \mathbf{0} & \alpha_2\alpha_1 H_k & \alpha_2^2 H_k & \cdots \alpha_2\alpha_{q-1}H_k \\ \vdots & \vdots & \vdots & \vdots \\ \mathbf{0} & \alpha_{q-1}\alpha_1 H_k & \alpha_{q-1}\alpha_2 H_k & \cdots & \alpha_{q-1}^2 H_k \end{pmatrix}. \qquad (13)$$

Let

$$P_k = \left((-1)^{\mathrm{Tr}(h_{ij})}\right)_{ij}, \quad \bar{P}_k = \left((-1)^{\mathrm{Tr}(\bar{h}_{ij})}\right)_{ij},$$

where h_{ij}, $i,j = 1,\ldots,\theta(q,k)$, and $\bar{h}_{ij}$, $i,j = 1,\ldots,q^k$, are the elements of the matrices H_k and $\bar{H}_k$, respectively. The core of the transform (9) is the multiplication of the matrix $\bar{P}_k$ by a characteristic vector (function). Our idea is to use only the matrix P_k, the characteristic vector χ, and the matrices $P_k^{(\alpha)} = \left((-1)^{\mathrm{Tr}(\alpha h_{ij})}\right)_{ij}$ for $\alpha \in \mathbb{F}_q^*$. Our algorithm uses these matrices to obtain not only the resulting vector of the generalized Walsh-Hadamard transform, but the weight distribution of the considered code. Since

$$\bar{P}_k = \begin{pmatrix} 1 & 1 & 1 & \cdots & 1 \\ 1 & P_k^{(\alpha_1^2)} & P_k^{(\alpha_1\alpha_2)} & \cdots & P_k^{(\alpha_1\alpha_{q-1})} \\ 1 & P_k^{(\alpha_2\alpha_1)} & P_k^{(\alpha_2^2)} & \cdots & P_k^{(\alpha_2\alpha_{q-1})} \\ \vdots & \vdots & \vdots & & \vdots \\ 1 & P_k^{(\alpha_{q-1}\alpha_1)} & P_k^{(\alpha_{q-1}\alpha_2)} & \cdots & P_k^{(\alpha_{q-1}^2)} \end{pmatrix}, \tag{14}$$

we have

$$\hat{\chi}^T = \bar{P}_k\bar{\chi}^T = \begin{pmatrix} 1 & 1 & 1 & \cdots & 1 \\ 1 & P_k^{(\alpha_1^2)} & P_k^{(\alpha_1\alpha_2)} & \cdots & P_k^{(\alpha_1\alpha_{q-1})} \\ 1 & P_k^{(\alpha_2\alpha_1)} & P_k^{(\alpha_2^2)} & \cdots & P_k^{(\alpha_2\alpha_{q-1})} \\ \vdots & \vdots & \vdots & & \vdots \\ 1 & P_k^{(\alpha_{q-1}\alpha_1)} & P_k^{(\alpha_{q-1}\alpha_2)} & \cdots & P_k^{(\alpha_{q-1}^2)} \end{pmatrix} \begin{pmatrix} 0 \\ \chi^T \\ \chi^T \\ \vdots \\ \chi^T \end{pmatrix}$$

$$= \begin{pmatrix} (q-1)\sum\chi \\ (P_k^{(\alpha_1^2)} + P_k^{(\alpha_1\alpha_2)} + \cdots + P_k^{(\alpha_1\alpha_{q-1})})\chi^T \\ (P_k^{(\alpha_2\alpha_1)} + P_k^{(\alpha_2^2)} + \cdots + P_k^{(\alpha_2\alpha_{q-1})})\chi^T \\ \vdots \\ (P_k^{(\alpha_{q-1}\alpha_1)} + P_k^{(\alpha_{q-1}\alpha_2)} + \cdots + P_k^{(\alpha_{q-1}^2)})\chi^T \end{pmatrix}.$$

Taking in mind that $\{\alpha_i\alpha_1,\ldots,\alpha_i\alpha_{q-1}\} = \mathbb{F}_q^*$ for all $i \in \{1,\ldots,q-1\}$, we obtain

$$\bar{P}_k\bar{\chi}^T = \begin{pmatrix} (q-1)\sum\chi \\ SP_k\chi^T \\ SP_k\chi^T \\ \vdots \\ SP_k\chi^T \end{pmatrix}, \tag{15}$$

where $SP_k = P_k^{(\alpha_1)} + P_k^{(\alpha_2)} + \cdots + P_k^{(\alpha_{q-1})}$. This means that we don't need the $2^k \times 2^k$ matrix $\bar{P}_k$ and the larger vector $\bar{\chi}$. To obtain $\hat{\chi}$ and the weight distribution of a code, it is enough to use the characteristic vector χ and the matrix P_k.

Using the recurrence relation (12), we obtain

$$
P_k = \begin{pmatrix}
P_{k-1} & P_{k-1} & \cdots & P_{k-1} & \mathbf{\Lambda_0} \\
P_{k-1} & \Lambda_{\alpha_1^2} P_{k-1} & \cdots & \Lambda_{\alpha_1 \alpha_{q-1}} P_{k-1} & \mathbf{\Lambda_{\alpha_1}} \\
P_{k-1} & \Lambda_{\alpha_2 \alpha_1} P_{k-1} & \cdots & \Lambda_{\alpha_2 \alpha_{q-1}} P_{k-1} & \mathbf{\Lambda_{\alpha_2}} \\
\vdots & \vdots & & \vdots & \vdots \\
P_{k-1} & \Lambda_{\alpha_{q-1} \alpha_1} P_{k-1} & \cdots & \Lambda_{\alpha_{q-1}^2} P_{k-1} & \mathbf{\Lambda_{\alpha_{q-1}}} \\
\mathbf{\Lambda_0} & \mathbf{\Lambda_{\alpha_1}} & \cdots & \mathbf{\Lambda_{\alpha_{q-1}}} & \Lambda_1
\end{pmatrix}
\tag{16}
$$

$$
= \begin{pmatrix}
& & & \mathbf{\Lambda_0} \\
\left(\Lambda_{\alpha_i \alpha_j} \right)_{ij} \otimes P_{k-1} & & & \mathbf{\Lambda_{\alpha_1}} \\
& & & \vdots \\
& & & \mathbf{\Lambda_{\alpha_{q-1}}} \\
\mathbf{\Lambda_0} \; \mathbf{\Lambda_{\alpha_1}} \; \cdots \; \mathbf{\Lambda_{\alpha_{q-1}}} & & & \Lambda_1
\end{pmatrix},
\tag{17}
$$

where $\Lambda_\alpha = (-1)^{\text{Tr}(\alpha)}$ for $\alpha \in \mathbb{F}_q$.

If the elements of $\mathbb{F}_q$ are ordered as in Section 2.3, we can apply (5) and provide a butterfly algorithm to compute $P_k \cdot \chi^{\text{T}}$. In order to explain better this algorithm, we split the characteristic vector χ (see (3)) as follows

$$
\chi = (\chi^{(0)} | \chi^{(1)} | \ldots | \chi^{(q-1)} | \chi_{\theta(q,k)})
\tag{18}
$$

where $\chi^{(0)}, \chi^{(1)}, \ldots, \chi^{(q-1)} \in \mathbb{Z}^{\theta(q,k-1)}$. Denote by Λ the matrix $\left(\Lambda_{\alpha_i \alpha_j} \right)_{ij}$. Thus

$$
P_k \cdot \chi^{\text{T}} = \begin{pmatrix}
(\Lambda \otimes P_{k-1}) \begin{pmatrix} \chi^{(0)T} \\ \chi^{(1)T} \\ \vdots \\ \chi^{(q-1)T} \end{pmatrix} + \chi_\theta \begin{pmatrix} \mathbf{\Lambda_0} \\ \mathbf{\Lambda_{\alpha_1}} \\ \vdots \\ \mathbf{\Lambda_{\alpha_{q-1}}} \end{pmatrix} \\
\Lambda_0 \sum \chi^{(0)} + \Lambda_{\alpha_1} \sum \chi^{(1)} + \cdots + \Lambda_{\alpha_{q-1}} \sum \chi^{(q-1)} + \Lambda_1 \chi_\theta
\end{pmatrix}
$$

$$
= \left(\begin{array}{c} (\Lambda \otimes I_{\theta(q,k-1)}) \left(\begin{array}{c} P_{k-1}\chi^{(0)T} \\ P_{k-1}\chi^{(1)T} \\ \vdots \\ P_{k-1}\chi^{(q-1)T} \end{array} \right) + \chi_\theta \left(\begin{array}{c} \Lambda_0 \\ \Lambda_{\alpha_1} \\ \vdots \\ \Lambda_{\alpha_{q-1}} \end{array} \right) \\ \Lambda_0\sum\chi^{(0)}+\Lambda_{\alpha_1}\sum\chi^{(1)}+\cdots+\Lambda_{\alpha_{q-1}}\sum\chi^{(q-1)}+\Lambda_1\chi_\theta \end{array} \right),
$$

where $\theta = \theta(q, k)$ for short and $\sum\chi^{(l)}$ means the sum of coordinates of $\chi^{(l)}$, $l = 0, 1, \ldots, q-1$.

Note that we obtain the same result excluding the last coordinate, if we add previously $\chi_{\theta(q,k)}$ to all coordinates of the block $P_{k-1}\chi^{(l)T}$, where $\alpha_l = 1$, and multiply it by the matrix $\Lambda \otimes I_{\theta(q,k-1)}$. For the last coordinate we need to add χ_θ once to the sum of the coordinates of $\chi^{(l)}$.

$$
P_k \cdot \chi^T = \left(\begin{array}{c} (\Lambda \otimes I_{\theta(q,k-1)}) \left(\begin{array}{c} P_{k-1}\chi^{(0)T} \\ P_{k-1}\chi^{(1)T} \\ \vdots \\ P_{k-1}\chi^{(l)T} + \chi_\theta \\ \vdots \\ P_{k-1}\chi^{(q-1)T} \end{array} \right) \\ \Lambda_0\sum\chi^{(0)}+\cdots+\Lambda_1\left(\chi_\theta+\sum\chi^{(l)}\right)+\cdots+\Lambda_{\alpha_{q-1}}\sum\chi^{(q-1)} \end{array} \right) \tag{19}
$$

Let split $\chi^{(l)}$ in the same way as in (18)

$$
\chi^{(l)} = (\chi^{(l,0)}|\chi^{(l,1)}|\cdots|\chi^{(l,q-1)}|\chi^{(l)}_{\theta(q,k-1)}) \tag{20}
$$

where $\chi^{(l,0)}, \chi^{(l,1)}, \ldots, \chi^{(l,q-1)} \in \mathbb{Z}^{\theta(q,k-2)}$. Further, we observe that

$$
P_{k-1}\chi^{(l)T} + \chi_\theta = \left(\begin{array}{c} (\Lambda \otimes I_{\theta(q,k-2)}) \left(\begin{array}{c} P_{k-2}\chi^{(l,0)T} + \chi_\theta \\ P_{k-2}\chi^{(l,1)T} \\ \vdots \\ P_{k-2}\chi^{(l,l)T} + \chi^{(l)}_{\theta(q,k-1)} \\ \vdots \\ P_{k-2}\chi^{(l,q-1)T} \end{array} \right) \\ \Lambda_0\left(\chi_\theta+\sum\chi^{(l,0)}\right)+\Lambda_{\alpha_1}\sum\chi^{(l,1)}+\cdots \end{array} \right)
$$

Thus one can proof by induction that it is enough to add χ_θ to the first coordinate of the vector $\chi^{(l)}$, previously multiplied by $P_1 = \Lambda_1$, and for the last coordinates of intermediate blocks add χ_θ to the first coordinate of $\chi^{(l)}$ before any summation.

Example 4.1. Let $q = 4$, $\mathbb{F}_4 = \{0, \alpha_1 = x, \alpha_2 = x+1, \alpha_3 = 1\}$, $k = 3$, and $\chi = (4, 5, 2, 8, 9, 3, 7, 4, 5, 3, 3, 5, 4, 7, 4, 5, 5, 8, 4, 3, 1)$. This is a characteristic vector of a quaternary linear code with length $n = 99$ and dimension 3. Then

$$
\Lambda = \begin{pmatrix} 1 & 1 & 1 & 1 \\ 1 & -1 & 1 & -1 \\ 1 & 1 & -1 & -1 \\ 1 & -1 & -1 & 1 \end{pmatrix}, \quad P_3\chi^T = \begin{pmatrix} (\Lambda \otimes I_5) \begin{pmatrix} P_2\chi^{(0)} \\ P_2\chi^{(1)} \\ P_2\chi^{(2)} \\ P_2\chi^{(3)} + 1 \end{pmatrix} \\ \sum \chi^{(0)} - \sum \chi^{(1)} - \sum \chi^{(2)} + \sum \chi^{(3)} + 1 \end{pmatrix}.
$$

Since $\Lambda \otimes I_1 = \Lambda$ and $P_1 = (1)$, we have

$$
P_2\chi^{(i)} = \begin{pmatrix} \Lambda \cdot (\chi_1^{(i)}, \chi_2^{(i)}, \chi_3^{(i)}, \chi_4^{(i)} + \chi_5^{(i)})^T \\ \chi_1^{(i)} - \chi_2^{(i)} - \chi_3^{(i)} + \chi_4^{(i)} + \chi_5^{(i)} \end{pmatrix}, \quad i = 0, 1, 2,
$$

$$
P_2\chi^{(3)} + \mathbf{1} = \begin{pmatrix} \Lambda \cdot (\chi_1^{(3)} + 1, \chi_2^{(3)}, \chi_3^{(3)}, \chi_4^{(3)} + \chi_5^{(3)})^T \\ \chi_1^{(3)} - \chi_2^{(3)} - \chi_3^{(3)} + \chi_4^{(3)} + \chi_5^{(3)} + 1 \end{pmatrix}.
$$

Hence

$$
(P_3\chi^T)^T = \chi P_3^T
$$
$$
= (99, -31, -23, 19, 19, 3, -33, -11, 19, 19, 1, -17, -1, 9, 9, 9, 3, -5, 9, 9, 9).
$$

Further, we consider the matrices $P_k^{(\alpha)}$ for $\alpha \in \mathbb{F}_q^*$. For $k = 1$ we have $P_1^{(\alpha)} = \Lambda_\alpha$. In the recurrence step we have

$$
P_k^{(\alpha)} = \begin{pmatrix} \Lambda^{(\alpha)} \otimes P_{k-1}^{(\alpha)} & \begin{matrix} \Lambda_0 \\ \Lambda_{\alpha\alpha_1} \\ \vdots \\ \Lambda_{\alpha\alpha_{q-1}} \end{matrix} \\ \Lambda_0 \ \Lambda_{\alpha\alpha_1} \ \cdots \ \Lambda_{\alpha\alpha_{q-1}} \ \Lambda_\alpha \end{pmatrix}, \tag{21}
$$

where $\Lambda^{(\alpha)} = (\Lambda_{\alpha\alpha_i\alpha_j})_{ij}$. Then

$$
P_k^{(\alpha)} \chi^{\mathrm{T}} =
\begin{pmatrix}
(\Lambda^{(\alpha)} \otimes I_{\theta(q,k-1)})
\begin{pmatrix}
P_{k-1}^{(\alpha)} \cdot \chi^{(0)\mathrm{T}} \\
P_{k-1}^{(\alpha)} \cdot \chi^{(1)\mathrm{T}} \\
\vdots \\
P_{k-1}^{(\alpha)} \cdot \chi^{(l)\mathrm{T}} + \chi_\theta \\
\vdots \\
P_{k-1}^{\alpha} \cdot \chi^{(q-1)\mathrm{T}}
\end{pmatrix} \\
\Lambda_0 \sum \chi^{(0)} + \Lambda_{\alpha\alpha_1} \sum \chi^{(1)} + \cdots + \Lambda_\alpha (\chi_\theta + \sum \chi^{(l)}) + \cdots
\end{pmatrix}. \quad (22)
$$

Let us consider the matrix $\Lambda^{(\alpha)}$ in more details. Since the multiplication by α induces a permutation of the elements of $\mathbb{F}_q$, the matrix above can be obtained from the matrix Λ by a suitable permutation of rows (and/or columns). Let $\pi_\alpha \in S_q$ be the permutation defined by $\pi_\alpha(i) = j$, where $\alpha_j = \alpha\alpha_i$, $i = 0, 1, \ldots, q-1$. The permutation π_α induces a permutation of block columns in the matrix $\Lambda \otimes I_{\theta(q,k-1)}$. We will obtain the same result as (22), if we multiply the matrix $\Lambda \otimes I_{\theta(q,k-1)}$ by the vector

$$
\begin{pmatrix}
P_{k-1}^{(\alpha)} \cdot \chi^{(\pi_\alpha^{-1}(0))\mathrm{T}} \\
P_{k-1}^{(\alpha)} \cdot \chi^{(\pi_\alpha^{-1}(1))\mathrm{T}} \\
\vdots \\
P_{k-1}^{(\alpha)} \cdot \chi^{(\pi_\alpha^{-1}(l))\mathrm{T}} + \chi_\theta \\
\vdots \\
P_{k-1}^{\alpha} \cdot \chi^{(\pi_\alpha^{-1}(q-1))\mathrm{T}}
\end{pmatrix}. \quad (23)
$$

With analogous considerations as in the case when $\alpha = 1$ one can conclude that for adding χ_θ to all coordinates of $P_{k-1}^{(\alpha)} \chi^{(l)\mathrm{T}}$ it is enough to add χ_θ to the first coordinate of the vector $\chi^{(l)}$, previously multiplied by $P_1^{(\alpha)} = \Lambda_\alpha$, and for the last coordinates of intermediate blocks to add χ_θ to the first coordinate of $\chi^{(l)}$ before any summation. The new element of the proof is that the permutation π preserves 0.

Example 4.2. Consider again the code from Example 4.1. But now we want to compute $P_3^{(x)} \chi^T$. Since $\pi_x = (1\ 2\ 3)$, we have

$$P_3^{(x)}\chi^T = \begin{pmatrix} (\Lambda^{(x)} \otimes I_5) \begin{pmatrix} P_2^{(x)}\chi^{(0)} \\ P_2^{(x)}\chi^{(1)} \\ P_2^{(x)}\chi^{(2)} \\ P_2^{(x)}\chi^{(3)} + 1 \end{pmatrix} \\ \sum \chi^{(0)} - \sum \chi^{(1)} + \sum \chi^{(2)} - \sum \chi^{(3)} - 1 \end{pmatrix}$$

$$= \begin{pmatrix} (\Lambda \otimes I_5) \begin{pmatrix} P_2^{(x)}\chi^{(0)} \\ P_2^{(x)}\chi^{(3)} + 1 \\ P_2^{(x)}\chi^{(1)} \\ P_2^{(x)}\chi^{(2)} \end{pmatrix} \\ \sum \chi^{(0)} - \sum \chi^{(1)} + \sum \chi^{(2)} - \sum \chi^{(3)} - 1 \end{pmatrix}.$$

Moreover, $\Lambda^{(x)} \otimes I_1 = \Lambda^{(x)}$ and $P_1^{(x)} = (-1)$, therefore

$$P_2^{(x)}\chi^{(i)} = \begin{pmatrix} \Lambda^{(x)} \cdot (-\chi_1^{(i)}, -\chi_2^{(i)}, -\chi_3^{(i)}, -\chi_4^{(i)} + \chi_5^{(i)})^T \\ \chi_1^{(i)} - \chi_2^{(i)} + \chi_3^{(i)} - \chi_4^{(i)} - \chi_5^{(i)} \end{pmatrix}$$

$$= \begin{pmatrix} \Lambda \cdot (-\chi_1^{(i)}, -\chi_4^{(i)} + \chi_5^{(i)}, -\chi_2^{(i)}, -\chi_3^{(i)})^T \\ \chi_1^{(i)} - \chi_2^{(i)} + \chi_3^{(i)} - \chi_4^{(i)} - \chi_5^{(i)} \end{pmatrix},$$

for $i = 0, 1, 2$, and

$$P_2^{(x)}\chi^{(3)} + \mathbf{1} = \begin{pmatrix} \Lambda^{(x)} \cdot (-\chi_1^{(3)} + 1, -\chi_2^{(3)}, -\chi_3^{(3)}, -\chi_4^{(3)} + \chi_5^{(3)})^T \\ \chi_1^{(3)} - \chi_2^{(3)} + \chi_3^{(3)} - \chi_4^{(3)} - \chi_5^{(3)} + 1 \end{pmatrix}$$

$$= \begin{pmatrix} \Lambda \cdot (-\chi_1^{(3)} + 1, -\chi_4^{(3)} + \chi_5^{(3)}, -\chi_2^{(3)}, -\chi_3^{(3)})^T \\ \chi_1^{(3)} - \chi_2^{(3)} + \chi_3^{(3)} - \chi_4^{(3)} - \chi_5^{(3)} + 1 \end{pmatrix}.$$

Hence

$$(P_3^{(x)}\chi^T)^T$$
$$= (-59, -13, 21, -5, -31, 9, -5, -7, 3, -19, 7, -11, -1, 5, -17, 3, -3, 3, -11, 3).$$

To obtain the vector $SP_k\chi^T = (P_k^{(\alpha_1)} + P_k^{(\alpha_2)} + \cdots + P_k^{(\alpha_{q-1})})\chi^T$, we do not use the matrices $P_k^{(\alpha)}$ for all $\alpha \in \mathbb{F}_q^*$. We apply the transformations described above to the characteristic vector χ. In other words, we take

$$(P_k^{(\alpha_1)} + P_k^{(\alpha_2)} + \cdots + P_k^{(\alpha_{q-1})})\chi^T = P_k(\chi^{(\pi_1)} + \chi^{(\pi_2)} + \cdots + \chi^{(\pi_{q-1})})^T,$$

where π_i are the corresponding permutations, $i = 1, 2, \ldots, q - 1$.

4.2. Precomputation

First, we define a map $\rho : \mathbb{Z}_{\theta(q,k)} \to \mathbb{Z}^{k-1}$ in the following way. If $0 \le t \le \theta(q,k) - 1$ and $\rho(t) = (\rho_1, \ldots, \rho_{k-1})$, then

$$\rho_1 = \lfloor \frac{t}{\theta(q,k-1)} \rfloor, \quad \rho_j = \lfloor \frac{t - \sum_{i=1}^{j-1} \rho_i \theta(q,k-i)}{\theta(q,k-j)} \rfloor \text{ for } j = 2, \ldots, k-1.$$

We will prove some properties of this map.

Lemma 4.1. *If* $0 \le t \le \theta(q,k) - 1$ *and* $\rho(t) = (\rho_1, \ldots, \rho_{k-1})$, *then*

$$t = \rho_1 \theta(q,k-1) + \rho_2 \theta(q,k-2) + \cdots + \rho_{k-2}\theta(q,2) + \rho_{k-1},$$

and $0 \le \rho_i \le q$, $i = 1, \ldots, k-1$.

Proof. Since $\rho_1 = \lfloor \frac{t}{\theta(q,k-1)} \rfloor$, then $t = \rho_1 \theta(q,k-1) + r_1$, where $r_1 < \theta(q,k-1)$ is the remainder. From $t \le \theta(q,k) - 1$ and $\theta(q,k) = q\theta(q,k-1) + 1$ we obtain

$$\rho_1 = \lfloor \frac{t}{\theta(q,k-1)} \rfloor \le \frac{q\theta(q,k-1)}{\theta(q,k-1)} = q.$$

Similarly, from $r_1 \le \theta(q,k-1) - 1$ and $\theta(q,k-1) = q\theta(q,k-2) + 1$ we have

$$\rho_2 = \lfloor \frac{r_1}{\theta(q,k-2)} \rfloor \le \frac{q\theta(q,k-2)}{\theta(q,k-2)} = q,$$

and $r_1 = \rho_2 \theta(q,k-2) + r_2$, $r_2 < \theta(q,k-2)$.

Continuing in this manner, we prove that $\rho_3 \le q, \ldots, \rho_{k-2} \le q$. In the last step,

$$\rho_{k-1} = \lfloor \frac{r_{k-2}}{\theta(q,1)} \rfloor = r_{k-2} \le \theta(q,2) - 1 = q.$$

Moreover,

$$t = \rho_1 \theta(q,k-1) + r_1 = \rho_1 \theta(q,k-1) + \rho_2 \theta(q,k-2) + r_2$$
$$= \cdots = \rho_1 \theta(q,k-1) + \rho_2 \theta(q,k-2) + \cdots + \rho_{k-2}\theta(q,2) + r_{k-2}$$
$$= \rho_1 \theta(q,k-1) + \rho_2 \theta(q,k-2) + \cdots + \rho_{k-2}\theta(q,2) + \rho_{k-1}.$$

As all considered integers are nonnegative, the quotients $\rho_1, \ldots, \rho_{k-1}$ are also nonnegative. $\square$

Corollary 4.1. *The map* ρ *is injective.*

Lemma 4.2. *If* $0 \le t \le \theta(q,k) - 1$, $\rho(t) = (\rho_1, \ldots, \rho_{k-1})$, *and there is an index* $i \in \{1, \ldots, k-1\}$ *such that* $\rho_i = q$, $\rho_j < q$ *for* $j = 1, \ldots, i-1$, *then* $\rho_{i+1} = \cdots = \rho_{k-1} = 0$.

Proof. Let $r_0 = t$, $r_{j-1} = \rho_j \theta(q, k-j) + r_j$, $r_j < \theta(q, k-j)$, $j = 1, \ldots, k-1$. If $\rho_i = q$ then

$$r_{i-1} = q\theta(q, k - i) + r_i < \theta(q, k - i + 1) = q\theta(q, k - i) + 1.$$

Hence $r_i < 1$ and so $r_i = 0$. This yields $r_{i+1} = \cdots = r_{k-1} = 0$ and $\rho_{i+1} = \cdots = \rho_{k-1} = 0$. $\qquad\square$

A small modification of the vector $\rho(t)$ is more convenient for our algorithm. We use the vector $\nu(t) = (\nu_1, \ldots, \nu_{k-1})$ defined by

$$\nu(t) = \begin{cases} \rho(t), & \text{if } \rho_i < q, \ i = 1, \ldots, k-1, \\ (\rho_1, \ldots, \rho_{i-1}, q, \ldots, q), & \text{if } \rho_i = q, \ i \le k-1. \end{cases}$$

Let $v_0, \ldots, v_{q^{k-1}-1}$ be the vectors in $\mathbb{F}_q^{k-1}$, ordered lexicographically with respect to the considered ordering of the elements in $\mathbb{F}_q$. Consider the map $\mu : \mathbb{F}_q^{k-1} \to \mathbb{Z}$, defined by $\mu(\alpha_{i_1}, \ldots, \alpha_{i_{k-1}}) = \rho^{-1}(i_1, \ldots, i_{k-1}) + 1$. Obviously, the images of the vectors from $\mathbb{F}_q^{k-1}$ are positive integers $\le \theta(q, k)$. Moreover, the different vectors have different images.

Similarly, we consider the maps $\rho_l : \mathbb{Z}_{\theta(q,l)} \to \mathbb{Z}^{l-1}$ and ν_l, $\mu_l : \mathbb{F}_q^{l-1} \to \mathbb{Z}$, defined in the same way as ρ, ν, and μ, respectively, but taking l instead of k, $l = 2, \ldots, k$. We denote by $v_0^{(l)}, \ldots, v_{q^{l-1}-1}^{(l)}$ the vectors in $\mathbb{F}_q^{l-1}$, ordered lexicographically with respect to the considered ordering of the elements in $\mathbb{F}_q$. Moreover, we need the map $\sigma_l : \mathbb{F}_q^{l-1} \to \mathbb{Z}^{l-1}$ defined by $\sigma_l(\alpha_{i_1}, \ldots, \alpha_{i_{l-1}}) = (i_1, \ldots, i_{l-1})$, $l = 2, \ldots, k$.

To prepare the butterfly algorithm, we use three arrays of length $\theta(q, k)$ named χ^0, χ^1 and S. The vectors χ^0 and χ^1 are modified copies of the characteristic vector χ. If $\nu(i-1)$ does not contain q as a coordinate, then $\chi^0[i] = \chi_i$ and $\chi^1[i] = -\chi_i$. Otherwise, the corresponding coordinate of χ will be added to suitable places in the copies. In this step we use the maps ν and ν^{-1}.

The array S is equal to the sum $\chi^{(\pi_1)} + \cdots + \chi^{(\pi_{q-1})}$ with a difference in the so-called 'inactive' coordinates (that are in position i if $\nu(i-1)$ contains q as a coordinate). Here we use the map μ.

Example 4.3. Consider again the code from Example 4.1 with $\chi = (4, 5, 2, 8, 9, 3, 7, 4, 5, 3, 3, 5, 4, 7, 4, 5, 5, 8, 4, 3, 1)$. Then the arrays obtained in Algorithm 4.1, are

$$\chi^0 = \left(4, 5, 2, 17, 0, 3, 7, 4, 8, 0, 3, 5, 4, 11, 0, 6, 5, 8, 7, 0, 0\right)$$
$$\chi^1 = \left(-4, -5, -2, 1, 0, -3, -7, -4, -2, 0, -3, -5, -4, -3, 0, -4, -5, -8, -1, 0, 0\right)$$
$$S = \left(-4, 4, -2, 10, 0, -4, 2, -4, -5, 0, -4, -5, -4, 2, 0, 0, -2, 1, -4, 0, 0\right).$$

Algorithm 4.1 Precomputations

Require: Integers q, k, and an array χ with length $\theta(q, k)$
Ensure: arrays χ^0 and S with length $\theta(q, k)$

1: Initialize arrays χ^0, χ^1, S with length $\theta(q, k)$ to 0;
2: **for** $i = 1$ **to** $\theta(q, k)$ **do**
3:     $v = \nu(i - 1)$; $w = v$; $j = k - 1$;
4:     **while** $v[j] = q$ and $j > 0$ **do**
5:         $w[j] = 0$; $j = j - 1$;
6:     **end while**
7:     **if** $j < k - 1$ **then**
8:         $w[j + 1] = \mathrm{Ord}(1)$; $t = \nu^{-1}(w) + 1$;
9:         $\chi^0[t] = \chi^0[t] + \chi[i]$; $\chi^1[t] = \chi^1[t] + \chi[i]$;
10:     **else**
11:         $\chi^0[i] = \chi[i]$; $\chi^1[i] = -\chi[i]$;
12:     **end if**
13: **end for**
14: **for** $i = 1$ **to** $q - 1$ **do**
15:     $s = \mathrm{Tr}(\alpha_i)$;
16:     **for** $j = 0$ **to** $q^{k-1} - 1$ **do**
17:         $t = \mu(v_j)$; $t_1 = \mu(\alpha_i v_j)$; $S[t_1] = S[t_1] + \chi^s[t]$;
18:     **end for**
19: **end for**

The value of the "inactive" coordinates in this step are 0's.

4.3. *Main algorithm*

Algorithm 4.2 realizes a butterfly algorithm over the sum S and the vector χ^0. Throughout the procedure, the algorithm finds the right places of the 'active' coordinates by applying suitable permutations. The butterfly algorithms for S and χ^0 are similar to Algorithm 3.1. Here we apply the maps σ_i, ν_i, ν^{-1}. We haven't defined the maps σ_1 and ν_1 that appear in the cycles when $k - l = 1$ and $l = 1$. In such a case we assume that the corresponding vectors $\sigma_1(v_j^{(1)})$ and $\nu_1(i - 1)$ are empty. The vectors $\mathbf{0}_i$ and $\mathbf{q}_i$ consist of i coordinates equal to 0 and q, respectively. Furthermore, $\mathbf{0}_0$ is the empty vector.

Algorithm 4.2 Main Algorithm

Require: Integers q, k, and the arrays χ^0 and S with length $\theta(q, k)$

Ensure: an array S with length $\theta(q, k)$

1: **for** $l = 1$ to $k - 1$ **do**
2:    **for** $j = 0$ to $q^{k-l-1} - 1$ **do**
3:      **for** $l_1 = 1$ to m **do**
4:        **for** $i_1 = 0$ to $2^{m-l_1} - 1$ **do**
5:          **for** $j_1 = 0$ to $2^{l_1-1} - 1$ **do**
6:            **for** $i = 1$ to $\theta(q, l)$ **do**
7:              $v = (\sigma_{k-l}(v_j^{(k-l)}), i_1 2^{l_1} + j_1, \nu_l(i - 1))$;
8:              $t_1 = \nu^{-1}(v) + 1$; $t_2 = t_1 + 2^{l_1-1}\theta(q, l)$;
9:              $A = S[t_1]$; $B = S[t_2]$;
10:              $S[t_1] = A + B$; $S[t_2] = A - B$;
11:            **end for**
12:            $v = (\sigma_{k-l}(v_j^{(k-l)}), i_1 2^{l_1} + j_1, \mathbf{0}_{l-1})$;
13:            $t_1 = \nu^{-1}(v) + 1$; $t_2 = t_1 + 2^{l_1-1}\theta(q, l)$;
14:            $A = \chi^0[t_1]$; $B = \chi^0[t_2]$;
15:            $\chi^0[t_1] = A + B$; $\chi^0[t_2] = A - B$;
16:          **end for**
17:        **end for**
18:      **end for**
19:      **for** $i = 1$ to $q - 1$ **do**
20:        $v = (\sigma_{k-l}(v_j^{(k-l)}), i, \mathbf{0}_{l-1})$; $w = (\sigma_{k-l}(\alpha_i v_j^{(k-l)}), \mathbf{q}_l)$;
21:        $t = \nu^{-1}(v) + 1$; $t_1 = \nu^{-1}(w) + 1$; $S[t_1] = S[t_1] + \chi_0[t]$;
22:      **end for**
23:    **end for**
24: **end for**

Example 4.4. For the code from Example 4.1 Algorithm 4.2 gives the following result:

$$S = (-19, -23, -7, 1, -35, 13, -27, -23, 21,$$
$$- 3, 13, -27, 5, -7, 1, 25, -3, -7, 1, -11, 13).$$

To obtain the weight distribution, we compute the array W with $W[i] = \dfrac{3n - S[i]}{4} = \dfrac{297 - S[i]}{4}$ for $i = 1, 2, \ldots, 21$, so

$$W = (79, 80, 76, 74, 83, 71, 81, 80, 69, 75, 71, 81, 73, 76, 74, 68, 75, 76, 74, 77, 71).$$

This means that the weight enumerator of this code is
$$W(y) = 1 + 3\big(y^{68} + y^{69} + 3y^{71} + y^{73} + 3y^{74} + 2y^{75}$$
$$+ 3y^{76} + y^{77} + y^{79} + 2y^{80} + 2y^{81} + y^{83}\big).$$

4.4. *Complexity analysis*

The algorithm for computation of χ^0 and χ^1 requires $2\theta(q,k)$ additions over $\mathbb{Z}$, so the complexity is $O(q^{k-1})$.

Step 2 of Algorithm 4.1 requires $(q-1)q^{k-1}.(k-1)$ multiplications in $\mathbb{F}_q$ to determine the new positions after the permutations and $(q-1)q^{k-1}$ additions over $\mathbb{Z}$. So the complexity of this step is $O(kq^k)$.

The main cycle in the main algorithm has $k-1$ iterations for l. About active positions we have $q^{k-l-1}.m.2^{m-1}.\theta(q,l).4 = O(mq^{k-1})$ additions in $\mathbb{Z}$. About calculations over χ^0 we have $q^{k-l-1}.m.2^{m-1}.4 = O(mq^{k-l})$ additions in $\mathbb{Z}$. About filling out the inactive positions we have $q^{k-l-1}.q.(k-l-1)$ multiplications in $\mathbb{F}_q$ and $q^{k-l-1}.q$ additions in $\mathbb{Z}$. So the summary complexity of Algorithm 4.2 is $O(kmq^{k-1})$.

It is easy to see that the complexity of Algorithm 3.1 is $O(kmq^k)$. Moreover, Algorithm 3.1 uses an array of length q^k, while Algorithm 4.2 uses three arrays but each of them of length $\theta(q,k) = (q^k - 1)/(q-1)$. Algorithm 4.2 is more effective than the previous algorithms for computing the weight distribution especially if the length n and/or the order of the considered field q are large numbers.

Acknowledgements

This research was supported by Grant DN 02/2/13.12.2016 of the Bulgarian National Science Fund.

References

[1] E. Berlekamp, R. McEliece and H. van Tilborg, On the inherent intractability of certain coding problems, *IEEE Trans. Inform. Theory* **24** (1978), 384–386.

[2] A. Vardy, The intractability of computing the minimum distance of a code, *IEEE Trans. Inform. Theory* **43** (1997), 1757–1766.

[3] A. Barg, *Complexity Issues in Coding Theory*, in *Handbook of Coding Theory*, eds. W. C. Huffman and V. Pless, Elsevier, 1998.

[4] M. G. Karpovsky, On the weight distribution of binary linear codes, *IEEE Trans. Inform. Theory* **25** (1979), 105–109.

[5] M. G. Karpovsky, Weight distribution of translates, covering radius, and perfect codes correcting errors of given weights, *IEEE Trans. Inform. Theory* **27** (1981), 462–472.

[6] E. Michielssen and A. Boag, A multilevel matrix decomposition algorithm for analysing scattering from large structures, *IEEE Trans. Antennas and Propagation* **44** (1996), 1086–1093.

[7] F. M. O'Neil and V. Rokhlin, An algorithm for the rapid evaluation of special function transforms, *Appl. Comput. Harmon. Anal.* **28** (2010), 203–226.

[8] W. C. Huffman and V. Pless, *Fundamentals of Error-Correcting Codes*, Cambridge University Press, Cambridge, UK, 2003.

[9] G. Seroussi and A. Lempel, Factorization of symmetric matrices and trace-orthogonal bases in finite fields, *SIAM J. Comput.* **9** (1980), 758–767.

[10] G. L. Mullen and D. Panario, *Handbook of Finite Fields*, Chapman and Hall/CRC, Boca Raton, FL 33487-2742, 2013.

[11] J. Seberry and M. Yamada, *Hadamard matrices, sequences and block designs*, in *Contemporary Design Theory: A Collection of Surveys*, 431–560, eds. J. H. Dinitz and D. R. Stinson, Wiley, New York, 1992.

[12] I. J. Good, The interaction algorithm and practical fourier analysis, *J. Roy. Statist. Soc. Ser. B (Methodological)* **20** (1958), 361–372.

[13] A. Joux, *Algorithmic Cryptanalysis*, Chapman and Hall/CRC, Boca Raton, FL 33487-2742, 2009.

Received January 22, 2019
Revised March 2, 2019

EXTRINSIC SHAPES OF TRAJECTORIES
ON REAL HYPERSURFACES OF TYPE (B)
IN A COMPLEX HYPERBOLIC SPACE

Tuya BAO*

*Colledge of Mathematics, Inner Mongolia University for the Nationalities,
Tongliao, Inner Mongolia, 028043, People's Republic of China*
E-mail: baotuya1981@126.com

Toshiaki ADACHI[†]

*Department of Mathematics, Nagoya Institute of Technology,
Nagoya 466-8555, Japan*
E-mail: adachi@nitech.ac.jp

We study the property that trajectories on real hypersurfaces of type (B) in a complex hyperbolic space are tangentially of order two. By investigating the image of the moduli space of such trajectories in the moduli space of circles on a complex hyperbolic space, we give some characterizations of homogeneous Hopf hypersurfaces.

Keywords: Sasakian magnetic fields; trajectories; real hypersurfaces; extrinsic shapes; circles; tangentially of order two; geodesic curvatures; complex torsions.

1. Introduction

When we study submanifolds it is a way to investigate how some typical curves on submanifolds can be seen from ambient spaces. If we take a standard sphere S^n in a Euclidean space $\mathbb{R}^{n+1}$, every geodesic on S^n can be seen as a circle in $\mathbb{R}^{n+1}$. Moreover, this property characterize S^n among hypersurfaces in $\mathbb{R}^{n+1}$. There are many results which characterize some submanifolds by such extrinsic properties of geodesics. For example, we have some characterizations of real hypersurfaces in a complex projective space and in a complex hyperbolic space (see [9, 11, 12], for example). On each real hypersurface in a Kähler manifold, we have almost contact metric structure, hence it is natural to consider that if we study more curves associated with this structure then we can refine these results.

---

*The first author is partially supported by National Natural Science Foundation of China (No. 11661062), Inner Mongolia Natural Science Foundation 2018MS01011, and Youth Technology Talents Program NJYT–19–A09.
†The second author is partially supported by Grant-in-Aid for Scientific Research (C) (No. 16K05126), Japan Society for the Promotion of Science.

In this context the authors study trajectories for Sasakian magnetic fields on real hypersurfaces which are induced from their almost contact metric structure. In a complex hyperbolic space, we have homogeneous Hopf hypersurfaces which are called of types (A_0), (A_1), (A_2) and (B). We studied hypersurfaces of types (A_0) and (A_1) in [5, 6] and those of type (A_2) in [2]. In these papers we paid attention to trajectories which can be seen as circles in the ambient space. To complete our study, we investigate hypersurfaces of type (B) in this paper. Being different from hypersurfaces of type (A), their shape operators and characteristic tensor fields are not necessarily simultaneously diagonalizable. This shows that "extrinsic shapes" of trajectories depend pointwisely. We study the relationship between geodesic curvatures and complex torsions of extrinsic shapes of trajectories at points where they can be seen "circle-like", and show some characterizations of some homogeneous Hopf hypersurfaces.

2. Homogeneous Hopf hypersurfaces

On a real hypersurface M in a Kähler manifold $\widetilde{M}$ with complex structure J and a Riemannian metric $\langle\ ,\ \rangle$, we have an almost contact metric structure $(\phi, \xi, \eta, \langle\ ,\ \rangle)$. With a unit normal local vector field $\mathcal{N}$ on M in $\widetilde{M}$, the characteristic vector field ξ is defined by $\xi = -J\mathcal{N}$, the 1-form η is given by $\eta(v) = \langle v, \xi \rangle$, the characteristic tensor field ϕ is defined by $\phi(v) = Jv - \eta(v)\mathcal{N}$, and $\langle\ ,\ \rangle$ is the induced metric. If we define a 2-form $\mathbb{F}_\phi$ by $\mathbb{F}_\phi(v, w) = \langle v, \phi(w) \rangle$, it is closed ([3]). Its constant multiple $\mathbb{F}_\kappa = \kappa\mathbb{F}_\phi$ ($\kappa \in \mathbb{R}$) is called a *Sasakian magnetic field* on this real hypersurface. A smooth curve γ on M which is parameterized by its arclength is said to be a *trajectory* for $\mathbb{F}_\kappa$ if it satisfies the differential equation $\nabla_{\dot\gamma}\dot\gamma = \kappa\phi\dot\gamma$. When $\kappa = 0$, which is the case without the influence of magnetic fields, trajectories are geodesics. We may hence consider that trajectories for Sasakian magnetic fields are generalizations of geodesics which are related with the contact structure.

For a real hypersurface M in a Kähler manifold $\widetilde{M}$, we decompose its tangent bundle TM into a complex subbundle T^0M of $T\widetilde{M}$ and the line bundle $\mathbb{R}\xi$, that is, $TM = T^0M \oplus \mathbb{R}\xi$. This hypersurface is said to be *Hopf* if its characteristic vector field ξ is principal at each point, that is, it is an eigenvector of its shape operator A_M at each point. It is known that the principal curvature δ_M of ξ of a Hopf hypersurface M in $\mathbb{C}H^n$ is locally constant ([10]). In this paper we study trajectories on a homogeneous Hopf hypersurfaces in a complex hyperbolic space $\mathbb{C}H^n(c)$ of constant holomorphic sectional curvature c. Such hypersurfaces are classified by Berndt [7]. They are

(A$_0$) a horosphere HS,
(A$_1$) a geodesic sphere $G(r)$ or a tube $T(r)$ around totally geodesic $\mathbb{C}H^{n-1}$,
(A$_2$) a tube $T_\ell(r)$ around totally geodesic $\mathbb{C}H^\ell$ $(1 \leq \ell \leq n-2)$,
(B) a tube $R(r)$ around totally geodesic $\mathbb{R}H^{n-1}$.

In the above list r's denote their radii. Real hypersurfaces $HS, G(r), T(r)$ and $T_\ell(r)$ are called of type (A), and $R(r)$ is called of type (B). When M is one of $HS, G(r)$ and $T(r)$, it has two principal curvatures δ_M and λ_M. All vectors orthogonal to ξ are principal and have common principal curvature λ_M. Their values are

$$\delta_M = \begin{cases} \sqrt{|c|}, & \\ \sqrt{|c|}\coth\sqrt{|c|}\,r, & \\ \sqrt{|c|}\coth\sqrt{|c|}\,r, & \end{cases} \qquad \lambda_M = \begin{cases} \sqrt{|c|}/2, & M = HS, \\ (\sqrt{|c|}/2)\coth(\sqrt{|c|}\,r/2), & M = G(r), \\ (\sqrt{|c|}/2)\tanh(\sqrt{|c|}\,r/2), & M = T(r), \end{cases}$$

When $M = T_\ell(r)$ or $M = R(r)$, it has three principal curvatures

$$\delta_M = \begin{cases} \sqrt{|c|}\coth\sqrt{|c|}\,r, & M = T_\ell(r), \\ \sqrt{|c|}\tanh\sqrt{|c|}\,r, & M = R(r), \end{cases}$$

$$\lambda_M = (\sqrt{|c|}/2)\coth(\sqrt{|c|}\,r/2), \qquad \mu_M = (\sqrt{|c|}/2)\tanh(\sqrt{|c|}\,r/2).$$

For real hypersurfaces of type (A), the tensor field ϕ preserves subbundles of principal curvature vectors, hence A_M and ϕ are simultaneously diagonalizable ($A_M\phi = \phi A_M$). On the other hand, for each real hypersurface M of type (B), we have $\phi(V_{\lambda_M}) = V_{\mu_M}$, $\phi(V_{\mu_M}) = V_{\lambda_M}$, where V_{λ_M}, V_{μ_M} are subbundles of principal curvature vectors associated with λ_M, μ_M, respectively.

3. Extrinsic shapes of trajectories

Let M be a real hypersurface in $\mathbb{C}H^n$. We denote by $\iota : M \to \mathbb{C}H^n$ an isometric embedding. For a smooth curve γ on M, we call the curve $\iota \circ \gamma$ on $\mathbb{C}H^n$ its *extrinsic shape*. For the sake of simplicity, we usually denote the extrinsic shape of γ also by γ.

We take a trajectory γ for a Sasakian magnetic field $\mathbb{F}_\kappa$ on M. We call the function $\rho_\gamma = \langle \dot{\gamma}, \xi \rangle$ the *structure torsion* of γ. As we have $\nabla_{\dot{\gamma}}\dot{\gamma} = \kappa\phi\dot{\gamma} = \kappa(J\dot{\gamma} - \rho_\gamma\mathcal{N})$, it is an important "character" of a trajectory. Since the Riemannian connections ∇ on M and $\widetilde{\nabla}$ on $\mathbb{C}H^n$ are related by Gauss and Weingarten formulae

$$\widetilde{\nabla}_X Y = \nabla_X Y + \langle A_M X, Y \rangle \mathcal{N}, \qquad \widetilde{\nabla}_X \mathcal{N} = -A_M X$$

for arbitrary vector fields X, Y on $\widetilde{M}$ tangent to M, we have $\nabla_X \xi = \phi A_M X$. Thus we obtain

$$\frac{d}{dt} \rho_\gamma = \langle \kappa \phi \dot{\gamma}, \xi \rangle + \langle \dot{\gamma}, \phi A_M \dot{\gamma} \rangle = \langle \dot{\gamma}, \phi A_M \dot{\gamma} \rangle = \frac{1}{2} \langle \dot{\gamma}, (\phi A_M - A_M \phi) \dot{\gamma} \rangle,$$

because ϕ is skew symmetric and A_M is symmetric. Therefore, the structure torsion of each trajectory on a real hypersurface of type (A) is a constant function. On the other hand, for a trajectory γ on a real hypersurface of type (B), if $\dot{\gamma}(t_0) \in (V_{\lambda_M})_{\gamma(t_0)} \oplus \mathbb{R}\xi_{\gamma(t_0)}$ or $\dot{\gamma}(t_0) \in (V_{\mu_M})_{\gamma(t_0)} \oplus \mathbb{R}\xi_{\gamma(t_0)}$, we find $\rho'_\gamma(t_0) = 0$ because $\phi(V_{\lambda_M}) = V_{\mu_M}$. But this does not guarantee the constancy of its structure torsion (cf. [4]).

Generally, a smooth curve σ on $\mathbb{C}H^n$ which is parameterized by its arclength is said to be a *circle* if it satisfies the equation $\widetilde{\nabla}_{\dot{\sigma}} \widetilde{\nabla}_{\dot{\sigma}} \dot{\sigma} + \|\widetilde{\nabla}_{\dot{\sigma}} \dot{\sigma}\|^2 \dot{\sigma} = 0$. The constant $k_\sigma = \|\widetilde{\nabla}_{\dot{\sigma}} \dot{\sigma}\|$ is called the *geodesic curvature* of σ. When a circle σ on $\mathbb{C}H^n$ is not a geodesic, that is, when $k_\sigma > 0$, we set $\tau_\sigma = \langle \dot{\sigma}, J\widetilde{\nabla}_{\dot{\sigma}} \dot{\sigma} \rangle / k_\sigma$. It is also constant along σ. We call this constant its *complex torsion*. A smooth curve on a real hypersurface M is said to be *extrinsic circular* if its extrinsic shape is a circle. Weakening the condition we say that a smooth curve γ on a real hypersurface M which is parameterized by its arclength is extrinsic circular at $\gamma(t_0)$ if $\widetilde{\nabla}_{\dot{\gamma}} \widetilde{\nabla}_{\dot{\gamma}} \dot{\gamma}(t_0) + \|\widetilde{\nabla}_{\dot{\gamma}} \dot{\gamma}\|^2 \dot{\gamma}(t_0) = 0$, and say that it is *tangentially of order two* at $\gamma(t_0)$ if $\widetilde{\nabla}_{\dot{\gamma}} \widetilde{\nabla}_{\dot{\gamma}} \dot{\gamma}(t_0) + \|\widetilde{\nabla}_{\dot{\gamma}} \dot{\gamma}\|^2 \dot{\gamma}(t_0)$ does not have a component tangent to M.

We now study extrinsic shapes of trajectories on real hypersurfaces. For differentials of the extrinsic shape of a trajectory γ for $\mathbb{F}_\kappa$, by using Gauss and Weingarten formulae, we have

$$\widetilde{\nabla}_{\dot{\gamma}} \dot{\gamma} = \kappa \phi \dot{\gamma} + \langle A_M \dot{\gamma}, \dot{\gamma} \rangle \mathcal{N}, \tag{3.1}$$

$$\widetilde{\nabla}_{\dot{\gamma}} \widetilde{\nabla}_{\dot{\gamma}} \dot{\gamma} = -\kappa^2 \dot{\gamma} - (\langle A_M \dot{\gamma}, \dot{\gamma} \rangle - \kappa \rho_\gamma)(A_M \dot{\gamma} + \kappa \xi)$$
$$+ \frac{d}{dt} (\langle A_M \dot{\gamma}, \dot{\gamma} \rangle - \kappa \rho_\gamma) \mathcal{N}. \tag{3.2}$$

Thus, its first geodesic curvature $k_\gamma = \|\widetilde{\nabla}_{\dot{\gamma}} \dot{\gamma}\|$ and its (first) complex torsion $\tau_\gamma = \langle \dot{\gamma}, J\widetilde{\nabla}_{\dot{\gamma}} \dot{\gamma} \rangle / k_\gamma$ are given as

$$k_\gamma^2 = \kappa^2 (1 - \rho_\gamma^2) + \langle A_M \dot{\gamma}, \dot{\gamma} \rangle^2, \quad \tau_\gamma = -\{\kappa(1 - \rho_\gamma^2) + \langle A_M \dot{\gamma}, \dot{\gamma} \rangle \rho_\gamma\} / k_\gamma.$$

By (3.2), we see that γ is tangentially of order 2 if and only if it satisfies

$$\{k_\gamma(t_0)^2 - \kappa^2\}\dot{\gamma}(t_0) = \big(\langle A_M\dot{\gamma}(t_0), \dot{\gamma}(t_0)\rangle - \kappa\rho_\gamma(t_0)\big)\big(A_M\dot{\gamma}(t_0) + \kappa\xi_{\gamma(t_0)}\big). \quad (3.3)$$

We therefore find the following:

Proposition 3.1 ([5]). *Let γ be a trajectory for $\mathbb{F}_\kappa$ on a Hopf real hypersurface M in $\mathbb{C}H^n$. Its extrinsic shape is as follows.*

(1) *If $\rho_\gamma(t_0) = \pm 1$, then γ is tangentially of order two at $\gamma(t_0)$. In this case $k_\gamma(t_0) = |\delta_M|$ and $\tau_\gamma(t_0) = \mp\mathrm{sgn}(\delta_M)$, where $\mathrm{sgn}(\delta_M)$ denotes the signature of δ_M.*

(2) *When $\rho_\gamma(t_0) \neq \pm 1$ and $\dot{\gamma}(t_0) - \rho_\gamma(t_0)\xi_{\gamma(t_0)}$ is principal, then γ is tangentially of order two at $\gamma(t_0)$ if and only if one of the following conditions holds:*

    i) $\alpha - \kappa\rho_\gamma(t_0) + (\delta_M - \alpha)\rho_\gamma(t_0)^2 = 0$,
    ii) $\kappa + (\delta_M - \alpha)\rho_\gamma(t_0) = 0$,

    where α denotes the principal curvature of $\dot{\gamma} - \rho_\gamma(t_0)\xi_{\gamma(t_0)}$. In the former case $k_\gamma(t_0) = |\kappa|$ and $\tau_\gamma = -\mathrm{sgn}(\kappa)$, and in the latter case $k_\gamma(t_0)^2 = \kappa^2 - 2\kappa\alpha\rho_\gamma(t_0) + \alpha^2$, $k_\gamma(t_0)\tau_\gamma(t_0) = \kappa(2\rho_\gamma(t_0)^2 - 1) - \alpha\rho_\gamma(t_0)$.

(3) *Suppose $k_\gamma(t_0) \neq |\kappa|$. If $\dot{\gamma} - \rho_\gamma(t_0)\xi_{\gamma(t_0)}$ is not principal, then γ is not tangentially of order two at $\gamma(t_0)$.*

(4) *If $k_\gamma(t_0) = |\kappa|$, then γ is tangentially of order two at $\gamma(t_0)$. In this case $\tau_\gamma(t_0) = -1$.*

The last assertion is added to the result in [5]. By (3.3), if $k_\gamma(t_0) = |\kappa|$, we have either $A_M\dot{\gamma}(t_0) = -\kappa\xi_{\gamma(t_0)}$ or $\langle A_M\dot{\gamma}(t_0), \dot{\gamma}(t_0)\rangle = \kappa\rho_\gamma(t_0)$. In the former case, as M is a Hope hypersurface, we see $\rho_\gamma(t_0) = \pm 1$, and in the latter case we have $\widetilde{\nabla}_{\dot{\gamma}}\dot{\gamma} = \kappa J\gamma$ and $\widetilde{\nabla}_{\dot{\gamma}}\widetilde{\nabla}_{\dot{\gamma}}\dot{\gamma} = -\kappa^2\dot{\gamma} + \frac{d}{dt}\big(\langle A_M\dot{\gamma}, \dot{\gamma}\rangle - \kappa\rho_\gamma\big)\mathcal{N}$ by (3.1) and (3.2). We hence get the assertion.

We here rewrite geodesic curvatures and complex torsions of extrinsic shapes of trajectories when they are tangentially of order two. In the case (2) ii) in Proposition 3.1, we have

$$k_\gamma(t_0)^2 = \alpha^2 + (\delta_M^2 - \alpha^2)\rho_\gamma(t_0)^2, \quad (3.4)$$

$$k_\gamma(t_0)\tau_\gamma(t_0) = \big\{\delta_M\big(1 - 2\rho_\gamma(t_0)^2\big) - 2\alpha\big(1 - \rho_\gamma(t_0)^2\big)\big\}\rho_\gamma(t_0). \quad (3.5)$$

4. Trajectories on real hypersurfaces of Type B

In this section we study the relationship between geodesic curvatures and complex torsions of extrinsic shapes of trajectories on a real hypersurface of type (B) when their extrinsic shapes are tangentially of order 2 and their complex torsions are not ± 1. Before go into our study we give a remark which is a consequence of Proposition 3.1.

Theorem 4.1. *For a trajectory γ for $\mathbb{F}_\kappa$ on a real hypersurface of type (B), under the assumption $k_\gamma(t_0) \neq |\kappa|$, it is tangentially of order two at $\gamma(t_0)$ if and only if it is extrinsic circular at that point.*

Proof. We are enough to show the "only if" part. Since $k_\gamma(t_0) \neq |\kappa|$, we find that $\dot\gamma(t_0) - \rho_\gamma(t_0)\xi_{\gamma(t_0)}$ is principal by Proposition 3.1. Hence we find $\rho'_\gamma(t_0) = 0$ (see §3). Let

$$\mathrm{Proj}_\lambda : TM = V_{\lambda_M} \oplus V_{\mu_M} \oplus \mathbb{R}\xi \to V_{\lambda_M}, \quad \mathrm{Proj}_\mu : TM \to V_{\mu_M}$$

denote projections onto subbundles of principal curvatures. We set $\eta_\gamma(t) = \|\mathrm{Proj}_\lambda\big(\dot\gamma(t)\big)\|$ and $\zeta_\gamma(t) = \|\mathrm{Proj}_\mu\big(\dot\gamma(t)\big)\|$. We then have $\eta_\gamma(t)^2 + \zeta_\gamma(t)^2 + \rho_\gamma(t)^2 = 1$ and

$$\langle A_M \dot\gamma(t), \dot\gamma(t)\rangle = \lambda_M \eta_\gamma(t)^2 + \mu_M \zeta_\gamma(t)^2 + \delta_M \rho_\gamma(t_0)^2.$$

When $\dot\gamma(t_0) - \rho_\gamma(t_0)\xi_{\gamma(t_0)} \in V_{\lambda_M}$, we see

$$\frac{d}{dt}\langle A_M \dot\gamma(t), \dot\gamma(t)\rangle\big|_{t=t_0}$$
$$= -2\lambda_M\big\{\zeta_\gamma(t_0)\zeta'_\gamma(t_0) + \rho_\gamma(t_0)\rho'_\gamma(t_0)\big\} + 2\delta_M \rho_\gamma(t_0)\rho'_\gamma(t_0) = 0,$$

because $\zeta_\gamma(t_0) = 0$. By the same argument we have $\frac{d}{dt}\langle A_M \dot\gamma(t), \dot\gamma(t)\rangle\big|_{t=t_0} = 0$ when $\dot\gamma(t_0) - \rho_\gamma(t_0)\xi_{\gamma(t_0)} \in V_{\mu_M}$. Hence we get the conclusion. $\square$

We take a real hypersurface $M = R(r)$ of type (B) in $\mathbb{C}H^n(c)$. If we put $\nu = \coth(\sqrt{|c|}\,r/2)$, then its principal curvatures are expressed as $\lambda_M = \sqrt{|c|}\nu/2$, $\mu_M = \sqrt{|c|}/(2\nu)$, $\delta_M = 2\sqrt{|c|}\nu/(\nu^2 + 1)$. We study a trajectory γ for $\mathbb{F}_\kappa$ in the case (2) ii) in Proposition 3.1. Its extrinsic shape is tangentially of order two at t_0.

When $\alpha = \lambda_M$, by (3.4), we have $k_\gamma(t_0) = \sqrt{3|c|}/2$ when $\lambda_M = \sqrt{3|c|}/2$, and

$$\rho_\gamma(t_0)^2 = \frac{\big(|c|\nu^2 - 4k_\gamma(t_0)^2\big)(\nu^2 + 1)^2}{|c|\nu^2(\nu^2 + 5)(\nu^2 - 3)} \tag{4.1}$$

when $\lambda_M \neq \sqrt{3|c|}/2$. The former case occurs when $\kappa = 0$, and in this case we have $\tau_\gamma(t_0) = -\rho_\gamma(t_0)$. In the latter case, by (3.5) and (4.1), we have

$$k_\gamma(t_0)^2 \tau_\gamma(t_0)^2$$
$$= \frac{4\big(|c|\nu^2 - 4k_\gamma(t_0)^2\big)\big\{|c|\nu^2(\nu^2 - 3) + 2k_\gamma(t_0)^2(\nu^2 + 1)^2\big\}^2}{c^2\nu^4(\nu^2 + 5)^3(\nu^2 - 3)}. \tag{4.2}$$

In order to study the dependency of $\tau_\gamma(t_0)$ on $k_\gamma(t_0)$, we set

$$g(K;\nu) = (|c|\nu^2 - 4K)\{|c|\nu^2(\nu^2-3) + 2K(\nu^2+1)^2\}^2/K.$$

Since we have

$$\frac{d}{dK}g(K;\nu) = -\frac{1}{K^2}\{|c|\nu^2(\nu^2-3) + 2K(\nu^2+1)^2\}$$
$$\times \{16(\nu^2+1)^2K^2 - 2K|c|\nu^2(\nu^2+1)^2 + |c|^2\nu^4(\nu^2-3)\},$$

we find that $\frac{d}{dK}g(K;\nu) = 0$ if and only if

$$K = \frac{|c|\nu^2(3-\nu^2)}{2(\nu^2+1)^2}, \quad \frac{|c|\nu^2(\nu^2-3)}{8(\nu^2+1)}, \quad \frac{|c|\nu^2}{2(\nu^2+1)}.$$

By (3.4), we have

$$k_\gamma(t_0)^2 = \frac{|c|\nu^2}{4} - \frac{|c|\nu^2(\nu^2+5)(\nu^2-3)}{4(\nu^2+1)^2}\rho_\gamma(t_0)^2.$$

Hence, $K = k_\gamma(t_0)^2$ varies $|c|\nu^2/4 \leq K < 4|c|\nu^2/(\nu^2+1)^2$ when $1 < \nu^2 < 3$, and varies $4|c|\nu^2/(\nu^2+1)^2 < K \leq |c|\nu^2/4$ when $\nu^2 > 3$. We here compare critical points and the end points of the interval where K varies:

- $\dfrac{|c|\nu^2(3-\nu^2)}{2(\nu^2+1)^2} < \dfrac{|c|\nu^2}{2(\nu^2+1)} < \dfrac{|c|\nu^2}{4}$ because $\nu^2 > 1$;

- $\dfrac{|c|\nu^2}{2(\nu^2+1)} < \dfrac{4|c|\nu^2}{(\nu^2+1)^2}$ if and only if $\nu^2 < 7$;

- $\dfrac{|c|\nu^2(\nu^2-3)}{8(\nu^2+1)} < \dfrac{|c|\nu^2}{2(\nu^2+1)}$ if and only if $\nu^2 < 7$.

We hence find the following.

1) $g(K;\nu)$ is monotone decreasing when $1 < \nu^2 < 3$ and when $3 < \nu^2 \leq 7$. By noticing (4.2), we have

$$\frac{4g(\frac{|c|\nu^2}{4};\nu)}{c^2\nu^4(\nu^2+5)^3(\nu^2-3)} = 0, \qquad \frac{4g(\frac{4|c|\nu^2}{(\nu^2+1)^2};\nu)}{c^2\nu^4(\nu^2+5)^3(\nu^2-3)} = 1.$$

2) When $\nu^2 > 7$, the function $g(K;\nu)$ is monotone decreasing in the intervals $\left(\frac{4|c|\nu^2}{(\nu^2+1)^2}, \frac{|c|\nu^2}{2(\nu^2+1)}\right]$ and $\left[\frac{|c|\nu^2(\nu^2-3)}{8(\nu^2+1)}, \frac{|c|\nu^2}{4}\right]$, and is monotone increasing in the interval $\left[\frac{|c|\nu^2}{2(\nu^2+1)}, \frac{|c|\nu^2(\nu^2-3)}{8(\nu^2+1)}\right]$. We additionally have

$$\frac{4g(\frac{|c|\nu^2}{2(\nu^2+1)};\nu)}{c^2\nu^4(\nu^2+5)^3(\nu^2-3)} = \frac{32(\nu^2-1)^3}{(\nu^2+5)^3(\nu^2-3)}, \qquad \frac{4g(\frac{|c|\nu^2(\nu^2-3)}{8(\nu^2+1)};\nu)}{c^2\nu^4(\nu^2+5)^3(\nu^2-3)} = 1.$$

We note that when $k_\gamma(t_0)^2 = |c|\nu^2(\nu^2-3)/\{8(\nu^2+1)\}$ $(\nu^2 > 7)$, we have $\rho_\gamma(t_0)^2 = 1/\{2(\nu^2-3)\}$ and $\kappa^2 = |c|\nu^2(\nu^2-3)/\{8(\nu^2+1)^2\}$. By these considerations, we obtain the behavior of $\tau_\gamma(t_0)^2$ with respect to $k_\gamma(t_0)^2$.

1) When $1 < \nu^2 < 3$, it is monotone increasing and takes all values in the interval $[0,1)$.
2) When $3 < \nu^2 \le 7$, it is monotone decreasing and takes all values in the interval $[0,1)$.
3) When $\nu^2 > 7$, it is monotone decreasing in the intervals $\left(\frac{4|c|\nu^2}{(\nu^2+1)^2}, \frac{|c|\nu^2}{2(\nu^2+1)}\right]$ and $\left[\frac{|c|\nu^2(\nu^2-3)}{8(\nu^2+1)}, \frac{|c|\nu^2}{4}\right]$, and is monotone increasing in the interval $\left[\frac{|c|\nu^2}{2(\nu^2+1)}, \frac{|c|\nu^2(\nu^2-3)}{8(\nu^2+1)}\right]$. It takes all values in the interval $[0,1]$. Moreover, it takes the value 1 when $k_\gamma(t_0)^2 = \frac{|c|\nu^2(\nu^2-3)}{8(\nu^2+1)}$, and takes the value 0 only when $k_\gamma(t_0)^2 = \frac{|c|\nu^2}{4}$.

Next we study the case $\alpha = \mu_M$. In this case we have

$$\rho_\gamma(t_0)^2 = \frac{\left(4k_\gamma(t_0)^2\nu^2 - |c|\right)(\nu^2+1)^2}{|c|(5\nu^2+1)(3\nu^2-1)} \tag{4.3}$$

and

$$k_\gamma(t_0)^2\tau_\gamma(t_0)^2$$
$$= \frac{4\nu^2\left(4k_\gamma(t_0)^2\nu^2 - |c|\right)\left\{|c|(3\nu^2-1) - 2k_\gamma(t_0)^2(\nu^2+1)^2\right\}^2}{|c|^2(5\nu^2+1)^3(3\nu^2-1)}. \tag{4.4}$$

If we set

$$h(K;\nu) = (4K\nu^2 - |c|)\left\{|c|(3\nu^2-1) - 2K(\nu^2+1)^2\right\}^2/K,$$

we have

$$\frac{d}{dK}h(K;\nu) = \frac{1}{K^2}\left\{2K(\nu^2+1)^2 - |c|(3\nu^2-1)\right\}$$
$$\times \left\{16\nu^2(\nu^2+1)^2K^2 - 2|c|(\nu^2+1)^2K - |c|^2(3\nu^2-1)\right\},$$

hence find that $\frac{d}{dK}h(K;\nu) = 0$ if and only if

$$K = \frac{|c|(3\nu^2-1)}{2(\nu^2+1)^2}, \quad \frac{|c|}{2(\nu^2+1)}, \quad -\frac{|c|(3\nu^2-1)}{8\nu^2(\nu^2+1)}.$$

By (3.4), we have

$$k_\gamma(t_0)^2 = \frac{|c|}{4\nu^2} + \frac{|c|(5\nu^2+1)(3\nu^2-1)}{4\nu^2(\nu^2+1)^2}\rho_\gamma(t_0)^2.$$

Hence, K varies $|c|/(4\nu^2) \le K < 4|c|\nu^2/(\nu^2+1)^2$. The critical points and the end points of the interval where K varies satisfy

$$\frac{|c|}{4\nu^2} < \frac{|c|}{2(\nu^2+1)} < \frac{|c|(3\nu^2-1)}{2(\nu^2+1)^2} < \frac{4|c|\nu^2}{(\nu^2+1)^2}.$$

Therefore, we obtain that $h(K;\nu)$ is monotone increasing in the intervals $\left[\frac{|c|}{4\nu^2}, \frac{|c|}{2(\nu^2+1)}\right]$ and $\left[\frac{|c|(3\nu^2-1)}{2(\nu^2+1)^2}, \frac{4|c|\nu^2}{(\nu^2+1)^2}\right)$, and is monotone decreasing in the interval $\left[\frac{|c|}{2(\nu^2+1)}, \frac{|c|(3\nu^2-1)}{2(\nu^2+1)^2}\right]$. By noticing (4.4), we moreover have

$$\frac{4\nu^2\, h\left(\frac{|c|}{4\nu^2};\nu\right)}{|c|^2(5\nu^2+1)^3(3\nu^2-1)} = 0, \qquad \frac{4\nu^2\, h\left(\frac{4|c|\nu^2}{(\nu^2+1)^2};\nu\right)}{|c|^2(5\nu^2+1)^3(3\nu^2-1)} = 1,$$

$$[5pt]\ \frac{4\nu^2\, h\left(\frac{|c|}{2(\nu^2+1)};\nu\right)}{|c|^2(5\nu^2+1)^3(3\nu^2-1)} = \frac{32\nu^2(\nu^2-1)^3}{(5\nu^2+1)^3(3\nu^2-1)} < 1,$$

$$\frac{4\nu^2\, h\left(\frac{|c|(3\nu^2-1)}{2(\nu^2+1)^2};\nu\right)}{|c|^2(5\nu^2+1)^3(3\nu^2-1)} = 0.$$

We note that when $k_\gamma(t_0)^2 = |c|(3\nu^2-1)/(2(\nu^2+1)^2)$ we have

$$\rho_\gamma(t_0)^2 = \frac{\nu^2-1}{3\nu^2-1}, \qquad \kappa^2 = \frac{|c|\,(\nu^2-1)(3\nu^2-1)}{4\nu^2(\nu^2+1)^2}.$$

By these considerations, we obtain the behavior of $\tau_\gamma(t_0)^2$ with respect to $k_\gamma(t_0)^2$. It is monotone increasing in the intervals $\left[\frac{|c|}{4\nu^2}, \frac{|c|}{2(\nu^2+1)}\right]$ and $\left[\frac{|c|(3\nu^2-1)}{2(\nu^2+1)^2}, \frac{4|c|\nu^2}{(\nu^2+1)^2}\right)$, and is monotone decreasing in the interval $\left[\frac{|c|}{2(\nu^2+1)}, \frac{|c|(3\nu^2-1)}{2(\nu^2+1)^2}\right]$. It takes all values in the interval $[0,1)$. Moreover it takes the value 0 when $k_\gamma(t_0)^2 = \frac{|c|}{4\nu^2}, \frac{|c|(3\nu^2-1)}{2(\nu^2+1)^2}$.

We here draw some graphs which show the relationship between $k_\gamma(t_0)$ and $|\tau_\gamma(t_0)|$. Unfortunately, we do not check whether the values of $|\tau_\gamma|$ corresponding to the case $\alpha = \mu_M$ are greater that those corresponding to the case $\alpha = \lambda_M$ or not when $\nu^2 < 3$ and $\lambda_M < k_\gamma < \delta_M$. We note that

- $\lambda_M = \sqrt{|c|}$ if and only if $\nu = 2$,
- $\delta_M = \sqrt{|c|}/2$ if and only if $\nu = 2+\sqrt{3}$,
- when $\nu > 7$ we have $|c|\nu^2(\nu^2-3)/\{8(\nu^2+1)\} > |c|$.

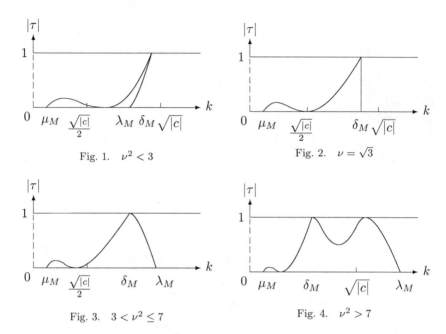

Fig. 1. $\nu^2 < 3$

Fig. 2. $\nu = \sqrt{3}$

Fig. 3. $3 < \nu^2 \le 7$

Fig. 4. $\nu^2 > 7$

5. Some characterizations of Hopf real hypersurfaces

In this section we give some characterizations of homogeneous Hopf hypersurfaces in a complex hyperbolic space $\mathbb{C}H^n$. We here pay attention to geodesic curvatures and complex torsions of extrinsic shapes of trajectories which are tangentially of order two.

Proposition 5.1 ([5]). *A real hypersurface M of $\mathbb{C}H^n$ is Hopf if and only if the following condition (TC) holds at each point $p \in M$:*

(TC) *There is κ_p such that the extrinsic shape of a trajectory γ_p for $\mathbb{F}_{\kappa_p}$ with $\dot{\gamma}_p(0) = \xi_p$ is tangentially of order two at $p = \gamma_p(0)$ and $k_{\gamma_p}(0) \ne |\kappa_p|$.*

For nonnegative constants k, τ with $\tau < 1$, we say a unit tangent vector $v \in U_pM$ at a point $p \in M$ of a real hypersurface M satisfy Condition (k, τ) if there exists a number κ with $|\kappa| \ne k$ such that the trajectory γ for $\mathbb{F}_\kappa$ of initial vector v is tangentially of order two at $p = \gamma(0)$ and satisfies $k_\gamma(0) = k$, $|\tau_\gamma(0)| = \tau$.

Theorem 5.1. *We put $r_0 = \left(1/\sqrt{|c|}\right) \log\left((\sqrt{3}+1)/(\sqrt{3}-1)\right)$. A real hypersurface M of a complex hyperbolic space $\mathbb{C}H^n(c)$ is locally congruent to either a tube $R(r)$ ($r \ge r_0$) around $\mathbb{R}H^n$ or a tube $T_\ell(r)$ ($r > r_0$) around $\mathbb{C}H^\ell$ if and only if the following conditions hold:*

i) *At each point $p \in M$, a trajectory γ_p for some $\mathbb{F}_{\kappa_p}$ with $\dot{\gamma}_p(0) = \xi_p$ is tangentially of order two at p and satisfies $k_{\gamma_p}(0) \neq |\kappa_p|$;*

ii) *There exist positive constants k, τ_1, τ_2 with $k < \sqrt{3|c|}/2$ and $\tau_1 < \tau_2 < 1$ such that at each point $p \in M$ we can choose linearly independent unit tangent vectors $v_1, \ldots, v_{2n-2} \in U_p M$ which satisfy the following;*

    a) *either Condition (k, τ_1) or Condition (k, τ_2) holds, but not all of them satisfy one of these conditions,*

    b) *the components $v_i - \langle v_i, \xi_p \rangle \xi_p$ $(i = 1, \ldots, 2n - 2)$ span the tangent subspace $T_p^0 M$.*

We here restrict ourselves to the cases that extrinsic shapes have null complex torsions and have complex torsions ± 1.

Theorem 5.2. *We put $r_1 = (2/\sqrt{|c|}) \log(\sqrt{5} + 2)$. A real hypersurface M of a complex hyperbolic space $\mathbb{C}H^n(c)$ is locally congruent to one of a tube $R(r)$ around $\mathbb{R}H^n$, a tube $T(r)$ $(r \geq r_1)$ around $\mathbb{C}H^{n-1}$ and a tube $T_\ell(r)$ $(r \geq r_1)$ around $\mathbb{C}H^\ell$ if and only if the following conditions hold:*

i) *At each point $p \in M$, a trajectory γ_p for some $\mathbb{F}_{\kappa_p}$ with $\dot{\gamma}_p(0) = \xi_p$ is tangentially of order two at p and satisfies $k_{\gamma_p}(0) \neq |\kappa_p|$;*

ii) *There exist positive constants k_1, k_2, τ with $\tau < 1$ such that at each point $p \in M$ we can choose linearly independent unit tangent vectors $v_1, \ldots, v_{2n-2} \in U_p M$ which satisfy the following;*

    a) *either Condition (k_1, τ) or Condition (k_2, τ) holds,*

    b) *the components $v_i - \langle v_i, \xi_p \rangle \xi_p$ $(i = 1, \ldots, 2n - 2)$ span the tangent subspace $T_p^0 M$;*

iii) *There exist a point $p_0 \in M$ and a positive constant k_0 with $k_0 \leq 3\sqrt{2|c|}/8$ such that we have a trajectory γ_0 for some $\mathbb{F}_{\kappa_0}$ with $\gamma_0(0) = p_0$ whose initial vector $\dot{\gamma}_0(0)$ is not orthogonal to ξ_p and that it is tangentially of order two at p_0 with $k_{\gamma_0}(0) = k_0$ and $\tau_{\gamma_0}(0) = 0$.*

Theorem 5.3. *We put $r_2 = (1/\sqrt{|c|}) \log((\sqrt{7}+1)/(\sqrt{7}-1))$. A real hypersurface M of a complex hyperbolic space $\mathbb{C}H^n(c)$ is locally congruent to a tube $R(r)$ $(r < r_2)$ around $\mathbb{R}H^n$ if and only if the following conditions hold:*

i) *At each point $p \in M$, a trajectory γ_p for some $\mathbb{F}_{\kappa_p}$ with $\dot{\gamma}_p(0) = \xi_p$ is tangentially of order two at p and satisfies $k_{\gamma_p}(0) \neq |\kappa_p|$;*

ii) *There exist positive constants k_1, k_2, τ with $\tau < 1$ such that at each point $p \in M$ we can choose linearly independent unit tangent vectors $v_1, \ldots, v_{2n-2} \in U_p M$ which satisfy the following;*

    a) *either Condition (k_1, τ) or Condition (k_2, τ) holds,*

    b) *the components $v_i - \langle v_i, \xi_p \rangle \xi_p$ $(i = 1, \dots, 2n-2)$ span the tangent subspace $T_p^0 M$;*

iii) *There exist a point $p_0 \in M$ and a positive constant k_0 such that we have a trajectory γ_0 for some $\mathbb{F}_{\kappa_0}$ $(|\kappa_0| \neq k_0)$ whose initial vector $\dot{\gamma}_0(0)$ is not parallel to ξ_p and that it is tangentially of order two at p_0 with $k_{\gamma_0}(0) = k_0$ and $|\tau_{\gamma_0}(0)| = 1$.*

In order to show our results, we need to recall extrinsic shapes of trajectories on real hypersurfaces of type (A). Since the structure torsion of each trajectory on these real hypersurfaces is constant, its extrinsic shape is tangentially of order two at some point if and only if it is extrinsic circular.

We only need to consider trajectories of the type (2) ii) of Proposition 3.1. In the case M is one of $G(r)$, HS and $T(r)$, when a trajectory is extrinsic circular, the geodesic curvature of its extrinsic shape takes the value in the interval $\lambda_M \leq k_\gamma < \delta_M$ by (3.4), and the complex torsion and the geodesic curvature of the extrinsic shape satisfies

$$\tau_\gamma^2 = \frac{(k_\gamma^2 - \lambda_M^2)(32\lambda_M^2 k_\gamma^2 + 4c\lambda_M^2 - c^2)^2}{|c|(8\lambda_M^2 + |c|)^3 k_\gamma^2} \tag{5.1}$$

by (3.5). When $M = G(r)$ or when $M = HS$, we see that $|\tau_\gamma|$ is monotone increasing with respect to k_γ, and takes all values in the interval $[0, 1)$. When $M = T(r)$, if we put $\nu = \coth(\sqrt{|c|}\, r/2)$, as we have $\lambda_M = \sqrt{|c|}/(2\nu)$ and $\delta_M = \sqrt{|c|}(\nu^2+1)/(2\nu)$, we find the following:

1) $|\tau_\gamma|$ is monotone increasing with respect to k_γ in the intervals $\left[\lambda_M, \sqrt{|c|(\nu^2+1)/(8\nu^2)}\,\right]$ and $\left[\sqrt{(|c|(\nu^2+1)/8}, \delta_M\right]$;
2) $|\tau_\gamma|$ is monotone decreasing in the interval $\left[\sqrt{(|c|(\nu^2+1)/(8\nu^2)}, \sqrt{(|c|(\nu^2+1)/8}\,\right]$;
3) $|\tau_\gamma|$ takes all values in the interval $[0, 1)$;
4) When $k_\gamma^2 = |c|(\nu^2+1)/(8\nu^2)$, we have
$\tau_\gamma^2 = (\nu^2-1)^3(\nu^2+1)/\{\nu^2(\nu^2+2)^3\} < 1$;
5) When $k_\gamma^2 = |c|(\nu^2+1)/8$, we have
$\tau_\gamma = 0$, $\rho_\gamma^2 = (\nu^2-1)/(2\nu^2)$, $\kappa^2 = |c|(\nu^2-1)/8$.

In the case $M = T_\ell(r)$, as it has two principal curvatures λ_M, μ_M for vectors orthogonal to ξ, and they satisfy $\lambda_M = \lambda_{G(r)}$ and $\mu_M = \lambda_{T(r)}$, we are enough to borrow those cases for $G(r)$ and for $T(r)$ at the same time. Thus the relationships between geodesic curvatures and complex torsions of extrinsic circular trajectories of the type (2) ii) of Proposition 3.1 on real hypersurfaces of type (A) are given like in Figs. 5–8 (see [2]).

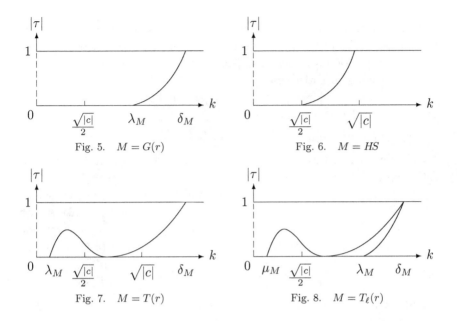

Fig. 5. $M = G(r)$

Fig. 6. $M = HS$

Fig. 7. $M = T(r)$

Fig. 8. $M = T_\ell(r)$

We are now in the position to prove our characterizations. Since Proposition 5.1 guarantees that M is a Hopf hypersurface if and only if the first condition of each of our results holds, We therefore need to check that Hopf hypersurfaces are characterized by the other conditions.

Proof of Theorem 5.1. First we check that the listed Hopf hypersurfaces satisfy the second condition. We note that $\nu = \coth(\sqrt{|c|}\, r/2)$ satisfies $\nu \le \sqrt{3}$ if and only if $r \ge r_0$.

When $M = R(r)$, we have $\delta_M \le \sqrt{3|c|}/2$ if and only if $\nu \le \sqrt{3}$. Therefore, if $r > r_0$, as we can see in Fig. 1, for k with $\lambda_M < k < \delta_M < \sqrt{3|c|}/2$ we can take τ_1, τ_2 so that they satisfy (4.2), (4.4), respectively. We define positive ρ_1, ρ_2 by (4.1), (4.3), respectively, and set

$$\kappa_1 = \frac{\sqrt{|c|}\,\nu(\nu^2-3)\rho_1}{2(\nu^2+1)} = -\sqrt{\frac{(4k^2-|c|\nu^2)(3-\nu^2)}{4(\nu^2+5)}},$$

$$\kappa_2 = -\frac{\sqrt{|c|}\,(3\nu^2-1)\rho_2}{2\nu(\nu^2+1)} = -\sqrt{\frac{(4k^2\nu^2-|c|)(3\nu^2-1)}{4\nu^2(5\nu^2+1)}},$$

according to the condition ii) in (2) of Proposition 3.1. We then find $|\kappa_1| \neq k$ and $|\kappa_2| \neq k$. Choose linearly independent tangent vectors $u_1, \ldots, u_{n-1} \in (V_{\lambda_M})_p$ with $\|u_i\|^2 = 1 - \rho_1^2$ and linearly independent tangent vectors $w_1, \ldots, w_{n-1} \in (V_{\mu_M})_p$ with $\|w_i\|^2 = 1 - \rho_2^2$ at each point

$p \in M$. We then find that trajectories for $\mathbb{F}_{\kappa_1}$ of initial vectors $u_i + \rho_1 \xi_p$ and those for $\mathbb{F}_{\kappa_2}$ of initial vectors $w_i + \rho_2 \xi_p$ are tangentially of order two at p by Proposition 3.1. The geodesic curvatures and complex torsions of extrinsic shapes of trajectories for $u_i + \rho_1 \xi_p$ are k and τ_1, and those for $w_i + \rho_2 \xi_p$ are k and τ_2. If $r = r_0$, regarding Fig. 2, we take arbitrary positive ρ_1, ρ_2 ($\rho_1 < \rho_2 < 1$), and put

$$\kappa_1 = \sqrt{|c|}\nu(\nu^2 - 3)\rho_1/(2(\nu^2+1)), \quad \kappa_2 = -\sqrt{|c|}(3\nu^2 - 1)\rho_2/(2\nu(\nu^2+1)).$$

We obtain that M satisfies the same property. Thus we see $R(r)$ ($r \geq r_0$) satisfies the second condition.

Next we study $M = T_\ell(r)$. For this real hypersurface we have $\lambda_M < \sqrt{3|c|}/2$ if and only if $r > r_0$ and $\delta_M > \sqrt{|c|}$. In this case, we take k with $\lambda_M < k \leq \sqrt{3|c|}/2$ and define τ_1, τ_2 with $0 < \tau_1 < \tau_2 < 1$ by the relation (5.1) and the relation obtained by replacing λ_M by μ_M in (5.1), That is,

$$\tau_1{}^2 = \frac{(4k^2 - |c|\nu^2)(8\nu^2 k^2 - |c|\nu^2 - |c|)^2}{4|c|^2(2\nu^2+1)^3 k^2}, \quad \tau_2{}^2 = \frac{(4k^2\nu^2 - |c|)(8k^2 - |c|\nu^2 - |c|)^2}{4|c|^2(\nu^2+2)^3 k^2}.$$

We define ρ_1, ρ_2 satisfying $0 < \rho_1, \rho_2 < 1$ by

$$\rho_1{}^2 = \frac{(4k^2 - |c|\nu^2)\nu^2}{|c|(2\nu^2+1)}, \quad \rho_2{}^2 = \frac{4k^2\nu^2 - |c|}{|c|\nu^2(\nu^2+2)},$$

and by noticing $\delta_M = \lambda_M + \mu_M$ we set $\kappa_1 = -\mu_M \rho_1 = -\sqrt{|c|}\rho_1/(2\nu)$, $\kappa_2 = -\lambda_M \rho_2 = -\sqrt{|c|}\nu\rho_2/2$ (see [2]). We then find that $|\kappa_1| \neq k$ and $|\kappa_2| \neq k$. We can check this also by the fourth assertion of Proposition 3.1. At each point $p \in M$, we choose linearly independent tangent vectors $u_1, \ldots, u_{2\ell} \in (V_{\lambda_M})_p$ with $\|u_i\|^2 = 1 - \rho_1{}^2$ and linearly independent tangent vectors $w_1, \ldots, w_{2n-2\ell-2} \in (V_{\mu_M})_p$ with $\|w_i\|^2 = 1 - \rho_2{}^2$. Then trajectories for $\mathbb{F}_{\kappa_1}$ of initial vectors $u_i + \rho_1 \xi_p$ and those for $\mathbb{F}_{\kappa_2}$ of initial vectors $w_i + \rho_2 \xi_p$ are tangentially of order two at p by Proposition 3.1. Since the geodesic curvatures and complex torsions of extrinsic shapes of trajectories for $u_i + \rho_1 \xi_p$ are k and τ_1, and those for $w_i + \rho_2 \xi_p$ are k and τ_2, we find that $M = T_\ell(r)$ satisfies the second condition.

On the other hand, we suppose that a Hopf hypersurface M satisfies the second condition. By Proposition 3.1, the trajectory γ_i for some $\mathbb{F}_{\kappa_i}$ of initial vector v_i is tangentially of order two at p, and $v_i - \langle v_i, \xi_p \rangle \xi_p$ is principal. Moreover, denoting the principal curvature of $v_i - \langle v_i, \xi_p \rangle \xi_p$ by α_i and putting $\rho_i = \langle v_i, \xi_p \rangle = \rho_{\gamma_i}(0)$ we find that $\alpha_i, \kappa_i, \rho_i$ satisfy

$$\begin{cases} \kappa_i + (\delta_M - \alpha_i)\rho_i = 0, \\ \kappa_i{}^2 - 2\kappa_i \alpha_i \rho_i + \alpha_i = k^2, \\ \kappa_i(2\rho_i{}^2 - 1) - \alpha_i \rho_i = k\tau_j, \end{cases}$$

$(j = 1, 2)$. If we regard this as a system of equations with unknown $\alpha_i, \kappa_i, \rho_i$, it has finite solutions. Since δ_M is locally constant, we see that principal curvatures of M are locally constant (cf. [8]). Therefore M is locally congruent to one of $HS, G(r), T(r), T_\ell(r)$ and $R(r)$. By the relationship between geodesic curvatures and complex torsions of extrinsic shapes of trajectories which are tangentially of order two that can be seen in Figs. 1–8, we get the conclusion. □

Proof of Theorem 5.2. In view of Figs. 1–8, we find that all homogeneous Hopf real hypersurfaces in $\mathbb{C}H^n$ satisfy the second condition along the similar lines as in the proof of Theorem 5.1. We note that we take k_1, k_2 as $k_1 = k_2$ for the case $M = G(r)$, HS, $T(r)$. We hence check the third condition. We set $\nu = \coth\sqrt{|c|}\,r/2$. When $M = R(r)$, if we take $k_0 = \sqrt{|c|(3\nu^2-1)/(2(\nu^2+1)^2)}$ $(\leq 3\sqrt{2|c|}/8)$, we saw in §4 that $R(r)$ satisfies the third condition at each point. When $M = T(r)$ or $M = T_\ell(r)$, if we take $k_0 = \sqrt{|c|(\nu^2+1)/8}$, which is not greater than $3\sqrt{2|c|}/8$ if and only if $\nu^2 \leq 5/4$, we saw that it satisfies the third condition at each point in this section.

On the other hand, if a Hopf hypersurface M satisfies the second condition, then by a similar argument as in the proof of Theorem 5.1 we find that all principal curvatures of M are locally constant. Hence M is locally congruent to one of $HS, G(r), T(r), T_\ell(r)$ and $R(r)$. Since $\tau_{\gamma_0}(0) = 0$, we find $|\kappa_0| \neq k_0$ by (4) in Proposition 3.1. When M is either HS or $G(r)$, extrinsic shapes of circular trajectory have null complex torsions if and only if their geodesic curvatures are λ_M, which is the case that trajectories have null structure torsion. When $M = T(r)$, extrinsic shapes of circular trajectory have null complex torsions if and only if their geodesic curvatures are either λ_M or the above k_0. If their geodesic curvatures are λ_M, then trajectories have null structure torsion. When M is either $T_\ell(r)$ or $R(r)$, extrinsic shapes of circular trajectory have null complex torsions if and only if their geodesic curvatures are one of λ_M, μ_M and the above k_0. If their geodesic curvatures are either λ_M or μ_M, then trajectories have null structure torsion. Thus, we get our conclusion. □

Proof of Theorem 5.3. As we pointed out in the proof of Theorem 5.2, all homogeneous Hopf real hypersurfaces in $\mathbb{C}H^n$ satisfy the second condition. We hence check the third condition. We set $\nu = \coth\sqrt{|c|}\,r/2$. When $M = R(r)$ with $\nu^2 > 7$, if we take $k_0 = \sqrt{|c|\nu^2(\nu^2-3)/(8(\nu^2+1))}$, we saw in §4 that it satisfies the third condition.

On the other hand, if a Hopf hypersurface M satisfies the second condition, M is locally congruent to one of $HS, G(r), T(r), T_\ell(r)$ and $R(r)$. When M is one of $HS, G(r), T(r)$ and $T_\ell(r)$, for each circular trajectory γ for $\mathbb{F}_\kappa$, under the assumption that $k_\gamma \neq |\kappa|$, we find $|\tau_\gamma| = 1$ if and only if $\rho_\gamma = \pm 1$. The situation for each trajectory which is tangentially of order two at some point on $R(r)$ with $\nu^2 \leq 7$ is the same. Thus, we get the conclusion.   □

Remark 5.1. The third condition in Theorem 5.3 is quite typical. This condition characterize $R(r)$ $(r < r_2)$ among homogeneous Hopf real hypersurfaces. Therefore, we can weaken the second condition in Theorem 5.3 as follows. There exist positive constants k_i, τ_i $(i = 1, \ldots, 2n - 2)$ with $\tau_i < 1$ such that at each point $p \in M$ we can choose linearly independent unit tangent vectors $v_1, \ldots, v_{2n-2} \in U_pM$ which satisfy the following;

  a) v_i satisfies Condition (k_i, τ_i),
  b) the components $v_i - \langle v_i, \xi_p \rangle \xi_p$ $(i = 1, \ldots, 2n - 2)$ span the tangent subspace $T_p^0 M$.

We here pay attention to Fig. 4. We compare the complex torsions at $k = \sqrt{|c|\nu^2/\{2(\nu^2+1)\}}$ and at $k = \sqrt{|c|/\{2(\nu^2+1)\}}$. Since we have

$$\frac{32(\nu^2-1)^3}{(\nu+5)^3(\nu^2-3)} < \frac{32\nu^2(\nu^2-1)^3}{(5\nu^2+1)^3(3\nu^2-1)}$$

if and only if $\nu > \nu_0$ with $\nu_0^2 = 91 + 12\sqrt{57} + 2\sqrt{6(687+91\sqrt{57})}$, we can say the following.

Proposition 5.2. *We take the positive r_4 satisfying $\coth(\sqrt{|c|}\, r_4/2) = \nu_0$. A real hypersurface M of a complex hyperbolic space $\mathbb{C}H^n(c)$ is locally congruent to a tube $R(r)$ $(r < r_4)$ around $\mathbb{R}H^n$ if and only if the following conditions hold:*

  i) *At each point $p \in M$, a trajectory γ_p for some $\mathbb{F}_{\kappa_p}$ with $\dot{\gamma}_p(0) = \xi_p$ is tangentially of order two at p and satisfies $k_{\gamma_p}(0) \neq |\kappa_p|$;*
  ii) *There exist positive constants $\tau, k_1, \ldots, k_6$ with $\tau < 1$ such that at each point $p \in M$ we can choose linearly independent unit tangent vectors $v_1, \ldots, v_{2n-2} \in U_pM$ which satisfy the following;*

      a) *each v_i satisfies one of Condition (k_j, τ) $(j = 1, \ldots, 6)$, and for each j at least one vector satisfies Condition (k_j, τ),*
      b) *the components $v_i - \langle v_i, \xi_p \rangle \xi_p$ $(i = 1, \ldots, 2n - 2)$ span the tangent subspace $T_p^0 M$.*

Next we give a characterization by paying attention to strengths of Sasakian magnetic fields. For constants κ, ρ with $|\rho| < 1$, we shall say that a unit tangent vector $v \in T_pM$ of a real hypersurface M in $\mathbb{C}H^n$ satisfies the condition (κ, ρ) at $p \in M$ if the extrinsic shape of the trajectory γ for $\mathbb{F}_\kappa$ of initial vector v is tangentially of order two at p and satisfies $|\tau_\gamma(0)| \neq 1$.

Proposition 5.3. *We put $r_0 = \left(1/\sqrt{|c|}\right) \log\left((\sqrt{3}+1)/(\sqrt{3}-1)\right)$. A real hypersurface M of a complex hyperbolic space $\mathbb{C}H^n(c)$ is locally congruent to a tube $R(r)$ $(r < r_0)$ around $\mathbb{R}H^n$ if and only if the following conditions hold:*

i) *At each point $p \in M$, a trajectory γ_p for some $\mathbb{F}_{\kappa_p}$ with $\dot{\gamma}_p(0) = \xi_p$ is tangentially of order two at p and satisfies $k_{\gamma_p}(0) \neq |\kappa_p|$;*
ii) *There exist a positive constant κ and constants ρ_1, ρ_2 with $-1 < \rho_1 < 0 < \rho_2 < 1$ such that at each point $p \in M$ we can choose linearly independent unit tangent vectors $v_1, \ldots, v_{2n-2} \in U_pM$ which satisfy*

    a) *either the condition (κ, ρ_1) or the condition (κ, ρ_2) holds, but not all of them satisfy one of these conditions,*
    b) *the components $v_i - \langle v_i, \xi_p \rangle \xi_p$ $(i = 1, \ldots, 2n-2)$ span the tangent subspace $T_p^0 M$.*

Proof. Since the first condition holds if and only if M is a Hopf hypersurface, we here study the second condition.

First we shall check that $R(r)$ $(r < r_0)$ satisfies the second condition. We put $\nu = \coth(\sqrt{|c|}\, r/2)$. When $M = R(r)$, we have

$$\delta_M - \lambda_M = \frac{\sqrt{|c|}\,\nu(3-\nu^2)}{2(\nu^2+1)}, \quad \delta_M - \mu_M = \frac{\sqrt{|c|}\,(3\nu^2-1)}{2\nu(\nu^2+1)}.$$

When $\nu^2 > 3$, which is the case $r < r_0$, we take a positive κ with $\kappa < \min\{|\delta_M - \lambda_M|, \delta_M - \mu_M\}$, and set $\rho_1 = -\kappa/(\delta_M - \mu_M)$, $\rho_2 = -\kappa/(\delta_M - \lambda_M)$. Then by Proposition 3.1 we find that $R(r)$ $(r < r_0)$ satisfies the second condition, because the complex torsion of the extrinsic shape of a trajectory which is tangentially of order two on this real hypersurface is equal to ± 1 if and only if its structure torsion is ± 1.

On the other hand, if a Hopf real hypersurface M satisfies the second condition, its principal curvatures are given by $(\kappa + \delta_M \rho_j)/\rho_j$ $(j = 1, 2)$ by Proposition 3.1. Since δ_M is locally constant, we find that principal curvatures of M are locally constant. Hence M is one of $HS, G(r), T(r), T_\ell(r)$ and $R(r)$. When M is one of $HS, G(r)$ and $T(r)$ we have one principal curvature corresponding to vectors in T^0M. Hence it does not satisfies the second condition by Proposition 3.1. When $M = T_\ell(r)$ we have $\delta_M = \lambda_M + \mu_M$ and $\lambda_M, \mu_M > 0$. Thus, we can not take ρ_1, ρ_2 so that their signatures are different. The situation is the same for $R(r)$ ($r \geq r_0$). We therefore get the conclusion.   □

At last, we make mention on Figs. 1–8. We say that two smooth curves σ_1, σ_2 on a Riemannian manifold N parameterized by their arclengths are *congruent* to each other if there exist an isometry φ of N and a real number t_c satisfying $\sigma_2(t) = \varphi \circ \sigma_1(t + t_c)$ for all t. When we can take $t_c = 0$, we say that they are congruent to each other in strong sense. On $\mathbb{C}H^n$, if two pairs $(\tilde{v}_1, \tilde{v}_2)$, $(\tilde{w}_1, \tilde{w}_2)$ of orthonormal tangent vectors with $\tilde{v}_1, \tilde{v}_2 \in T_{\tilde{p}}\mathbb{C}H^n$ and $\tilde{w}_1, \tilde{w}_2 \in T_{\tilde{q}}\mathbb{C}H^n$ satisfy $|\langle \tilde{v}_1, J\tilde{v}_2 \rangle| = |\langle \tilde{w}_1, J\tilde{w}_2 \rangle|$, then there exists an isometry φ satisfying $\varphi(\tilde{p}) = \tilde{q}$ and $d\varphi(\tilde{v}_i) = \tilde{w}_i$ ($i = 1, 2$). Therefore, two circles σ_1, σ_2 on $\mathbb{C}H^n$ are congruent to each other in strong sense if and only if they have the same geodesic curvatures and the same absolute values of complex torsions. Hence, the moduli space $\mathcal{C}_2(\mathbb{C}H^n)$ of circles of positive geodesic curvatures, which is the set of all congruence classes of these curves, is set theoretically identified with $(0, \infty) \times [0, 1]$.

When M is one of $HS, G(r)$ and $T(r)$, if two unit tangent vectors $v \in T_pM$ and $w \in T_qM$ satisfy $|\langle v, \xi_p \rangle| = |\langle w, \xi_q \rangle|$, then there is an isometry ψ of M satisfying $\psi(p) = q$ and $d\psi(v) = w$. When M is either $T_\ell(r)$ or $R(r)$, if two unit tangent vectors $v \in T_pM$ and $w \in T_qM$ satisfy $\|\mathrm{Proj}_\lambda(v)\| = \|\mathrm{Proj}_\lambda(w)\|$, $\|\mathrm{Proj}_\mu(v)\| = \|\mathrm{Proj}_\mu(w)\|$ and $|\langle v, \xi_p \rangle| = |\langle w, \xi_q \rangle|$, then there is an isometry ψ of M satisfying $\psi(p) = q$ and $d\psi(v) = w$. Here, $\mathrm{Proj}_\lambda : T^0M \to V_{\lambda_M}$, $\mathrm{Proj}_\mu : T^0M \to V_{\mu_M}$ denote the projections onto subbundles of principal curvature vectors. Hence two trajectories γ_1, γ_2 for $\mathbb{F}_\kappa$ on these homogeneous Hopf hypersurfaces are congruent to each other in strong sense if their initial vectors $\dot{\gamma}_1(0), \dot{\gamma}_2(0)$ satisfy the above corresponding conditions. We note that these real hypersurfaces are equvariant in $\mathbb{C}H^n$, that is, for each isometry ψ of such a real hypersurface M there is an isometry φ of $\mathbb{C}H^n$ satisfying $\varphi \circ \iota = \iota \circ \psi$ for the embedding $\iota : M \to \mathbb{C}H^n$. Hence, if two trajectories are congruent to each other, then their extrinsic shapes are congruent to each other.

When M is a real hypersurfaces of type (A), the structure torsion of each trajectory is constant. Moreover, when $M = T_\ell(r)$, it is known that $\|\mathrm{Proj}_\lambda(\dot{\gamma})\|$ and $\|\mathrm{Proj}_\mu(\dot{\gamma})\|$ for each trajectory γ are also constant along γ (see [2]). Therefore, denoting by $\mathcal{E}(M)$ the set of all congruence classes of extrinsic circular trajectories of the type (2) ii) in Proposition 3.1 on a real hypersurface M of type (A), we find that Figs. 5–8 show images of embeddings given by $\mathcal{E}(M) \ni [\gamma] \mapsto (k_\gamma, \tau_\gamma) \in \mathcal{C}_2(\mathbb{C}H^n)$. On the other hand, when M is a real hypersurface of type (B), as the structure torsion of a trajectory is not necessarily constant, we can not say that we can define such a map. Hence Figs. 1–4 only show the relationship between geodesic curvatures and complex torsions at points where trajectories are tangentially of order two. Thus, we are interested in the following problem: On a real hypersurface of type (B), is there a trajectory γ which is tangentially of order two at $\gamma(t_1)$ and $\gamma(t_2)$ such that $\dot{\gamma}(t_1) - \rho_\gamma(t_1)\xi_{\gamma(t_1)}$ is a principal curvature vector corresponding to λ_M and $\dot{\gamma}(t_2) - \rho_\gamma(t_2)\xi_{\gamma(t_2)}$ is that corresponding to μ_M?

References

[1] T. Adachi, Foliation on the moduli space of extrinsic circular trajectories on a complex hyperbolic space, *Topology Appl.* **196** (2015), 311–324.

[2] _____, Trajectories on real hypersurfaces of type (A$_2$) which can be seen as circles in a complex hyperbolic space, *Note Math.* **37** suppl. 1 (2017), 19–33.

[3] T. Bao and T. Adachi, Circular trajectories on real hypersurfaces in a nonflat complex space form, *J. Geom.* **96** (2009), 41–55.

[4] _____, Trajectories for Sasakian magnetic fields on real hypersurfaces of type (B) in a complex hyperbolic space, *Differential Geom. Appl.* **29** (2011), S28–S32.

[5] _____, Characterizations of some homogeneous Hopf real hypersurfaces in a nonflat complex space form by extrinsic shapes of trajectories, *Differential Geom. Appl.* **48** (2016), 104–118.

[6] _____, Extrinsic circular trajectories on totally η-umbilic real hypersurfaces in a complex hyperbolic space, *Kodai Math. J.* **39** (2016), 615–631.

[7] J. Berndt, Real hypersurfaces with constant principal curvatures in complex hyperbolic space, *J. Reine Angew. Math.* **395** (1989), 132–141.

[8] T. Kato, *Peerturbation theory for linear operators*, Springer, 1966.

[9] T. Kajiwara and S. Maeda, Sectional curvatures of geodesic spheres in a complex hyperbolic space, *Kodai Math. J,* **38** (2015), 604–619.

[10] U-H. Ki and Y.J. Suh, On real hypersurfaces of a complex space form, *Math. J. Okayama Univ.* **32** (1990), 207–221.

[11] S. Maeda and T. Adachi, Sasakian curves on hypersurfaces of type (A) in a nonflat complex space form, *Reults Math.* **56** (2009), 489–499.

[12] S. Maeda, T. Adachi and H.K. Kim, Characterizations of geodesic
     spheres in a nonflat complex space form, *Glasgow Math. J.* **55** (2013),
     217–227.

Received December 26, 2018
Revised   March 29, 2019

PERIODIC SURFACES OF REVOLUTION in $\mathbb{R}^3$
AND CLOSED PLANE CURVES

Yasuhiro NAKAGAWA*

*Faculty of Advanced Science and Technology, Kumamoto University,
2-40-1 Kurokami, Kumamoto, 860-8555, Japan
E-mail: yasunaka@educ.kumamoto-u.ac.jp*

Hidekazu SATO

*Department of Mathematics, Saga University,
1 Honjo-machi, Saga, 840-8502, Japan
E-mail: analysis.d.e@gmail.com*

In this paper, we give a very simple method to construct closed plane curves. Furthermore, we prove that the space of curvature functions of closed plane curves is dense in the space of periodic continuous functions on $\mathbb{R}$.

Keywords: Closed plane curve; surfaces of revolution.

1. Introduction

In [2] and [3], Kenmotsu studied the mean curvature of a surface of revolution in $\mathbb{R}^3$. In particular, he proved that when a closed plane curve γ of curvature κ is given one can then obtain the periodic surface of revolution in $\mathbb{R}^3$ whose mean curvature is $\kappa/2$. Therefore, if we can construct closed plane curves, then we obtain periodic surfaces of revolution in $\mathbb{R}^3$.

In this paper, by taking smooth periodic functions with period 2π and some rational numbers, we construct closed plane curves in quite an elementary way. In view of our method in our construction, we can show that the space of curvature functions of closed plane curves is dense in the space of periodic continuous functions on $\mathbb{R}$.

After the authors had finished writing this paper, Kenmotsu informed them of the paper [1]. In that paper, Arroyo, Garay and Mencía had already achieved more general results ([1, Corollary 2.2]). However, the proof in this paper is extremely simple and elementary. This is a reason why the authors present this paper.

In concluding this introduction, the authors wish to thank Professor Katsuei Kenmotsu for his interest in this work and valuable suggestions.

---

*Partly supported by JSPS Grant-in-Aid for Scientific Research (C) No. 15K04848.

2. Closed plane curves and periodic surfaces

In order to explain our interest in this paper, we shall start by recalling some results due to Kenmotsu ([2, 3]) on mean curvatures of surfaces of revolution in $\mathbb{R}^3$.

Theorem 2.1 (Kenmotsu [3]). *Let $H(s)$ be a continuous periodic function on $\mathbb{R}$ of period $L > 0$. We set a function η_H by $\eta_H(u) = 2 \int_0^u H(s)ds$. Then the function $H(s)$ is the mean curvature of a periodic surface of revolution S in $\mathbb{R}^3$ with period L if and only if it satisfies one of the following three conditions:*

(i) $\sin \eta_H(L) \neq 0$ *and*

$$\frac{\int_0^L \cos \eta_H(u)du}{\sin \eta_H(L)} = \frac{\int_0^L \sin \eta_H(u)du}{1 - \cos \eta_H(L)};$$

(ii) $\sin \eta_H(L) = 1 - \cos \eta_H(L) = 0$ *and*

$$\int_0^L \cos \eta_H(u)du = \int_0^L \sin \eta_H(u)du = 0;$$

(iii) $\sin \eta_H(L) = 0$, $1 - \cos \eta_H(L) \neq 0$ *and*

$$\int_0^L \cos \eta_H(u)du = 0, \qquad \int_0^L \sin \eta_H(u)du \neq 0.$$

We say a C^2-plane curve $\gamma \colon \mathbb{R} \to \mathbb{R}^2$ to be closed if it satisfies

$$\gamma(s + L) = \gamma(s) \qquad (s \in \mathbb{R})$$

with some positive L. When a function H satisfies the condition (ii) in Theorem 2.1, if we put

$$F_H(s) := \int_0^s \sin \eta_H(u)du, \qquad G_H(s) := \int_0^s \cos \eta_H(u)du,$$

then, $\Gamma_H(s) := (G_H(s), F_H(s))$ is a closed plane curve parametrized by its arc length s. Conversely, Kenmotsu also proved the following:

Theorem 2.2 ([3]). *Let Γ be a closed plane curve parametrized by its arc length s, and let $k(s)$ denote its curvature. Then there exists a periodic surface of revolution in $\mathbb{R}^3$ whose mean curvature is $k(s)/2$.*

Therefore, if we can construct closed plane curves, then we also obtain periodic surfaces of revolution in $\mathbb{R}^3$. The following is the main result in this paper:

Theorem 2.3. *Given a periodic C^1-function $p(s)$ with period 2π, an arbitrary natural number $k \in \mathbb{N}$ and an arbitrary integer $l \in \mathbb{Z}$, we define a function $H(s)$ on $\mathbb{R}$ by*

$$H(s) := p'(s) + \frac{2l+1}{4k}.$$

Then, it satisfies the condition (ii) *in Theorem 2.1 with $L = 4k\pi$. In particular, the corresponding plane curve Γ_H is closed, and H is the mean curvature of a periodic surface of revolution in $\mathbb{R}^3$.*

Let $\mathcal{P}$ be the set of continuous periodic functions on $\mathbb{R}$ with period 2π. We put

$$\mathcal{P}_0 := \{p \in \mathcal{P} \mid p \text{ is the curvature of some closed plane curve}\}.$$

As a corollary of our result, we can also prove the following:

Corollary 2.1. *For an arbitrary $p \in \mathcal{P}$ and $\epsilon > 0$, there exists δ satisfying $-\epsilon < \delta < \epsilon$ and $p + \delta \in \mathcal{P}_0$. In particular, $\mathcal{P}_0$ is dense in $\mathcal{P}$.*

3. Proofs of our results

In this section we give proofs of our results by checking that the corresponding function η_H is periodic and satisfies the condition (ii) in Theorem 2.1.

Proof of Theorem 2.3. Since $H(s) = p'(s) + \dfrac{2l+1}{4k}$, we find

$$\eta_H(u) = 2p(u) - 2p(0) + \frac{2l+1}{2k}u.$$

As p is a periodic function with period 2π, we have

$$\eta_H(4k\pi) = (2l+1)2\pi \quad \text{and} \quad \sin\eta_H(4k\pi) = 1 - \cos\eta_H(4k\pi) = 0.$$

Furthermore, since we have

$$\eta_H(u + 2k\pi) = 2p(u + 2k\pi) - 2p(0) + \frac{2l+1}{2k}(u + 2k\pi)$$

$$= 2p(u) - 2p(0) + \frac{2l+1}{2k}u + (2l+1)\pi$$

$$= \eta_H(u) + (2l+1)\pi,$$

we obtain

$$
\begin{aligned}
\int_0^{4k\pi} \sin\eta_H(u)du &= \int_0^{2k\pi} \sin\eta_H(u)du + \int_{2k\pi}^{4k\pi} \sin\eta_H(u)du \\
&= \int_0^{2k\pi} \sin\eta_H(u)du + \int_0^{2k\pi} \sin\eta_H(t+2k\pi)dt \\
&= \int_0^{2k\pi} \sin\eta_H(u)du + \int_0^{2k\pi} \sin\left(\eta_H(t)+(2l+1)\pi\right)dt \\
&= \int_0^{2k\pi} \sin\eta_H(u)du - \int_0^{2k\pi} \sin\eta_H(t)dt = 0.
\end{aligned}
$$

Similarly, we can conclude that $\int_0^{4k\pi} \cos\eta_H(u)du = 0$. Therefore, $H(s)$ satisfies the condition (ii) in Theorem 2.1. This completes the proof of Theorem 2.3 with Theorem 2.1. $\qquad\square$

Proof of Corollary 2.1. For $p \in \mathcal{P}$, we put $A_p := \int_0^{2\pi} p(s)ds$. Then the function $\widetilde{p}$ defined by

$$
\widetilde{p}(u) := \int_0^u \left(p(s) - \frac{A_p}{2\pi}\right)ds
$$

is a periodic function with period 2π, because, for each $u \in \mathbb{R}$ we have

$$
\begin{aligned}
\widetilde{p}(u+2\pi) &= \int_0^{u+2\pi} \left(p(s) - \frac{A_p}{2\pi}\right)ds \\
&= \int_0^{2\pi} \left(p(s) - \frac{A_p}{2\pi}\right)ds + \int_{2\pi}^{u+2\pi} \left(p(s) - \frac{A_p}{2\pi}\right)ds \\
&= \int_0^u \left(p(t+2\pi) - \frac{A_p}{2\pi}\right)dt \\
&= \int_0^u \left(p(t) - \frac{A_p}{2\pi}\right)dt \\
&= \widetilde{p}(u).
\end{aligned}
$$

We can approximate the real number $A_p/(2\pi)$ by rational numbers of the form $(2l+1)/(2k)$ with $l \in \mathbb{Z}$ and $k \in \mathbb{N}$, that is, for an arbitrary $\epsilon > 0$, there exist $l \in \mathbb{Z}$ and $k \in \mathbb{N}$ satisfying

$$
0 \leqq \left| \frac{A_p}{2\pi} - \frac{2l+1}{2k} \right| < \epsilon.
$$

If we put $\delta := \{\pi(2l + 1) - kA_p\}/(2\pi k)$, it clearly satisfies $-\epsilon < \delta < \epsilon$. Since Theorems 2.2 and 2.3 allow us to conclude that

$$p(s) + \delta = \tilde{p}'(s) + \frac{A_p}{2\pi} + \delta = 2\left(\frac{1}{2}\tilde{p}'(s) + \frac{2l + 1}{4k}\right) \in \mathcal{P}_0.$$

This completes the proof of Corollary 2.1. □

4. Examples

In this section, we give some examples of periodic functions and their corresponding plane curves. We here give a periodic function H of the form in Theorem 2.3 and then drow the graph of the corresponding closed plane curve $\Gamma_H(s) = (G_H(s), F_H(s))$ obtained as

$$F_H(s) := \int_0^s \sin \eta_H(u)du, \qquad G_H(s) := \int_0^s \cos \eta_H(u)du,$$

with $\eta_H(u) = 2\int_0^u H(s)ds$.

Once we obtain such a closed curve satisfying the condition (ii) in Theorem 2.1, according to [2, 3] we can construct a surface of revolution in the following manner. We define a profile curve $C_H(s) = (x_H(s), y_H(s))$ ($s \in \mathbb{R}$) by

$$x_H(s) := \int_0^s \frac{(G_H(u) - c_2)F_H'(u) - (F_H(u) - c_1)G_H'(u)}{\sqrt{(F_H(u) - c_1)^2 + (G_H(u) - c_2)^2}} du,$$

$$y_H(s) := \sqrt{(F_H(s) - c_1)^2 + (G_H(s) - c_2)^2},$$

with some suitable constants $c_1, c_2 \in \mathbb{R}$. By using this curve we can construct a periodic surface of revolution in $\mathbb{R}^3$. Unfortunately, the figure of this surface of revolution is very complicated, in general. We therefore show only graphs of plane curves.

We note that figures of these curves are all programmed by the second named author by using the symbolic manipulation program Mathematica.

Example 4.1. When $H(s) = \cos s + \dfrac{1}{4}$, it is periodic with $L = 4\pi$. In this case we have

$$\eta_H(u) = 2\int_0^u H(s)ds = 2\sin u + \frac{1}{2}u \qquad (u \in \mathbb{R}).$$

The closed plane curve $\Gamma_H(s) = (G_H(s), F_H(s))$ corresponding to this is like Fig. 1.

Example 4.2. When $H(s) = (5/2)\sin s + (1/4)$, it is periodic with $L = 4\pi$. In this case we have $\eta_H(u) = -5\cos u + 5 + (u/2)$, and the corresponding closed plane curve $\Gamma_H(s)$ is like Fig. 2.

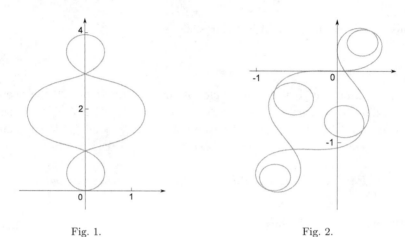

Fig. 1.   Fig. 2.

Example 4.3. When $H(s) = \sin s + (3/2)\cos s + (1/8)$, it is periodic with $L = 8\pi$. In this case we have $\eta_H(u) = -2\cos u + 3\sin u + 2 + (u/4)$, and the corresponding closed plane curve $\Gamma_H(s)$ is like Fig. 3.

Example 4.4. When $H(s) = \sin s + (1/2)\cos s + (5/4)$, it is periodic with $L = 4\pi$. In this case we have $\eta_H(u) = -2\cos u + \sin u + 2 + (5u/2)$, and the corresponding closed plane curve $\Gamma_H(s)$ is like Fig. 4.

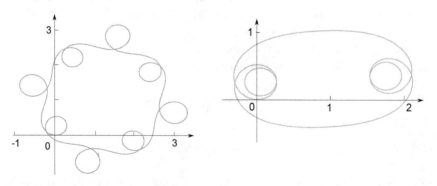

Fig. 3.   Fig. 4.

Example 4.5. When $H(s) = \sin^3 s + (1/12)$, it is periodic with $L = 12\pi$. In this case we have $\eta_H(u) = -2\cos u + (2/3)\cos^3 u + (4/3) + (u/6)$, and the corresponding closed plane curve $\Gamma_H(s)$ is like Fig. 5.

Example 4.6. When $H(s) = \cos^3 s + (7/16)$, it is periodic with $L = 16\pi$. In this case we have $\eta_H(u) = 2\sin u - (2/3)\sin^3 u + (7u/8)$, and the corresponding closed plane curve $\Gamma_H(s)$ is like Fig. 6.

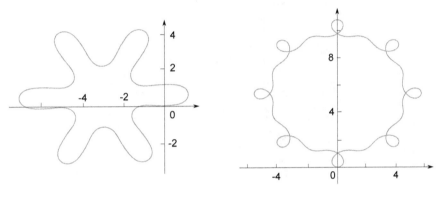

Fig. 5.   Fig. 6.

References

[1] J. Arroyo, O. J. Garay and J. J. Mencía, *When Is a Periodic Function the Curvature of a Closed Plane Curve?*, Amer. Math. Monthly **115** (2008), 405–414.

[2] K. Kenmotsu, Surfaces of revolution with prescribed mean curvature, *Tohoku Math. J.* **32** (1980), 147–153.

[3] K. Kenmotsu, Surfaces of revolution with periodic mean curvature, *Osaka J. Math.* **40** (2003), 687–696.

Received November 8, 2018
Revised February 14, 2019

CPSIA information can be obtained
at www.ICGtesting.com
Printed in the USA
BVHW041358141119
563788BV00006B/74/P

9 789811 206689